Cosmic Challenge

Listing more than 500 sky targets, both near and far, in 187 challenges, this observing guide will test novice astronomers and advanced veterans alike. Its unique mix of Solar System and deep-sky targets will have observers hunting for the Apollo lunar landing sites, searching for satellites orbiting the outermost planets, and exploring hundreds of star clusters, nebulae, distant galaxies, and quasars. Each target object is accompanied by a rating indicating how difficult the object is to find, an in-depth visual description, an illustration showing how the object realistically looks, and a detailed finder chart to help you find each challenge quickly and effectively. This guide introduces objects often overlooked in other observing guides, and challenges are provided for the naked eye, through binoculars, to the largest backyard telescopes. This paperback edition has updated charts and data tables to challenge observers for many years to come.

Philip S. Harrington is the author of eight previous books for the amateur astronomer, including *Touring the Universe through Binoculars*, *Star Ware*, and *Star Watch*. He is also a contributing editor for *Astronomy* magazine, where he has authored the magazine's monthly "Binocular Universe" column and "Phil Harrington's Challenge Objects," a quarterly online column on Astronomy.com. He is an adjunct professor at Dowling College, New York, where he teaches courses in stellar and planetary astronomy.

Cosmic Challenge

The Ultimate Observing List for Amateurs

PHILIP S. HARRINGTON
Dowling College, New York

CAMBRIDGE
UNIVERSITY PRESS

CAMBRIDGE
UNIVERSITY PRESS

University Printing House, Cambridge CB2 8BS, United Kingdom

One Liberty Plaza, 20th Floor, New York, NY 10006, USA

477 Williamstown Road, Port Melbourne, VIC 3207, Australia

314–321, 3rd Floor, Plot 3, Splendor Forum, Jasola District Centre, New Delhi – 110025, India

79 Anson Road, #06-04/06, Singapore 079906

Cambridge University Press is part of the University of Cambridge.

It furthers the University's mission by disseminating knowledge in the pursuit of education, learning, and research at the highest international levels of excellence.

www.cambridge.org
Information on this title: www.cambridge.org/9781108710756
DOI: 10.1017/9781108695558

First published 2011
First paperback edition 2019
Reprinted 2019

Printed in the United Kingdom by TJ International Ltd. Padstow Cornwall

A catalogue record for this publication is available from the British Library.

ISBN 978-1-108-71075-6 Paperback

For my wife Wendy, the center of my universe

Contents

Preface

Surely there is not another field of human contemplation so wondrously rich as astronomy! It is so easy to reach, so responsive to every mood, so stimulating, uplifting, abstracting, and infinitely consoling. Everybody may not be a chemist, a geologist, a mathematician, but everybody may be and ought to be, in a modest, personal way, an astronomer, for star-gazing is a great medicine of the soul.

With those words, Garret Serviss embarked on his book *Round the Year with the Stars*. Published in 1910, *Round the Year with the Stars* brought readers to sights that few had ever seen before. Serviss was one of his generation's best-known astronomical authors, with several previous titles to his credit. Indeed, he was almost apologetic for writing this latest work. "The writer's only real excuse for appearing again in this particular field is that he has never yet finished a book, and seen it go forth, without feeling that he had overlooked, or cast aside, or of necessity omitted a multitude of things quite as interesting and important as any he had touched upon."

That is my excuse, as well. In the 100 years since Serviss's book first appeared, there have been hundreds, if not thousands, of observing guides published. I have written a couple of them myself. Some were general guides intended to introduce the reader to the sky's finest objects. Others paid homage to only certain classes of objects, restricting their discussion to only deep-sky objects or perhaps members of the Solar System. Many were geared toward newcomers to the hobby and to science, while others were intended for veterans who had been around the block many times.

Many of the published guides, my own included, have overlooked some fascinating objects, perhaps in part because the author felt those objects were too difficult for the intended audience.

The book you hold before you is a little different. *Cosmic Challenge* focuses on a wide variety of sky targets, including some old favorites and some that you probably have never even heard of before. Each object included will have been selected not because it is easy, but because it is difficult to spot in some way. The type of challenge posed will vary from one target to the next. An object might be very faint, or very small, or tough to spot for any of a number of other reasons.

Of course, what's challenging to one person might be an easy catch for another. So much depends on each person's level of experience, the clarity and darkness of the observing site, and the telescope used. A tough test for a 4-inch telescope should be quite easy through a 14-inch. To help level the playing field, each chapter is devoted to one of six instrument categories based on aperture: naked eye, binoculars, 3- to 5-inch telescopes, 6- to 9.25-inch telescopes, 10- to 14-inch telescopes, and 15-inchers and up. Each chapter is then further segmented by season.

Although the book mainly covers deep-sky objects that may require dark skies regardless of telescope size, each chapter also includes targets of interest to city dwellers. Many lunar and planetary features, visible year-round, as well as some close-set double and multiple star systems, are included, since they are equally challenging regardless of the observing site.

I would very much enjoy hearing from you, the reader, as you attempt to view the objects outlined in this book. Feel free to email me at phil@philharrington.net. And be sure to check for additions and addenda in the "Cosmic Challenge" section of my website, www.philharrington.net.

Acknowledgments

I wish to pass on my sincere appreciation to those dedicated amateur astronomers who reviewed each chapter to correct errors, offer suggestions, and in general polish the product you hold before you. Those proofreaders include Glenn Chaple, Phil Creed, Rod Mollise, and Sam Storch. I am very fortunate to have had this skilled set of veteran amateur astronomers – all among the most knowledgeable amateurs in the world – review the final manuscript. Thank you all for your comments and your suggestions.

I also wish to thank John Boudreau, Bill Bradley, Kevin Dixon, Bob King, Larry Landolfi, Frank Melillo, and Dan Wright, the astrophotographers whose work adorns some of the pages to come. I find it truly amazing to see the results being produced by accomplished backyard astronomers these days. These seasoned amateurs are among today's most talented astrophotographers.

My thanks to Christian Legrand and Patrick Chevalley, creators of Virtual Moon Atlas software, for allowing me to use their program as the basis for the lunar charts found in the later chapters.

Many thanks also to my editor Vincent Higgs, Abigail Jones, Megan Waddington, and Claire Poole at Cambridge University Press for their diligent guidance and help throughout the production phase of this book. Without their input, the book as it exists simply would not.

I would also like to single out four people who were very influential in my early days as a teenage amateur astronomer. The first is George Clark, my 6th-grade science teacher in Norwalk, Connecticut. It was his homework assignment to watch the total lunar eclipse of April 13, 1968, that sparked my interest, and ultimately lifelong passion, in astronomy in the first place.

My thanks also to Russ Harding, former director of the Robert B. Oliver Planetarium for the Norwalk school system. With the patience of a saint, he allowed this high schooler to putter around the planetarium nearly every day after school, very likely getting in the way and causing trouble rather than helping. But I loved it.

Another influential teacher in my life was Fred Bump from the neighboring Westport, Connecticut, schools. Fred was the power behind resurrecting Rolnick Observatory and the Westport Astronomical Society, where I was an active member "back in the day."

Finally, thanks to Charles Scovil, curator of Stamford Observatory in Stamford, Connecticut. He fostered my interest, and that of many other young astronomers. My parents would drive me there every Friday night, as the observatory became my "hang out."

And, of course, were it not for my parents, Frank and Dorothy Harrington, none of that would have happened. I love them dearly for that gift.

Last, but certainly not least, my deepest thanks, love, and appreciation go to my ever-patient family. My wife Wendy, daughter Helen, and mother-in-law Helen Hunt, have continually provided me with boundless love and encouragement over the years. Wendy also looked everything over with her eagle eyes one final time before I shipped the manuscript off to the publisher. Writing a book such as this entails more work and longer hours than most people realize. Were it not for their support and patience, allowing me time away from the family to assemble the work you hold in your hands, this book would not have come to pass.

Photo credits

Unless otherwise credited, all illustrations are by the author.

Meeting the challenge

"It's going to be clear tonight," you think to yourself as you stare out of the window. "I can't wait for the Sun to go down!" Your mind immediately leaves whatever it is you're doing at the time, be it work, chores, school, or something else, and flies into the cosmos.

Shhh, did you hear that? That's the universe calling you.

As amateur astronomers, we aren't content to sit home at night, watch television, or drive to the local multiplex cinema to take in the latest movie. We won't have any of that. We're explorers. That's what drew us into astronomy in the first place: the idea that we can explore this marvelous universe of ours right from our backyards.

But then, hours later, as you're gathering your observing gear for a night under the stars, you stop dead in your tracks. "What am I going to look at?"

Has that ever been you? Probably. In fact, that may have been you just last night. Let's face it, if you've been involved with observational astronomy long enough, there is bound to come a time when you've seen "everything" your telescope can show you. All of a sudden, that show on television is beginning to sound tempting.

Okay, now stop right there! I'm not going to let you do it. You're an explorer. You're a pioneer. As amateur astronomers, we share a unique perspective on life. We enjoy searching for sights that few others in the entire course of human history have ever witnessed first hand. We blaze a trail that few have trod before. We won't miss a clear night just because "there's nothing new to observe."

That's where this book comes in. You and I are about to go observing together. No, not for the "same old stuff" that many observing guides run through time and again. We're hunting for bigger game. We're going

to challenge our telescopes, our observing site, and our powers of observation.

Each object included in the later chapters has been selected not because it is easy, but because it is difficult to see in some way. The type of challenge posed will vary from one target to the next. An object might be very faint, or very small, or tough to spot for any of a number of other reasons.

Of course, what might be a challenge for one person could be an easy catch for another. So much depends on each observer's level of experience, clarity and darkness of the observing site, and the telescope used. A tough test for a 4-inch telescope, for instance, will probably prove easy for a 14-inch – although not necessarily.

To help level the playing field, each chapter will be segmented further into six instrument categories based on aperture: naked eye, binoculars, 3- to 5-inch telescopes, 6- to 9.25-inch telescopes, 10- to 14-inch telescopes, and 15-inchers and up. Although many of the included deep-sky objects require dark skies regardless of telescope size, each chapter also includes targets of interest to city dwellers. Many close-set double and multiple star systems are equally challenging regardless of the observing site.

Since many of those deep-sky objects will be unfamiliar to most readers, each listing is accompanied by an eyepiece field chart. These charts are not intended to supersede the need for a separate star atlas, however. In order to locate these objects, you will also need one of several star atlases that are currently in print (Appendix B lists some of the best). Use the atlas to find the target's general location, and then zero in on it using the eyepiece field chart.

Closer to home, we also try to see tough targets in the Solar System, visible through the year. For example, one challenge is to see several small craters within the prominent lunar crater Plato. Another is spying the major moons orbiting Uranus. Both of these tests are notoriously difficult, but can also be attempted with equal effort from within a city or out in the country, given the right equipment.

Each entry in the chapters to come is rated on a 1-to-4 star scale alongside the challenge header according to the level of the challenge that it presents. A "Challenge Factor" of 1 star represents the easiest level, while a rating of 4 stars means that extra effort will be needed.

You will find that some of the 4-star challenges are not doable unless all conditions are perfect. Consider

those to be "hors catégorie," a term used in bicycle racing to designate an uphill climb that is "beyond categorization."

Okay, the stage is set, but before we can dive right into the challenges, we need to do a pre-check. In order to meet the Cosmic Challenge, your observing gear must be selected to eke out every photon possible from each target. Your observing site must also be matched to the test. Finally, you must be prepared, both mentally and physically. How do we go about ensuring these? Read on!

OPTIMIZING YOUR EQUIPMENT

Your eyes

Let's begin on common ground. Whether you own a telescope, binoculars, or no optical device at all, you have the greatest optical device on Earth: the human eye (Figure 1.1). Have you ever stopped and thought about it? Unless you are suffering from an eye-related malady, you probably take your sense of sight for granted. But consider that upwards of 90% of the information our brain receives and processes comes from just your two eyes.

As amateur astronomers, we are most interested in optimizing our so-called *night vision*, since that will allow us to push our observing skills to their greatest extent. To understand how to do that, it is important to understand just how the eye adapts to different lighting conditions.

We have all had occasion to wake up in the middle of the night and turn on a light, instantly blasting our eyes with a huge quantity of photons. It can be pretty painful at first, but slowly our eyes become accustomed to the bright light and our vision returns to normal.

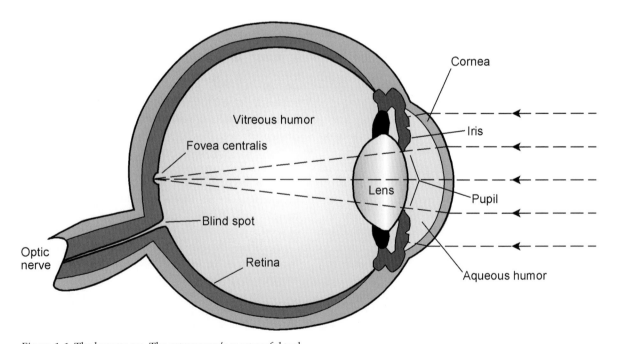

Figure 1.1 The human eye. The astronomer's most useful tool

Then, as we shut the light off again and our eyes plunge into darkness, we are once again blinded for a while until our eyes acclimate to the new conditions.

To appreciate our ability to perceive objects under varying lighting conditions, it is first important to have a basic understanding of how the human eye works. The human eye measures about an inch in diameter and is surrounded by a two-part protective layer: the transparent, colorless cornea and the white, opaque *sclera*. The *cornea* acts as a window to the eye and lies in front of a pocket of clear fluid called the *aqueous humor* and the eye's *iris*. Besides giving the eye its characteristic color, the iris regulates the amount of light entering the eye and, more importantly, varies its focal ratio. Under low-light conditions, the iris relaxes, dilating the pupil (the circular opening in the center of the iris), while, in bright light, the iris will tense, constricting the pupil, increasing the focal ratio, and masking lens aberrations to produce sharper views.

From the pupil, light passes through the eye's lens and across the eyeball's interior, which is filled with fluid called the *vitreous humor*. Both the lens and cornea act to focus the image onto the *retina*. The retina is composed of ten layers of nerve cells, including photo-sensitive receptors called *rods* and *cones*. Cones are concerned with brightly lit scenes, color vision, and resolution. Rods are low-level light receptors but cannot distinguish color. There are more cones towards the *fovea centralis* (the center of the retina and our perceived field of view), while rods are more numerous toward the edges. There are neither rods nor cones at the junction with the optic nerve, the eye's so-called *blind spot*.

In order to perceive images under dim lighting, the eye experiences a two-step process to adapt to the changing conditions. First, after being plunged into darkness, the eye's pupil quickly dilates to between 4 and 7 mm in diameter, doubling the pupil's normal, daytime aperture of approximately 2.5 mm. Of course, your numbers may vary, especially with age. While the irises of a 20-year-old may dilate to 7 mm, they may only expand to 4 or 5 mm for someone 50 or 60 years old.

A shift in the eye's chemical balance also occurs, but much more slowly. The build-up of a chemical substance called rhodopsin, or visual purple, increases the sensitivity of the rods. Most people's eyes become adjusted to the dark in 20 to 30 minutes, although some require as few as 10 minutes or as long as one hour.

The center of the retina, the *fovea centralis*, is only made up of bright-light receptors, or cones. As a result, apart from the blind spot itself, the center of our view turns out to be the least sensitive area of the eye to dim light. Knowing this is especially important when looking for faint, diffuse objects, like comets and many deep-sky objects. By averting our vision, looking a little to one side or the other, rather than staring directly at a faint object, the target's feeble light is directed onto the peripheral area of the retina, which is rich in dim-lighting sensing rods. Figure 1.2 shows just how big a difference averted vision can make.

There's more to averted vision than just looking to one side or the other. Studies show that the retina is most sensitive to dim light in the direction of your nose, and least sensitive toward your ear, in the direction of your eye's blind spot. The areas above and below your center of vision are also not quite as sensitive.

Have you ever caught yourself holding your breath as you search for a difficult target through your telescope? I know that I have. Oxygen deprivation, even if for only a few seconds, can actually desensitize your eyes, so keep breathing. In fact, some observers find that by breathing deeply for 10 to 15 seconds before peering into an eyepiece, and continuing to breathe normally once in position, actually accentuates faint objects. Be sure to keep this mind if you are observing at altitude.

Sometimes, averted vision alone is not enough. Another way to detect difficult objects is to tap the side of the telescope tube very gently. The eye's peripheral vision is also very sensitive to motion, so a slight side-to-side motion will often reveal marginally visible objects, even if only for a moment.

Our ability to perceive color involves the complex interaction of the wavelengths across the visible spectrum and the human visual perception. If our visible window was restricted to only one particular wavelength, then our perception would be restricted to that one color. For example, if the human eye was only sensitive to energy at 550 nanometers ("nm" is the abbreviation for nanometer, a very small unit of measure: one nanometer is equal to 10^{-9} meters or 10 angstrom (Å) units), then our world would appear only as varying intensities of yellow. If it were stimulated only at 485 nm, then our world would appear blue, and so on. Our eyes' ability to perceive the wavelength composition of light is critical to the sensation of color perception.

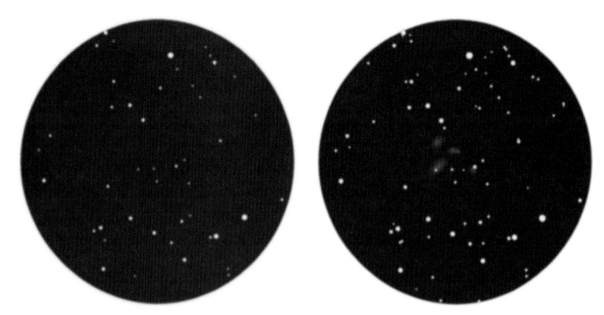

Figure 1.2 Direct vision (left) versus averted vision (right). Looking to one side rather than directly at a target can make all the difference in the world

While the eye's sensitivity to dim lighting increases dramatically during the dark adaptation process, it becomes nearly color-blind. That's why, at night, everything looks gray, varying only in light and dark intensity. Sadly, despite what some books and telescope promotional materials may imply, we visual observers will never enjoy the vibrant colors that are captured in astronomical photographs.

Most extended deep-sky objects display little color, apart from the greenish and bluish tints of some brighter planetary nebulae, and the blue, yellow, and reddish-orange tinges of some star clusters. The most notorious deep-sky objects to spot visually are emission nebulae. For instance, take a look at a photograph of M42, the Great Nebula in Orion, and you will immediately see a labyrinth of red filaments excited into fluorescence by a colorful brood of blue-white stars that are snared within. But when we look at M42 through our telescopes, all we see is a grayish image, perhaps with a hint of green through 8-inch and larger apertures.

Where's all that red? True, M42 will show the most subtle ruddy tinge along its fringe in 12-inchers and up, but that is about it. That's because, unfortunately, of all the colors in the visible portion of the electromagnetic spectrum, our eyes are least sensitive to red. Even the greenish tint to M42 is an exception to the rule. The vast majority of emission nebulae appear as vague grayish blurs to the eye, barely above the blackness of the surrounding sky. Cameras do not have the restricted color perception that plagues our eyes, and so can record nebulae in all of their colorful splendor given the proper exposure. So, while we can marvel at these glorious targets in books and magazines, we can only imagine their true magnificence when trying to spot them for ourselves.

Your binoculars

I've long preached that, when it comes to stargazing, two eyes are better than one. Using both eyes is not only more relaxing than squinting through a conventional telescope, it also has a more natural feel to it. Beyond the aesthetic appeal, ophthalmological studies prove that binocular vision increases the observer's perception of challenging objects. This effect is referred to as *binocular summation*. Depending on the type of object being viewed, the brain processes up to 40% more information using both eyes than just one.

The studies show that the improvement stems from how our brains process the information received from our eyes. The reason for this is actually two-fold. First, with two signals, any "noise" occurring along one stream will be canceled out by the second stream, thereby improving perception. The second advantage is that, when some of the cells in the brain's visual cortex receive two almost identical signals simultaneously,

Table 1.1 *Binocular versus monocular vision*

Binocular aperture	Equivalent telescope aperture
30	36
35	42
40	47
42	50
50	59
56	66
63	75
70	83
80	95
90	107
100	119
125	148
150	178

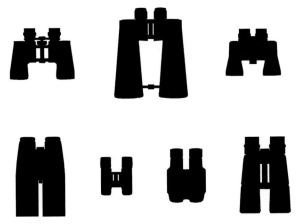

Figure 1.3 *Binocular silhouettes. Binoculars come in different shapes and sizes, though all share one of two basic designs. The top three silhouettes show typical Porro-prism binoculars, while the bottom four illustrate the roof prism design*

they show a greater level of activity than simply doubling that from one eye alone.

How much of an advantage does binocular vision offer over a one-eyed telescope? Despite the term, *binocular summation* does not simply mean adding the apertures of a pair of binoculars together. An 80-mm binocular is not equivalent to a 160-mm telescope. Instead, we must look at the total light-gathering area. To calculate the true advantage of using two eyes over one, plug the binocular's aperture (A) into the following formula:

$$\text{Equivalent telescope aperture} = \sqrt{(A^2 \times 1.41)}$$

Table 1.1 summarizes findings for most popular binocular apertures. Therefore, a pair of 80-mm binoculars is equivalent to a single 95-mm telescope operating at the same magnification. Magnification is key, since different values, and the different exit pupils that result, will affect performance greatly. The *exit pupil* is the circle of light exiting each eyepiece. This topic will be addressed in the section discussing eyepieces later in this chapter.

While it is true that just about any pair of binoculars (Figure 1.3) will show the night sky in greater depth than viewing by eye alone, some binoculars are more suitable to stargazing than others.

Let's begin with the optics. Are the lenses coated? Optical coatings reduce lens flare and improve light transmission, two desirable characteristics for astronomical binoculars. A plain, uncoated lens reflects about 4% of the light striking it, reducing image contrast and causing the black background to appear grayish. By applying a microscopically thin coating of magnesium fluoride (abbreviated MgFl) to both sides of a lens, reflection is reduced to about 1.5%. The optics in top-end binoculars receive multiple anti-reflection coatings. These reduce reflection to less than 0.5%, producing even finer views.

Manufacturers usually state on the binocular tailstock the type of optical coating, but don't just take their word for it. To check for yourself, hold your binoculars at arm's length and look into the objective lenses at a narrow angle. What color do you see? A lens coated with a single layer of magnesium fluoride will show a bluish or purplish tint, while multicoated lenses have a greenish tint. Uncoated lenses have a whitish glint.

Unfortunately, some low-end binoculars that state "fully multicoated optics" are not fully multicoated. Often, the outer surfaces of the objectives are multicoated, but internal optical surfaces are only single coated. Yes, this is false advertising, but unfortunately the only way to check for sure is to disassemble the binoculars and look for yourself. This is not recommended!

Next, let's talk about binocular prisms. Binoculars use one of two different types of prism assemblies to flip images right-side-up. Porro prisms are more common than roof prisms, especially for astronomical viewing. Low-end roof prisms usually generate dimmer images than similarly priced Porro prisms owing to their need to have one prism face aluminized.

The best Porro-prism binoculars use prisms made of BaK-4 glass, while less expensive models use BK-7 glass. The index of refraction of BaK-4 allows total internal reflection of all light entering the prisms. That is, all the light entering the prisms exits into the eyepieces. The reflection properties of BK-7 Porro prisms require that one face of the prisms be aluminized in order to reflect light. The result, as with inexpensive roof-prism binoculars, is dimmer images.

Again, the manufacturer should state the type of prism glass used on the binocular's tailpiece, but to see for yourself hold the binoculars at arm's length and look at the circle of light (the exit pupil) exiting each eyepiece. Do you see a clear circle, or is there a grayish diamond shape within? Clear circles indicate that the pair's prisms are made of BaK-4, while the diamond effect is caused by the light fall-off from using BK-7 glass.

While it is important to look for the right features when buying a pair of binoculars for stargazing, it is just as critical that the binoculars match the person using them. As was discussed earlier in this chapter, the amount of light entering the eye is controlled by the pupil's iris. To be perfectly matched to an observer for nighttime sky watching, the exit pupil should be no larger than the dilated diameter of the observer's eyes. Too large an exit pupil will diminish image contrast by washing out the background.

The exit pupil for a particular pair of binoculars is determined by dividing the objective's diameter by the power. For example, a 7×50 binocular has a 7.1-mm exit pupil, while 10×50 glasses have a 5-mm exit pupil.

The binoculars' objective diameter and resulting exit pupil should closely match the conditions under which they will most frequently be used. If you spend most of your observing time under dark rural skies, then your eyes' pupils should dilate fully to about 7 mm. Therefore, a diameter/power combination producing a matching 7-mm exit pupil should be chosen.

Depending on your age as well as the level of light pollution at your observing site, your pupils may never dilate beyond 4 or 5 mm. In these instances, 10×50 binoculars, with their 5-mm exit pupils, will produce superior results than similar quality 7×50s, even though their apertures are the same.

Even the mythical "finest binocular in the world" is of little value if the user cannot hold it steadily. That is

why, when trying to get the most out of any binocular (even those that are image-stabilized), always place it on some sort of external support. Not only will this steady the view, it will also allow the observer to go back and forth between binocular and star chart when hunting for a particularly challenging target.

The traditional choice has always been a conventional camera tripod, but not all tripods are suitable for the task at hand. Tripods are typically designed to aim a camera toward a subject that is, more or less, the same height above the horizon as the camera itself. As binocular astronomers, however, we have higher aspirations than that. Because we spend most of our time looking skyward, most tripods prove too short to view the sky comfortably, especially when aiming near the zenith.

Instead of, or perhaps in combination with, a tripod, use a mount that offsets the binoculars away from the tripod legs. The favorite design, called a "flexible parallelogram" mount, works on the same basic principle as swing-arm desk lamps. The binoculars are attached to the end of a pivoting beam that allows the glasses to be pointed anywhere between horizon and zenith. Once the binoculars are aimed toward a target, the height of the eyepieces is raised or lowered to a comfortable position without affecting aim.

Many binocularists prefer mirror-based mounts. With these, the binoculars are mounted on a tabletop bracket that is tilted down toward a large flat mirror. The mirror is angled skyward, allowing it to be aimed toward any part of the sky without the observer craning his or her neck. While I admit the design has some attractiveness, these binocular mounts are compromises at best. For one, all commercial mirror-style binocular mounts that I know of use only standard aluminizing, which reflects between 85% and 88% of the light striking them. This effectively reduces the binoculars' aperture. They also flip images around, negating one of the greatest appeals of binocular astronomy: that the binoculars act as direct extensions of our eyes. They are prone to dewing over in damp environments.

Your telescope

Telescopes are like some children. Both need to be coddled and fussed over all the time. You have to spoil them if you want to get the most out of them. That goes for when you are viewing through them as well as when you are not. In fact, the latter is actually more important than the former.

Unless a telescope is stored correctly, dust, grime, and other contaminants will impair (indeed, ruin) the telescope in surprisingly little time. Unless a telescope is stored in a cool, dry place, with its optics dry and properly sealed, its optics will quickly turn into an expensive Petri dish cultivating all sorts of biological activity. On the other hand, a telescope that is properly cared for between uses can easily outlast its owner.

It sounds obvious, but when you are not using your telescope, keep it sealed. This is usually just a matter of putting the dust cap that came with the telescope over the front of the tube and capping the focuser. If your telescope's dust cap is long gone, use a plastic shower cap. Further, if your telescope or binoculars came with a case, use it as a second seal against dust as well as a cushion against bumps.

Never cap and seal a telescope, however, before it has had time to dry. If you live in a humid environment, remember that a dark, dank telescope is the perfect breeding ground for mold and mildew. If the outside of the tube is damp after a night under the stars, then the insides probably are, as well. To help the telescope dry without condensation puddling on the mirror or lens, tilt the tube at about 45° and leave it that way, uncapped, until the morning.

The same advice applies to eyepieces. I recently received an email from a new amateur astronomer named Terry, who was concerned that his eyepieces were damaged. He wrote, "I was looking at the Moon the other night when I noticed what appeared to be something on or in the eyepiece. After careful examination, I found what I can only describe as something growing on one of the glass surfaces. I always keep them capped in 'bolt cases' and inside a carrying case. Do I need to let them dry out more before I put them away?" The quick answer to Terry's question is "yes!" Never store a telescope, binoculars, eyepieces, or any other optical equipment unless you know that it is dry. Even then, it's not a bad idea to keep a small pouch of desiccant in your eyepiece or binocular case.

Even if you are diligent about storing your optics correctly, dust is bound to accumulate. Many amateurs are surprised to learn that even a moderate amount of dust will have very little effect on a telescope's performance. Mildew, however, will result in dimmer, hazier images.

Collimation

Many amateurs complain that their telescope optics just don't have what it takes. Instead, they curse the manufacturer for selling them a lemon. But I have found that, more often than not, the optics are perfectly fine. The only thing that is wrong is that they are out of collimation.

Some telescope designs are more easily knocked out of collimation than others. Refractors should never go out of collimation unless the telescope tube has become bent or warped, usually as a result of mishandling by the user. Catadioptric telescopes should also stand up well, but reflectors should be checked before each session.

To see if your telescope's optics are properly collimated, place a bright star in the exact center of a high-power eyepiece (say, 200× or more). Defocus the image so that the star expands into a disk. If the telescope is properly collimated, then the disk should appear perfectly round. The silhouette of the secondary mirror should be smack dab in the middle of things through reflectors and catadioptric instruments.

If, however, the defocused star appears oval or lop-sided – and it is perfectly centered in view – then the instrument is in need of adjustment.

Collimating a Newtonian

To collimate a telescope accurately, you're going to need the proper tools. At a minimum, you will need a sight tube, which consists of a long empty tube with a set of crosshairs at one end and a peephole at the other.

Check your Newtonian's collimation by aiming the telescope toward a brightly lit wall or the daytime sky (far from the Sun, please). Rack the focuser in as far as it will go, insert the sight tube, and take a look. Ideally, you should see the secondary mirror centered in the crosshairs. If you do, skip to the next step; if not, then the secondary must be adjusted. Move to the front of the telescope tube and loosen the nut holding the secondary's central bolt in place. Slide the secondary assembly in and out along the telescope's optical axis until it appears centered[1] in the sight tube. Before

---

[1] Technically, the secondary should be offset ever-so-slightly toward the primary, rather than centered directly under the focuser. To learn more about this fine point of Newtonian collimation, review the discussion in D. Kriege and R. Berry, *The Dobsonian Telescope* (Willmann-Bell, 1997).

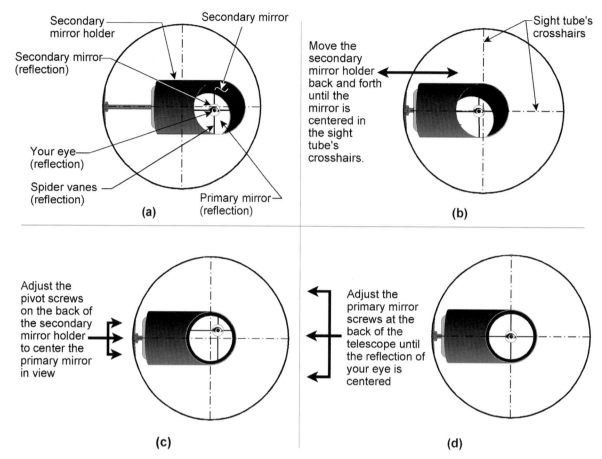

Figure 1.4 Newtonian collimation. Collimating a Newtonian reflector is as easy as 1–2–3. (a) The view through an uncollimated telescope. (b) Adjust the secondary mirror's central post until the mirror is centered under the focuser tube. (c) Turn the secondary mirror's three (possibly four) adjustment screws until the reflection of the primary mirror is centered. (d) Finally, adjust the primary mirror until the reflected image of your eye is centered in view

tightening the nut, check to make sure that the diagonal is not rotated left or right; it should appear perfectly circular. When done with this step, the view through the sight tube should look like Figure 1.4b.

With the diagonal centered under the focuser, look through the sight tube at the reflection of the primary mirror. You should see the end of the telescope tube and at least part of the mirror. Most secondary mirror mounts have three equally spaced screws that, when turned, pivot the secondary's angle. Alternately loosen and tighten these three screws until you see the end of the telescope tube perfectly centered in the secondary. Don't worry if the primary isn't centered; we will take care of that in a moment. When properly aligned, the view after this step should look like Figure 1.4c.

Finally, it is time to adjust the primary mirror's tilt, as in Figure 1.4d. Look at the back of the primary mirror's cell. There should be three, sometimes six, screws. These adjust the tilt of the primary. Go back and forth between the sight tube and these adjustment screws, turning only one at a time slowly, until the primary's reflection is centered in the sight tube's crosshairs. Many companies place a center dot or ring in the exact middle of the primary to make this step easier. (If your primary mirror mount has three sets of two screws, one in each pair adjusts the mirror, while the other presses against the mirror's cell to keep it from rocking. If your telescope uses this type of mirror mount, the three locking screws must be loosened slightly before any adjustment can be made.)

If your Newtonian's focal ratio is *f*/7 or slower (that is, has a higher *f* number), you're done. If, however, your Newtonian is faster than *f*/7, a second tool, called a Cheshire eyepiece, is strongly recommended for more precise alignment. Like a sight tube, a Cheshire eyepiece contains no optics. Instead, a hole in the side of the Cheshire's barrel reveals a polished metal surface cut at a 45° angle. A small hole centered in that surface opens to a peephole at the top of the tool.

To understand how a Cheshire eyepiece works, insert it into your telescope's focuser and shine a flashlight beam into the eyepiece's side opening while you look through the peephole. Centered in the dark silhouette of the diagonal, you should see a bright donut of light. That's the reflection of the Cheshire's polished surface off the primary mirror. The dark center is actually the hole in that surface. Adjust the primary's tilt until its black reference spot is centered in the Cheshire eyepiece's donut. That's it.

Collimating a Schmidt–Cassegrain

Unlike a Newtonian reflector, where both the primary and secondary mirrors can be readily accessed for collimation, commercially made Schmidt–Cassegrain telescopes (SCTs) have their primary mirrors set at the factory. Fortunately, SCTs are rugged enough to put up with the minor bumps that might occur during set-up without affecting collimation. The secondary mirror, however, should be checked often to see if all is well.

The best way to check an SCT for collimation is to aim at a star and slightly defocus the image. Ideally, the diffraction rings surrounding the defocused star should be perfectly centered with the star perfectly centered in view, as in Figure 1.5a. If not, then something needs to be done.

A Schmidt–Cassegrain's secondary mirror floats in the middle of the telescope's corrector plate, held in place by three equally spaced adjustment screws. (Some models hide the adjustment screws with a plastic cover. You'll need to remove this cover, very carefully, to find the screws.) You change the tilt of the secondary by turning those screws in and out.

Slowly turn one of the adjustment screws no more than a quarter turn. Return the star to the center of view and take a look. Are the rings centered? If not, check which direction they are off and turn the corresponding screw slightly in or out, then recheck. Did that help or

hurt? If collimation is still off, return and try turning another screw. Keep going back and forth until everything is centered, but do not just work one screw. Turn two, possibly all three, to avoid loosening or tightening one screw too much.

If the image is still not correctly aligned even after repeated attempts, then there is a distinct possibility that the primary is not square to the secondary. Focus on a rich star field. If any coma (ellipticity) is evident around the stars at the center of view, then chances are good that the primary is angled incorrectly. In this case, your only alternative is to contact either the dealer where the telescope was purchased or the manufacturer.

Baffling/flocking

While ensuring that a telescope's optics are collimated correctly is the single best thing that an observer can do to get optimal performance out of an instrument, it's really just the beginning. Most commercial telescopes can be made to outperform themselves by adding a few enhancements.

First, let's talk baffling. By baffling your telescope properly, you can enhance image contrast greatly, improving the odds of seeing dim, low-contrast objects. Most refractors and catadioptric telescopes have internal baffles (Figure 1.6) that are designed to keep stray light from washing out the in-view target. The dew shield that sticks out in front of a refractor's objective does double duty as a shield against lens fogging as well as a baffle to keep errant light from a foreign source off the objective. The vast majority of catadioptric scopes do not come supplied with a dew shield, even though they need one just as much as any refractor. Newtonian and Cassegrain reflectors do not come with these as standard equipment either, but both would benefit from one, as well.

To be of maximum benefit, an external dew/light shield should extend in front of a telescope a distance equal to the aperture, as in Figure 1.7. For example, an 8-inch telescope should have an 8-inch long dew/light shield protruding in front of it in order to block stray light properly. Let's say that 8-inch telescope is a Schmidt–Cassegrain, with its corrector plate set back just a few millimeters from the front edge of the tube. That telescope's dew/light shield needs to measure a full 8 inches in length, plus enough extra material for the shield to slip over the tube and be held in place.

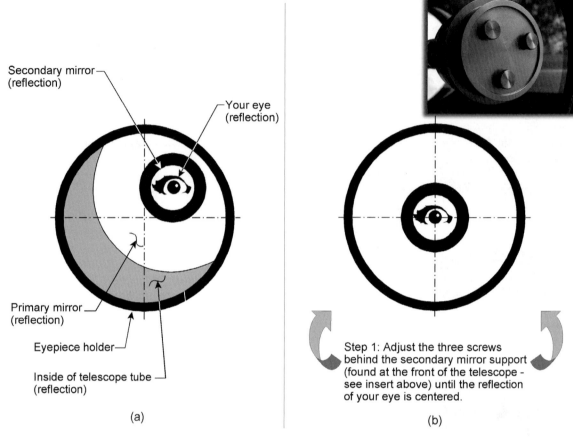

Secondary mirror
(reflection)

Your eye
(reflection)

Primary mirror
(reflection)

Eyepiece holder

Inside of telescope tube
(reflection)

(a)

Step 1: Adjust the three screws
behind the secondary mirror support
(found at the front of the telescope -
see insert above) until the reflection
of your eye is centered.

(b)

Figure 1.5 SCT collimation. Collimating a Schmidt–Cassegrain involves adjusting the secondary mirror (a) until its reflection appears centered in the primary as you look through the empty focuser (b). The insert at upper right shows an example of the secondary mirror's three adjustment knobs. Insert photo courtesy of Bob's Knobs

Now, let's say the 8-inch telescope is a Newtonian reflector. Measure how far back the focuser is located behind the front edge of the tube. For this example, let's say that it is located 2 inches back from the front of the tube. The light/dew shield here only needs to project 6 inches in front of the tube to meet our criterion.

There are many aftermarket light shields available for sale. The most convenient models are foldable for easy storage when not in use. Alternatively, you can make your own from several readily available materials. I have a shield on my own 18-inch Newtonian made from Kydex, a plasticized material commonly used for upper cage assemblies of truss-style Newtonian reflectors. Strips of adhesive-backed Velcro® hook-and-loop material holds the shield in place.

Just as light can sneak into a Newtonian's focal plane from the front, it can also enter through the tail

end of the instrument as well, around the primary mirror. To combat this problem, some Newtonians come with a metal plate sealing the end of the tube. While this certainly blocks light, the plate also blocks air from entering, slowing the time it takes the mirror to acclimate to the cool night air. Instead, I recommend fashioning a cover of opaque black cloth, such as nylon, that fits over the tube. Nylon will allow the tube to breathe while effectively preventing light from entering.

Most refractors, Cassegrain reflectors, and catadioptric telescopes have internal light baffles in order to maximize image contrast, which is crucial to seeing dim objects. Surprisingly few observers realize that Newtonian reflectors can also benefit from internal baffling.

To evaluate your own Newtonian, take a look at the inside of the optical tube assembly. Every mechanical

Figure 1.8 *Flocking opposite focuser. By adding flocking paper at critical interior spots within a telescope tube, such as opposite the focuser, image contrast can be improved*

Figure 1.6 *Internal baffles. A properly baffled telescope, such as this Cassegrain reflector, is a must for observing many of the challenging objects described in later chapters*

Figure 1.7 *Dew shield. A dew shield can not only slow fogging of a telescope's optics, it can also help block extraneous light from washing out the field of view*

component, from the inside of the tube itself to every nut and bolt, should be painted flat black to minimize any stray reflections bouncing onto the optical components. Be sure to look through the empty focuser at the secondary mirror holder, as well. If you see anything that is not painted flat black, touch it up, taking great care not to drip paint where you don't want it!

Flat black paint still reflects some light that strikes it. To minimize internal reflections and maximize image contrast further, many amateurs *flock* the insides of their Newtonian tube assemblies. The term "flock" in this case has nothing to do with gathering sheep. Instead, flocking a telescope tube means altering the inside surface in some manner to eliminate stray light from bouncing around and through the instrument's optical path and into the focal plane.

Flocking the entire interior of a telescope tube is not necessary to enjoy the benefits of improved contrast. In a Newtonian reflector, the critical areas are the tube wall directly opposite the focuser (Figure 1.8) and immediately forward of the primary mirror. The best material to use is adhesive-backed flock paper. Flock paper has a velvety feel to it and reflects less than 1% of the light striking it. Alternatively, low-pile adhesive-backed velvet or felt can be used, although it is not necessarily as effective at squelching unwanted light. Flock paper is available from several specialized sources, including Protostar (www.fpi-protostar.com; 866−227−6240), while velvet and felt can be found at most arts and crafts stores.

To begin, measure the area of the tube that you want to cover. Lay the material out on a smooth cutting surface, flock side down, and transfer your measurement onto the peel-away paper backing. When the material has been cut to size, lay it inside the tube to check the fit, but do not remove the backing. Once all is correct, peel back about an inch of the paper backing along one "tube length" edge and slide the flock paper back into the tube, taking care not to let the adhesive touch the tube surface before it is in position. If you are flocking to the edge of the tube, allow a little paper to protrude out of the tube itself; trim back this overlap after the paper is in place. Slowly and carefully peel away more of the paper backing and press the adhesive into place. Start in the center of the sheet and work your way out toward the two edges to prevent wrinkling. Avoid the temptation to stretch the paper into place, as that will also cause wrinkles. If one sheet of paper is not long enough to line the circumference of your telescope tube, begin a second sheet in the same manner. Be sure to allow enough of an overlap seam to ensure good adhesion.

Many people have successfully flocked the inside of their catadioptric telescopes, as well. Since this operation requires the removal of the corrector plate, however, I would caution that doing so probably voids the telescope's warranty. It also carries some inherent risks if the telescope is not reassembled correctly. If you still want to do it, make sure to mark the exact orientation of the corrector plate relative to the tube. The orientation of the corrector plate is usually matched precisely with the primary mirror on an optical bench at the factory. If the corrector is put back in place, but rotated left or right relative to its original position, telescope performance will most likely suffer noticeably.

Limiting magnitude

A telescope's ability to reveal faint objects depends primarily on the area of its objective lens or primary mirror. Quite simply, the larger the aperture, the more light gathered. Recall from school that the area of a circle is equal to its radius squared multiplied by pi (approximately 3.14). For example, the prime optic in a 6-inch telescope has a light-gathering area of 28.3 square inches (since $3 \times 3 \times 3.14 = 28.3$). Doubling the aperture to 12 inches expands the light-gathering area to 113.1 square inches, an increase of 400%. Tripling it to 18 inches nets an increase of 900%, to 254.5 square inches.

A telescope's "limiting magnitude" is a measure of how faint a star or deep-sky object the instrument will show. Table 1.2 lists limiting magnitudes for several common telescope apertures, but trying to quantify this

Table 1.2 *Limiting magnitudes*

Telescope aperture		Faintest magnitude
inches	millimeters	
2	51	10.6
3	76	11.5
4	102	12.1
6	152	13.0
8	203	13.6
10	254	14.1
12.5	318	14.6
14	356	14.8
16	406	15.1
18	457	15.4
20	508	15.6
24	610	16.0
30	762	16.5

value is anything but precise. And also keep in mind that just because, say, an 18-inch telescope might see 15th-magnitude stars, it may not necessarily be able to reveal 15th-magnitude galaxies, owing to their extended size.

A deep-sky object's visibility is more dependent on its surface brightness, or magnitude per unit area, rather than its total integrated magnitude, as catalog values represent. Other factors affecting limiting magnitude include the quality of the telescope's optics, seeing conditions, light pollution, excessive magnification, and the observer's vision and experience. Some may find these numbers are conservative estimates; experienced observers under dark crystalline skies can better these by half a magnitude or more.

Resolving power

The ability to distinguish closely spaced objects is referred to as the telescope's resolving power, and is especially important when viewing subtle planetary detail, small surface markings on the Moon, or tight double stars.

It has often been said that the stars at night are so far away that even the world's largest telescopes can only show them as points of light. That's not entirely true. Regardless of the size, quality, or location of a

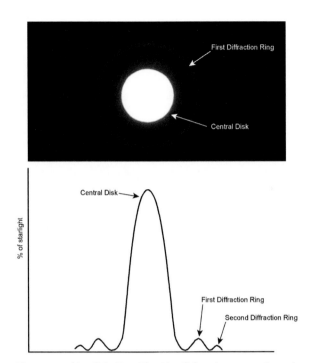

Figure 1.9 Airy disk. Top: The Airy disk as it appears through a highly magnified telescope. Bottom: Graph showing the distribution of light

telescope, stars will never appear as perfect points of light. This is partially due to atmospheric interference, and partially due to the fact that light is emitted in waves rather than mathematically straight beams. Even with perfect atmospheric conditions, each star will appear through a telescope as a tiny disk, referred to as the *Airy disk* in honor of its discoverer, Sir George Biddell Airy (1801–1892).

Owing to the wave property of light, rays from different parts of a telescope's prime optic (be it a mirror or a lens) alternately interfere with and enhance each other, producing a series of dark and bright concentric rings around the Airy disk (Figure 1.9). The whole display is known as a diffraction pattern.

Ideally, through a telescope without a central obstruction (that is, without a secondary mirror, which impacts light distribution), 85% of the starlight remains concentrated in the central Airy disk, with the rest distributed among progressively fainter rings.

The apparent diameter of the Airy disk plays a direct role in determining an instrument's resolving power. This becomes especially critical for observations of close binary stars. Just like determining a telescope's

limiting magnitude, how close a pair of stars will be resolved in a given aperture depends on many variables, but especially optical quality of the telescope and sky conditions.

How close can two stars appear and still be resolvable as two? The single most important factor that influences the result is a telescope's aperture. The larger the aperture, the finer the level of detail resolved. Of the many observational experiments that have been conducted to determine the resolution limits for telescopes, the two most often cited are the Rayleigh Criterion and the Dawes' Limit.

The Rayleigh Criterion, devised by John William Strutt, the third Baron Rayleigh, in 1878, predicts how close two stars can be to each other and still be distinguishable as two separate points. Based on empirical data, the Rayleigh Criterion for any telescope can be calculated using the formula:

Rayleigh Criterion $= 138 \div D$

where D is the aperture in millimeters, and the result is expressed in arc-seconds.

The nineteenth-century English astronomer William Dawes derived a formula for calculating just how close a pair of 6th-magnitude yellow stars can be to each other, and appear elongated, but not separately resolved (Figure 1.10). His formula, known as Dawes' Limit, is:

Dawes' Limit $= 114 \div D$

Again, D is the aperture in millimeters and the result is in arc-seconds.

Table 1.3 lists both values for some common telescope apertures. Some amateurs can readily exceed Dawes' Limit when using telescopes 6 inches in aperture or less, while others will never reach it. That's because a telescope's actual performance can be adversely affected by many factors, including turbulence in our atmosphere, a great disparity in the test stars' colors and/or magnitudes, misaligned or poor-quality optics, and the observer's visual acuity.

A large-aperture telescope (i.e., greater than about 10 inches) rarely reaches its Dawes' Limit. Even the largest backyard instruments can almost never show detail finer than between 0.5 arc-seconds and 1 arc-second because of our atmosphere. In other words, a 16- to 18-inch telescope will offer little additional detail on a planet or better resolve a tight double star than an 8- to 10-inch when used under

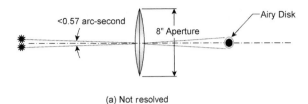

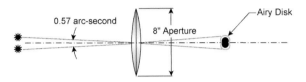

(a) Not resolved

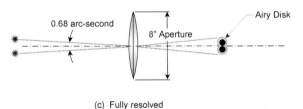

(b) Barely resolved (Dawes' Limit for an 8-inch)

(c) Fully resolved

Figure 1.10 Dawes' Limit. The resolving power of an 8-inch telescope. (a) Not resolved; (b) Barely resolved, or the Dawes' Limit for the aperture; (c) Fully resolved

Table 1.3 *Rayleigh Criterion and Dawes' Limit for common apertures*

Aperture		Rayleigh Criterion	Dawes' Limit
inches	millimeters	arc-seconds	arc-seconds
2	50	2.76	2.32
3	76	1.82	1.53
4	102	1.35	1.14
6	152	0.91	0.76
8	203	0.68	0.57
9.25	235	0.59	0.49
10	254	0.54	0.46
12	305	0.45	0.38
14	356	0.39	0.33
16	406	0.34	0.29
18	457	0.30	0.25
20	508	0.27	0.23
25	635	0.22	0.18
30	762	0.18	0.15

most observing conditions (although the larger telescope will certainly enhance an object's color and brightness). Interpret Dawes' Limit as a telescope's equivalent to the projected gas mileage of an automobile: "your mileage may vary!"

At the other side of the aperture scale, we have binoculars. While it would appear from Table 1.3 that a pair of, say 10×50 binoculars (i.e., 10 power, 50-mm objective lenses) should be able to resolve binary stars that are separated by only 2.3″, in reality, they don't even come close. That's because Dawes' Limit assumes high magnifications far beyond the values of most common binoculars. As a result, binoculars will never approach their theoretical Dawes' Limit. To estimate how tightly a pair of stars can still be resolved with binoculars, it's more appropriate to consider magnification than aperture.

That value, in turn, depends on how closely a pair of stars can be resolved with the unaided eye, assuming 20/20 far vision. In his classic reference *Amateur Astronomer's Handbook*, author J. B. Sidgwick suggests

that 3 arc-minutes is a reasonable value, although he goes on to note that some especially keen observers can better that by nearly a factor of 2. Other sources state that naked-eye observers can expect to resolve two stars no closer than 5 arc-minutes apart.

For our discussion here, let's draw a happy medium and assume, given 20/20 vision, that the human eye can resolve two stars separated by 4 arc-minutes, or 240 arc-seconds. Using this as our basis, we can now estimate the minimum value for a given binocular by dividing its magnification into 240. Table 1.4 lists some common values, rounded to the nearest whole number. Can you do better? Chapters 3 and 4 will offer you a chance to prove it.

Your eyepieces

Just as the correct telescope is critical for seeing a challenging object, so too the choice of eyepiece (Figure 1.11) can either make or break the test. Most people select an eyepiece based solely on the resulting magnification, but I suggest that this is going about it the wrong way. While the correct magnification certainly plays a role when hunting for a particular

Table 1.4 *Binocular resolving limits*

Binocular magnification	Resolution threshold (arc-seconds)
6	40
7	34
8	30
9	27
10	24
11	22
12	20
14	17
15	16
16	15
18	13
20	12
25	10
30	8

Table 1.5 *Recommended exit pupils for various sky targets*

Target	Exit pupil (mm)
Wide star fields under the best dark-sky conditions (e.g., large star clusters, diffuse nebulae, and galaxies)	5 to 7
Smaller deep-sky objects; complete lunar disk	3 to 5
Small, faint deep-sky objects (especially planetary nebulae and smaller galaxies); double stars, lunar detail, and planets on nights of poor seeing	1 to 3
Double stars, lunar detail and planets on exceptional nights	0.5 to 1

or

$$\text{exit pupil} = \text{aperture} \times \text{fl}_{\text{eyepiece}}/\text{fl}_{\text{telescope}}$$

or

$$\text{exit pupil} = \frac{\text{fl}_{\text{eyepiece}}}{\text{telescope focal ratio}}$$

The pupil of the human eye dilates to about 7 mm when acclimated to dark conditions (although, as stated earlier, this varies from one person to the next and shrinks as we age). If an eyepiece's exit pupil exceeds 7 mm, then the observer's eye will be incapable of taking in all of the light that the ocular has to offer. Further, contrast between the target and the background sky will be diminished, causing a washed-out look. At the other extreme, too small an exit pupil will probably dim the view dramatically, making it nearly impossible to focus the image sharply.

There is no one exit pupil value that is best for every object in the sky, or for that matter every challenge discussed in this book. You might come to your own conclusions that are quite different from mine. Table 1.5 summarizes my personal preferences.

We should not ignore magnification altogether, however. Many observing handbooks cite the maximum-magnification rule of $60\times$ per inch ($23\times$ per centimeter) of aperture. By this thinking, a 6-inch telescope has a magnification ceiling of $360\times$ before image quality degrades due to atmospheric and optical imperfections. This is not a hard and fast law of physics, however. There will be nights when a high-quality telescope will be able to surpass this limit by perhaps a factor of 2. But just as often, and perhaps more often, atmospheric conditions will dictate that no more than

Figure 1.11 Eyepieces. A selection of eyepieces

object, practice shows that the diameter of the exit pupil produced by a given telescope–eyepiece combination is just as critical, if not even more so. As discussed previously, the exit pupil is the diameter of the beam of light that exits an eyepiece and enters the observer's eye.

To determine which eyepiece in your collection is best for a specific target, you must first know the resulting exit pupil. This is easily calculated using any of the following three formulae:

$$\text{exit pupil} = \frac{\text{aperture}}{\text{magnification}}$$

Table 1.6 *Telescope aperture versus maximum magnification*

Telescope aperture		Magnification	
inches	centimeters	Theoretical (60×/inch)	Practical
2.4	6	144	100
3.1	8	186	125
4	10	255	170
6	15	360	240
8	20	480	300
10	25	600	300
12	30.5	720	300
14	36	840	300
16	41	960	300
18	46	1080	300
20	50	1200	300
25	64	1500	300
30	76	1800	300

30× per inch of aperture, possibly even less, be used to maintain image integrity.

Just how much magnification is too much and how much is just right? The choice should be based largely on your personal experience. Table 1.6 summarizes mine.

The table purposely tops out at 300× for all telescopes beyond 8 inches. My personal experience across more than 40 years as an observer shows that little is usually gained by using more than 300× to view an object, regardless of aperture. Although some of the challenges in this book require more than 300×, especially a few in the later chapters, a good rule to follow is to use only as much magnification as needed to see what you are trying to see.

Eyepiece design plays a large role in all this, as well. Just because two eyepieces happen to have the same focal length, and therefore yield the same magnification and exit pupil, does not mean that they will work equally well on all targets. That's where an eyepiece's optical design comes into play. Today, we have a wide range of designs from which to choose. Which is best for one class of object might be quite different from which is best for another class of object.

Based again on my own personal experiences as well as those of other veteran observers, I will offer the following suggestions. But first, a disclaimer: these

choices are quite subjective and, as such, can differ quite dramatically from person to person. Therefore, if you find that a different eyepiece design works best for you, by all means continuing using it.

Thus, for objects that require the highest possible sharpness and contrast, such as fine planetary detail or a tight binary star, use premium orthoscopic and Plössl eyepieces. Top-end monocentric eyepieces are also strong contenders in this category, although personally I find their very narrow fields of view aesthetically unpleasant. Image contrast is also quite good in some hybrid eyepiece designs, although their wider fields of view are not normally needed for small details. Unless, of course, your telescope is not motorized; then, the large fields are welcome, since an object will stay in view longer before the scope has to be nudged to keep up with Earth's rotation.

When it comes to hunting for faint galaxies, planetary nebulae, and other small-scale deep-sky objects, it's tough to beat premium hybrid eyepieces that have apparent fields of view in the 60°–70° range. Larger-scale objects, such as many of the bright and dark nebulae discussed later, benefit from super-wide eyepieces with apparent fields ranging from 80° to 100°. Again, consider both the magnification as well as the exit pupil. You'll find that, if given a choice between a 25-mm eyepiece with a 40° apparent field of view and an 18-mm eyepiece with a 56° apparent field, the 18-mm will have better image contrast even though their real fields of view are approximately the same. That's thanks to the 18-mm's smaller exit pupil.

Many of the challenges to follow pit a faint target against the overwhelming glare of a nearby star or planet. In that case, clean optics and exceptional skies are prerequisites. Beyond that, however, seeing the target will probably require that you do something to lower or eliminate the glare from the detractor. To do that, many observers move the offender off the edge of the eyepiece field. That may be fine but, in many cases, offsetting the glare will also bring the target to just inside the field edge where, depending on the eyepiece, distortion may blur it out of existence.

Instead, use an eyepiece fitted with an occulting bar across the center to block the detractor. Although no company sells an occulting eyepiece, it is easy enough to make one at home. To work correctly, the edge of the occulting bar must appear sharp in view, which means that it must lie at the eyepiece's focal plane. This usually coincides with the metal field stop near the field lens. Begin by cutting a thin strip of opaque black photographic tape, no more than a third of the width of the field lens and just long enough to span the inside diameter of the eyepiece barrel. (Recommendation: Use a low-power eyepiece with a relatively large field lens, and then insert it into a Barlow lens to get the high magnification needed for the task.) Holding the tape gently with tweezers, carefully lay it into place.

Your filters

Just as photographers use different filters on their cameras to create different effects, amateur astronomers can also use filters when viewing the night sky. Using the correct filter will help reveal the object you're trying to see, but using the wrong filter could well make it disappear.

Before we delve deeper into the subject, let's understand what filters will do and what they won't. Filters will not add to the view; they only subtract. Depending on its tinting or coating, a filter will transmit those specific wavelengths of light, while suppressing others. The net result is that subtle detail on the surface of a planet or perhaps the dim glow of a distant nebula will become visible by isolating its wavelengths of light from among the electromagnetic throng.

Color filters

Color filters are the simplest and least expensive filters available to amateur astronomers. These are also called lunar and planetary filters by some, since that is their true strength. Viewing planetary surfaces through color filters can greatly reduce the distortion of the boundary between a lighter area and a darker region on the Moon's or a planet's surface, an effect called *irradiation*. They also help increase subtle contrasts between two adjacent regions of a planet or the Moon by transmitting one color while absorbing the other.

While we can certainly refer to a color filter by its inherent tint, it is much more accurate to speak of filters by their Wratten number, a system originally devised by Frederick Wratten a century ago. For example, when talking about a red filter, we might be referring to a Wratten #23A (light red), #24 (red), or #25 (deep red) filter. The complete spectrum of Wratten filters includes over 100 individual shades; however, we can distill it down to only a handful that prove worthwhile for our astronomical needs. Table 1.7 gives a broad overview of their application.

Bear in mind that even these color filters are not magical incantations summoned up by some mystical power. While filters will change the appearance of an object dramatically (a #58 green filter will make you think the Moon really is made out of green cheese!), the impact they have on enhancing specific features is actually quite subtle. There are many experienced planetary observers who actually frown on using color filters at all. Other, equally talented planet-watchers swear by them.

If you are interested in experimenting with color filters, start out slowly. Choose four or five basic colors, such as deep yellow (#15), red (#23), green (#58), and medium blue (#80A), but watch those transmission percentages. As you increase magnification to see fine planetary details, keep in mind that image brightness is going to fall off quickly. Introducing a filter is only going to exacerbate the situation. In general, if you are using a 6-inch or smaller instrument, avoid planetary filters with transmission percentages that are lower than 40%. In that case, substitute another filter. A #82A might be more suitable than a #80A in a 6-inch.

Some filters do double duty, perhaps even better than their manufacturers realize. For instance, I actually prefer narrowband deep-sky filters, discussed below, to color filters when viewing Mars and Jupiter. I find that these filters bring out subtle layering in Jupiter's bands as well as Martian surface detail better than the more traditional orange and red filters listed in the previous table.

At the same time, a #82A light blue filter can also be used as a poor man's broadband light-pollution reduction filter, since it suppresses light in the yellow portion of the visible spectrum, the wavelength where many artificial lights, including the ubiquitous high-pressure sodium streetlight, shine most strongly.

My best advice is never to stick to convention and take someone's word for something (even the author of a book!). Sure, use the advice that you garner from others, but never be afraid to try using a piece of astronomical equipment in an unconventional way. You just may stumble upon something unexpected.

Deep-sky object filters

As amateur astronomers everywhere are painfully aware, the problem of light pollution is rapidly swallowing up our skies. In regions across the country and around the world, the onslaught of civilization has reduced the flood of starlight to a mere trickle of what it

Table 1.7 *Color filters: a comparison*

Filter				
Wratten number	Color	Transmission percentage	Object	Benefit
Moon filter	Neutral density	Note 1	Moon	Reduces brightness of Moon evenly across the spectrum
Polarizer	Neutral density	Note 2	Moon	Like the neutral density filter, reduces brightness without introducing false colors
			Mercury	Darkens sky background to increase contrast of planet; helpful for determining phase of Mercury
			Venus	Reduces glare without adding artificial color (especially helpful for viewing planet through larger telescopes)
11	Yellow-green	78%	Jupiter	Reveals fine details in Jovian cloud bands
15	Deep yellow	67%	Moon	Enhances contrast of lunar surface
			Saturn	Helps to reveal Saturnian cloud bands
21	Orange	46%	Mercury	Helps to see planet's phases
			Mars	Penetrates Martian atmosphere to reveal reddish areas and highlight surface features such as plains; best for 6-inch and smaller scopes
			Jupiter	Accentuates Jovian cloud bands
			Saturn	Helps to reveal Saturnian cloud bands
23A	Light red	25%	Mercury	Increases contrast of planet against blue sky, aiding in daytime or bright twilight observation
			Mars	Penetrates Martian atmosphere to reveal reddish areas and highlight surface features such as plains; best for 8- to12.5-inch scopes
25	Deep red	14%	Mercury	Increases contrast of planet against blue sky, aiding in daytime or bright twilight observation
			Venus	Darkens background to reduce glare; some say it also helps reveal subtle cloud markings
			Mars	Penetrates Martian atmosphere to reveal reddish areas and highlight surface features such as plains; best for scopes larger than 12.5 inches
38A	Deep blue	17%	Mars	Brings out dust storms on surface of Mars
56	Light green	53%	Jupiter	Accentuates reddish Jovian features such as the Red Spot
58	Green	24%	Moon	Enhances contrast of lunar surface
			Mars	Accentuates "melt lines" around Martian polar caps
			Jupiter	Accentuates reddish Jovian features such as the Red Spot
80A	Medium blue	30%	Moon	Reduces glare
			Mercury	Improves view of Mercury against bright orange twilight sky
			Venus	Improves view of Venus against bright orange twilight sky
			Mars	Brings out Martian polar caps and high clouds, especially near the planet's limb
			Jupiter	Highlights details in orange and purple belts as well as white ovals in Jovian atmosphere
			Comets	Increases contrast of some comets' tails
82A	Very light blue	73%	Jupiter	Highlights details in orange and purple belts as well as white ovals in Jovian atmosphere

Note 1. Neutral density filters are available in several densities and are usually labeled with their transmission percentage. An ND 50 filter transmits 50% of the light passing through, while an ND 25 filter transmits 25%. Some manufacturers use a logarithmic scale instead; here, an ND 0.3 filter transmits 50%, while an ND 0.6 transmits 25%.
Note 2. Polarizing filters are, effectively, a matched pair of neutral density filters that rotate relative to one another. As the filters are rotated, their combined neutral density filtering changes from low to high.

Figure 1.12 Filters. A collection of light-pollution reduction (LPR) filters and nebula filters

once was. Are we powerless against this beast? Not entirely.

While the dilemma of light pollution is discussed in more specific detail later in this chapter when evaluating observing sites, this section will explain how filters can be used to help counteract part of the problem.

First, let's briefly define light pollution. Light pollution is unwanted illumination of the night sky caused largely by poorly designed or poorly aimed artificial lighting fixtures. Rather than illuminating only their intended terrestrial targets, many fixtures scatter their light in all directions, including up.

There are two kinds of light pollution: local and general. Local light pollution shines directly into the observer's eyes, and may be caused by anything from a nearby streetlight to an inconsiderate neighbor's porch light. There isn't a filter made that will counteract this sort of interference, although this problem is also addressed later under "Strategies and Techniques."

Light-pollution reduction (LPR) filters, such as those shown in Figure 1.12, are much more effective against general light pollution, or "sky glow." Sky glow, the most destructive type of light pollution, is the collective glare from untold hundreds or even thousands of distant lighting fixtures. It can turn a clear blue daytime sky into a yellowish, hazy night sky of limited usefulness.

Although modern technology caused the problem in the first place, it also offers partial redemption. Many sources of light pollution, including high-pressure sodium streetlights, shine in the yellow region of the visible spectrum, between 550 nm and 630 nm. At the same time, planetary nebulae and emission nebulae emit most of their light in the blue-green portion of the

spectrum.[2] Emission nebulae, for example, glow primarily in the hydrogen-beta (486 nm) and oxygen-III (496 nm and 501 nm) regions of the spectrum. In theory, if the yellow wavelengths could somehow be suppressed while the blue-green wavelengths were allowed to pass, then the effect of light pollution would be greatly reduced.

What exactly do light-pollution reduction filters do? A popular misconception is that LPR filters make faint objects look brighter. Not true! LPR filters are designed to block specific wavelengths of "bad" light while letting "good" light pass. The observer need only attach the filter to his or her telescope, usually by screwing the filter into the field end of an eyepiece. The net result is increased contrast between the object under observation and the filter-darkened background.

Light-pollution reduction filters (also called nebula filters) come in three varieties: broadband, narrowband, and line filters. The biggest difference is in their application. As their name implies, broadband filters pass a wide swath of the visible spectrum, from about 430 nm to around 550 nm. Narrowband filters limit their transmission to between about 480 and 520 nm. Line filters have extremely narrow transmission windows, allowing only one or two specific wavelengths of light to pass.

Filters do a very good job at increasing contrast between certain types of deep-sky objects and the background sky, but no one type of filter is best for everything. Some heighten an object's visibility greatly, some have little or no effect, while others can actually make a target fade or disappear completely! Which LPR filter is best under which circumstances remains one of the hottest topics of debate among amateur astronomers today.

If you live under extreme light pollution, consider adding a broadband filter to your equipment arsenal. And if you enjoy searching for tough planetaries or challenging emission nebulae, then either a narrowband or an oxygen-III (O-III) line filter − or both! − is a must-have. Finally, for deep-sky diehards who already own the others, a hydrogen-beta (Hβ) filter will prove handy to have, even though it will

---

[2] Many deep-sky objects also shine in the deep-red hydrogen-alpha (656 nm) portion of the spectrum, a region that is all but invisible to the eye under dim light conditions. Still, most light-pollution reduction filters transmit these wavelengths as well for photographic purposes.

Table 1.8 *Deep-sky filters: a comparison*

Filter	Benefit
Broadband	• Limited usefulness from observing locations that suffer from severe light pollution. No value for those observing from suburban and rural sites.
Narrowband	• Very effective at helping to isolate faint emission and planetary nebulae by suppressing surrounding sky glow. • Accentuate surface features on Mars as well as cloud structure on Jupiter.
Oxygen-III	• Strongly recommended for telescopes of 8 inches aperture and larger. Excellent for isolating faint emission and planetary nebulae.
Hydrogen-beta	• Because of spectral characteristics of these filters, they are of limited value. But, for the few objects that they work well with, including the elusive California and Horsehead Nebulae, they far exceed other filters.

probably see less use than other filters in your collection. Take a look at Table 1.8.

Your observing site

Time was when, to find an observing site, a stargazer would simply step out into the backyard and look up. Sadly, for most of us, those days are a distant, fading memory. Unless you live in the desert southwest of the United States, in the middle of a dense forest, or perhaps at the summit of a mountain, odds are that your backyard may not be the best site for spotting many of the challenging objects to come. Today's amateur astronomer has to take many factors into account when choosing a site. Concerns such as light pollution, overhead air turbulence, security, and accessibility all impact our ability to see the sky in all its glory. To make matters worse, the chances of finding one spot that is perfect for all of the challenges presented in this book are very slim.

If you have been an amateur astronomer for a long time, then you know that a sky forecasted as "clear" by a TV or radio meteorologist may not be clear at all for stargazing. There's much more to it than that. We need to consider temperature, relative humidity, upper-level wind speed and direction, sources of pollution both near and far, suspended particulates, and a host of

other issues that are never discussed on the evening news.

From a stargazer's perspective, sky conditions can be divided into three categories: transparency, seeing, and sky darkness. "Transparency" refers to how clear the sky is, while "seeing" refers to the steadiness of the air mass overhead. Clouds, haze, humidity, and artificial and natural air pollutants all adversely affect both in different ways. Finally, "sky darkness" speaks to the ambient level of background light. Light pollution raises this level. People often confuse the terms transparency and sky darkness. It is certainly possible to have a city sky that is more transparent than a rural sky, but because of the lower level of sky darkness (due to urban light pollution), fainter stars will still be visible from the country site, even with its poorer transparency.

There's much more in the air than just air. There are also small particles, called aerosols, suspended in the air. Atmospheric aerosols come from natural sources such as volcanoes, dust storms, forest fires, and sea spray, as well as artificial sources, including the burning of fossil fuels, and plowing or digging of soil. Individual particles are invisible to the eye, but their combined effect is evident in air that looks dirty or hazy.

Concentrations of aerosols vary significantly with location and season. Most aerosols are in the troposphere, although volcanic aerosols can rise into the stratosphere. The latter may affect a large portion of the globe and can persist for years.

We need only look along the horizon during the day to see their effect. Nautical and aviation forecasts will often quote visibility in terms of miles or kilometers. In the western portions of North America, especially in the southwestern corner of the United States, visibilities in excess of 100 miles (160 km) are not uncommon. Move east of the Mississippi River, however, and the numbers plummet. Further, the seasonal average varies significantly in eastern North America, from only 25–30 miles (32–48 km) during the hazy days of summer to more than 60 miles (97 km) in the winter months.

These numbers only refer to visibility along the horizon. Upper-level aerosols from distant forest fires or volcanic ash may have little adverse effect on the transparency of the lower atmosphere, but still cause a milky sky.

One of the most popular sources for determining just how transparent the North American sky will be at any given time over the next 48 hours is the Clear Sky

Chart website. Each of the hundreds of observing sites listed in the Clear Sky Chart directory uses meteorological data from the Canadian Meteorological Centre to predict if and when it will be clear in the near future. Created by Canadian amateur astronomer Attilla Danko, each Clear Sky Chart graphically predicts cloud cover, transparency, and darkness for that particular site. Check www.cleardarksky.com for more information. For the rest of the world, Chinese amateur Ye Quanzhi offers his 7Timer! website. Pronounced "Tri-Timer", which means "bell of clear sky" in Chinese, the 7Timer! site offers predictions of transparency, cloud cover, and darkness over the next 72 hours. Visit 7timer.y234.cn for details.

The best way to gauge the true air transparency between ground and outer space is by the Aerosol Optical Thickness (AOT) and Aerosol Optical Depth (AOD). Both terms, used synonymously, refer to how much small particles in the atmosphere will affect the passage of light through our atmosphere. The larger the optical thickness, the less light of that wavelength reaches Earth's surface. A typical Aerosol Optical Thickness value in clear air is roughly 0.1. A very clear sky may have an AOT of 0.05 or less. Very hazy skies can have AOT values of 0.5 or greater. There are a number of Internet websites that plot AOT and AOD data. A select few can be found in Appendix B.

Transparency and sky darkness are critical for finding the vast majority of deep-sky challenges described in this book, but not for spotting fine details on the Moon and planets, or splitting tight binary stars. For these tests, higher magnifications are a must. But, as soon as we increase magnification, we also increase the negative impact that air turbulence will have on the view. To see the greatest detail on the Moon or a planet, steady seeing conditions (Figure 1.13) are a must.

Seeing is dependent on many factors, few of which overlap with transparency. Ironically, the circumstances that bring very transparent skies often create poor seeing conditions. Transparent skies are often ushered in by the passage of a cold front, which is often characterized by a turbulent air stream. Steady seeing conditions, evident by the stars' lack of twinkling (or *scintillation*, as the effect is properly known), require a very smooth, laminar of air overhead. "Twinkle, twinkle little star" may be a fondly remembered childhood rhyme, but it's the bane of planetary observers everywhere.

To judge seeing conditions as objectively as possible, the Harvard College Observatory astronomer

Figure 1.13 Seeing. Waiting for the steadiest atmospheric conditions can turn a blurry, boiling target into a sharp object ripe with great detail

William H. Pickering (1858–1938) devised the scale shown in Table 1.9 based on observations he made through a 5-inch (130-mm) refractor. Use this scale as a general guide only. Unless you are also observing through a 5-inch telescope, the statements about the sizes and appearance of diffraction disks and rings must be modified for larger or smaller apertures.

Pickering's scale was derived from the appearance of a star's Airy disk and is a good general way of quantifying the atmospheric impact on a challenging binary star. It is not however, necessarily, a legitimate system for judging planetary seeing. For that, we turn to another seeing scale devised by the acclaimed Greek planetary observer Eugène Michel Antoniadi (1870–1944). Antoniadi, remembered mostly for being one of the first to dispute the existence of Martian canals, created the scale in Table 1.10 that is still used by most lunar and planetary observers to this day.

The air often seems steadiest when the sky is slightly hazy. While any cloudiness may make faint objects invisible, the presence of a thin haze may indicate a smooth, or laminar, air mass overhead. At those times, we can increase our telescopes' magnification and make out finer details in double stars and the planets that are missed at other times because of atmospheric turbulence.

This turbulence is caused by temperature differences in the air. Some of those differences occur miles above Earth's surface, where boundary layers of air masses with different temperatures meet. As previously

Table 1.9 *Pickering seeing scale*

1	Very poor	Star image 2× the diameter of the third diffraction ring – star image 13 arc-seconds in diameter.
2	Very poor	Star image occasionally 2× the diameter of the third ring.
3	Poor to very poor	Star image about the same diameter as the 3rd ring (6.7 arc-seconds) and brighter at the centre.
4	Poor	Airy disk often visible. Arcs of diffraction rings sometimes seen.
5	Fair	Airy disk always visible. Arcs frequently seen.
6	Fair to good	Airy disk always visible. Short arcs constantly seen.
7	Good	Disk sometimes sharply defined. Diffraction rings seen as long arcs or complete circles.
8	Good to excellent	Disk always sharply defined. Rings seen as long arcs/complete circles, always in motion.
9	Excellent	Inner diffraction ring stationary. Outer rings occasionally stationary.
10	Perfect	The complete diffraction pattern is stationary.

Table 1.10 *Antoniadi seeing scale*

I	Perfect seeing, without a quiver
II	Slight undulations, with moments of calm lasting several seconds
III	Moderate seeing, with large tremors
IV	Poor seeing, with constant troublesome undulations
V	Very bad seeing, scarcely allowing the making of a rough sketch

Figure 1.14 Jet stream. The jet stream passing across North America in the summer (top path) and winter (bottom path)

mentioned, this often occurs immediately before or after the passage of a cold front, such as when cold, dry air replaces moist, typically cloudy conditions. These frontal passages may bring outstandingly clear, but very turbulent, skies. Poor seeing may also result from windy or unseasonably cold weather conditions, while good seeing conditions typically occur after a high-pressure system causes clear skies for several days in a row.

These generalizations do not necessarily apply for every location on planet Earth, however. Keep your own weather records over the course of a full year, or even two, comparing conditions, such as temperature, humidity, wind speed and direction, and cloud cover, to seeing steadiness. For consistency, be sure to check these factors using the same equipment.

As with transparency, seeing can also vary quite dramatically from one season to the next. The path of the jet stream (Figure 1.14) can make all the difference. During the summer, the jet stream usually hugs the polar regions, allowing tropical air to infiltrate into middle latitudes. In winter, the jet stream droops toward the middle latitudes, bringing with it colder, but far more turbulent conditions.

Time and location can also either make or break seeing. In many locations, the steadiest conditions occur right after sunset, before radiational cooling begins to throw the atmosphere into chaos. Therefore, look for the challenge objects that require good seeing as soon as you can spot them in evening twilight. Waiting even an hour after sunset may be enough time for atmospheric calmness to worsen. As the evening wears on, nighttime cooling may slow, allowing turbulence to slow. Seeing often improves after midnight, although this is not a hard-and-fast worldwide rule.

Local topography also plays a major part in seeing conditions. Most major observatories are located high atop mountains near an ocean for good reason. Not only does their altitude get them out from under low-lying clouds, but, by facing the telescopes into the prevailing wind direction, the air has been calmed after crossing a cool, flat ocean. Flat plains, including deserts, can also spawn good seeing conditions.

One of the worst places for seeing is in a valley downwind from a mountain. As the wind flows over and around the peak, the air will be churned up before it dives into the valley, causing poor seeing, and even clouds. That's exactly what happened to me during the 1991 July 11 total eclipse from Baja California, Mexico. The tour leader misjudged where we should set up, choosing a site to the east of a local mountain range. As the temperature dropped during the partial phases, clouds formed in this otherwise parched environment, and in the process, cut a 6-minute 58-second total phase in half. Even before clouds did us in, it was clear that seeing conditions were faltering.

The surface that you set a telescope on can also influence seeing conditions, since different materials absorb heat at different rates. Concrete and blacktop are the worst offenders because they readily retain heat. Grass, although also requiring a cool-down period, is better since it does not retain as much heat. It is also always best to choose a location far from buildings, since observing over a roof, or worse still, near an operating chimney, can be ruinous.

Other sources of air turbulence lie even closer to home. The most dramatic temperature conditions are often found only a few millimeters in front of your telescope's objective lens or mirror. If the optics are not at thermal equilibrium with the ambient air temperature, they will be engulfed in an undulating envelope of air that is either warmer or cooler. To help squelch this problem, be sure to allow enough time between set-up and observing for a telescope to become acclimated to the night. For small telescopes, this means a minimum of 30 minutes after bringing them out from a warm (or cool) storage location. The larger the telescope, the longer it takes to acclimate – perhaps significantly longer.

As the night continues to cool, the air often loses its heat faster than your telescope. In these cases, a telescope acclimated to the evening may unexpectedly fall out of thermal balance if the air temperature drops suddenly in the middle of the night. That becomes apparent when, quite suddenly, steady images begin to dance and shimmer. This can be especially destructive in Newtonian reflectors, since their long, open-ended

Figure 1.15 Open mirror cell. By allowing air to circulate around a reflector's primary mirror, a telescope will acclimate to the nighttime temperature more quickly

tubes can act like chimneys channeling warm air currents up into the cool night. The closed tubes of refractors and catadioptric telescopes are not as prone to the chimney effect, but they can still be affected by so-called tube currents.

To help combat the chimney effect, the primary mirror's cell should be an "open design." That means the back end of the telescope is not covered with a metal plate or other obstruction but, rather, shows the underside of the primary, as in Figure 1.15.

An open mirror cell promotes the cooling process, but to speed it along even more, many amateurs have installed small muffin-style cooling fans behind their Newtonian's primary mirrors. In fact, many of the mirror cells in today's Newtonians have predrilled mounting holes for this very purpose. If yours does, check with your telescope manufacturer to see if they sell a specially designed fan for your instrument. If not, muffin fans are commonly available from electronic and computer hardware supply stores, such as Radio Shack.

While placing a fan behind the mirror helps to promote air flow in a Newtonian's tube, many amateurs add a second fan to the side of their primaries to move air across the mirror's surface.

Figure 1.16 Light pollution color map. Where do you observe from? The darker the location on this worldwide light-pollution map, the better

Conversely, if the optics are cooler than the surrounding air, then dew or frost will form on them during the evening. While we can't halt the formation of dew, there are ways to slow the whole process down dramatically. The simplest way is to install a dew shield on the telescope. As previously discussed, a dew shield is a tube extension that protrudes in front of a telescope tube to shield the optics from wide exposure to the cold air, thus slowing radiational cooling. Newtonian reflectors usually do not need a dew shield since their primary mirrors lie at the bottom of the tube, unless the secondary mirror is prone to dewing over (only in exceptionally damp conditions) or the reflector has an open truss "tube." The former situation can be slowed by installing battery powered heater elements around the secondary. For the latter, many amateurs wrap the truss with cloth, effectively shielding the mirror from radiational cooling as well as from the observer's body heat.

Heater straps are popular accessories for refractors and catadioptric telescopes. These electrically powered heaters wrap around the front end of the telescope tube to heat the objective or corrector plate by conduction to just above the air's dew point. Although these may be used independently of a dew shield, straps are more effective with one in place.

Even if the air overhead is clear, dry, and steady as a rock, runaway light pollution can have a disastrous effect on observing. How big an impact does light pollution have on your observing site? In 2001, veteran amateur astronomer and comet observer John Bortle introduced his now-famous nine-point scale for determining sky darkness (Figure 1.16). Quite honestly, many amateurs took offense at Bortle's scale when it first came out because it shed light, if you'll pardon the pun, on just how pervasive the problem was. What was thought to be a truly dark site was now suddenly revealed to be only mediocre at best.

To judge your own observing site for yourself, Table 1.11 summarizes Bortle's criteria.

If you are troubled by light pollution, don't just sit there. Do something about it! The International Dark-Sky Association (IDA) can provide essential facts, strategies, and resources in your quest to enact light-pollution legislation. Contact them at 3545 North Stewart, Tucson, Arizona 85716, or on the Internet at www.darksky.org. Just don't make the same mistake that so many people do. They expect the IDA to ride into town on a white horse and rescue them. That's not likely to happen. They can't do it alone. You need to become your own activist and contact your elected officials.

My own town of Brookhaven, New York, enacted light-pollution legislation in 2006 after I simply sent an email to my councilman a year earlier expressing concern that a farm just south of me was about to be

Table 1.11 *Bortle scale of sky darkness*

Bortle class	Color on light-pollution maps	Criteria
1	Black	**Excellent dark-sky site** • Naked-eye limiting magnitude: 7.6 to 8.0 • Zodiacal light, Gegenshein, and zodiacal band are all visible • Galaxy M33 is obvious to the naked eye • Milky Way in Scorpius and Sagittarius region casts obvious diffuse shadows • Terrestrial objects (trees, vehicles, people) are invisible
2	Gray	**Typical truly dark site** • Naked-eye limiting magnitude: 7.1 to 7.5 • M33 easily visible with direct vision • Zodiacal light still bright enough to cast weak shadows just before dawn and after dusk • Earthly clouds are visible only as dark holes or voids • Terrestrial objects visible only vaguely, except where they project against the sky
3	Blue	**Rural sky** • Naked-eye limiting magnitude: 6.6 to 7.0 • Some light pollution evident along the horizon • Milky Way structure appears complex • Brighter globular clusters such as M5, M15, and M22 are all distinct naked-eye objects • M33 easy with averted vision • Terrestrial objects vaguely apparent at distances of 20 to 30 feet
4	Green	**Rural/suburban transition** • Naked-eye limiting magnitude: 6.1 to 6.5 • Fairly obvious light-pollution domes apparent over population centers in several directions • Zodiacal light is clearly evident but doesn't even extend halfway to the zenith • Milky Way well above the horizon is still impressive but lacks all but the most obvious structure • M33 is a difficult averted-vision object and is detectable only when at an altitude higher than 50° • Terrestrial objects can be seen clearly at a distance
5	Orange	**Suburban sky** • Naked-eye limiting magnitude: 5.6 to 6.0 • Faint hints of the zodiacal light only on the best spring and autumn nights • Milky Way looks rather washed out overhead and is very weak or invisible near the horizon • Artificial light sources are evident in most if not all directions • Clouds are noticeably brighter than the sky itself
6	Red	**Bright suburban sky** • Naked-eye limiting magnitude: 5.5 • Zodiacal light invisible • Milky Way only visible when near the zenith • Clouds anywhere in the sky appear fairly bright • Andromeda Galaxy is only faintly apparent to the unaided eye
7	Red	**Suburban/urban transition** • Naked-eye limiting magnitude: 5.0 • Entire sky background has a grayish-white hue • Milky Way is invisible • Clouds anywhere in the sky appear bright
8	White	**City sky** • Naked-eye limiting magnitude: 4.5 • Sky appears whitish gray or orangish • Newspaper headlines can be read without a flashlight • Some familiar constellation patterns are difficult to see
9	White	**Inner-city sky** • Naked-eye limiting magnitude: 4.0 or less • Entire sky is brightly lit • Many familiar constellation figures are invisible

developed into a multitude of athletic fields. I was very concerned about the inevitable lighting that would result. The problem captured his attention. Creating a committee of five citizens, I being one of them, he fashioned stringent light-pollution legislation that was subsequently ratified and enacted. You can find information about the legislation on my personal website, www.philharrington.net.

Some final strategies and techniques

It takes more than a first-rate observing site and top-notch equipment to spot the challenges in this book. It also takes an experienced eye. Of course, that experience can only come with time, but, by using some of the tricks and tips already mentioned as well as some included below, it is hoped that process can be accelerated.

It is common knowledge that, in order to preserve our night vision, we should only use red-filtered lights when observing. Many amateurs, however, make the mistake of using far too bright a light. Filtered or not, a light that is too bright will harm night vision. I always have at least two flashlights with me when observing. The first is my "walking light," which is bright enough to illuminate the ground, but far too intense for looking at charts. My second light is my "chart light," which shines just brightly enough for me to see my references.

Keeping your eyes dark-adapted once they are acclimated to low light is important to the success of every observer. In addition to using a chart light, you may find it helpful to observe with one eye and read charts with the other. Keep the one eye closed when using the other, or, better still, use an eye patch. Eye patches can be found at most pharmacies.

You know the adage "don't drink and drive," right? The same can be said for observing. Studies show that alcohol weakens both visual acuity as well as night vision. Smoking and drinking too much coffee will also impair your night vision, to say nothing of fouling your telescope's optics. If you're at a star party and someone is smoking nearby, ask him or her to move away, both for the sake of your own health as well as your telescope's. Finally, and this is something that I never have to worry about, always observe on a full stomach.

But don't eat junk, which will likely make you sleepy. Instead, eat healthily. A vitamin shortage, especially vitamin A, will diminish your night-vision acuity as well. But, at the same time, don't overdose on multivitamins before heading out. Vitamins are not the kind of thing that if a little is good, a lot is better.

As we age, ultraviolet radiation from the Sun both reduces the sensitivity of our eyes' retinas and increases the odds of contracting cataracts, age-related macular degeneration, and other ocular disorders. To slow their onset if you wear eyeglasses, make sure that your glasses have an anti-ultraviolet coating on the lenses. If possible, also have your eyeglasses coated with an anti-reflection coating.

Of course, if your eye is distracted by the glare of a nearby streetlight or porch light, then a challenging object will be extinguished quickly. To help shield our eyes from peripheral light, many eyepieces and binoculars come with rubber eyecups. While they prove adequate under most conditions, eyecups alone may not block out all extraneous light. The earliest photographers faced the same problem when taking a picture, which is why they used to drape their cameras and themselves with opaque, black shrouds to keep stray light from fouling their sensitive photographic plates. These "personal cloaking devices," as I call them, are very effective at the eyepiece, as well. Choose a material that is lightweight, but also sufficiently dense to block outside light. I prefer an old, black turtleneck sweater from the dark days of the 1970s. I stick my head through the opening, but upside-down, so that my head ends up inside the sweater. The rest of the material acts as the shield. It works very well, except that it also traps body heat, which can cause eyepieces to fog over quickly when humidity is high.

The greatest tool in anyone's arsenal to help find and see difficult objects is experience. As your eyes and brain become more experienced at looking for faint, diffuse deep-sky objects, you will find that seeing them will actually become easier. Unfortunately, the only way to speed up this process is to get out on as many clear nights as possible and test your abilities.

With all these tricks and tips now discussed and detailed, it's now time to put them into action. The universe awaits us in the next six chapters. Let's go!

2
Naked-eye challenges

Of all the challenges presented in this book, this chapter should be considered the great equalizer. All of the objects to come in later chapters require some level of auxiliary equipment – binoculars or a telescope – in order to be seen. But not these. The targets described below require only one piece of optical equipment: your eyes. That's right, we're about to go nude stargazing together!

While some readers will enjoy distinct advantages over others in the chapters to come in terms of equipment, here we are as close to on a level playing field as possible. That's the fun of it. It's amazing just what can be seen by eye alone when you really make a concerted effort under dark skies.

1   M81

Target	Type	RA	Dec.	Constellation	Magnitude	Size	Chart
M81	Galaxy	09 55.6	+69 03.8	Ursa Major	7.9b (note 1)	27.1′×14.2′	2.1

Note 1. Magnitude numerical values followed by "p" denote photographic magnitudes based on photographic studies. Owing to the photographic emulsions used, a target's photographic magnitude may be slightly higher or lower than its visual magnitude.

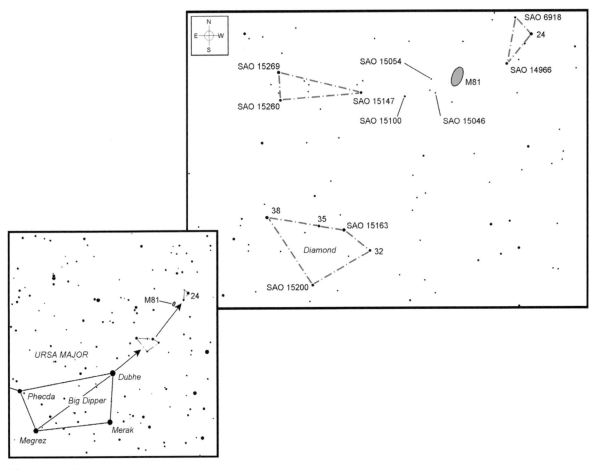

Chart 2.1 M81

Has anyone ever asked you "What is the most distant object visible to the naked eye?" Probably? It could have been someone at a club star party or maybe a family member at a reunion who is bemused by your interest in astronomy, but most astronomers have been asked that question before. What did you tell them? More likely than not, most city slickers would give the stock answer that we have all used: M31, the

Andromeda Galaxy (naked-eye Challenge 11), which by modern reckoning is 2.5 million light years away. Those who live farther from civilization, without the pall of light pollution shrouding their sky, may have answered M33, the Triangulum Galaxy (naked-eye Challenge 12), which is 2.9 million light years away. As discussed later in this chapter, M33 is often used as the telltale sign of a dark autumn sky.

Although these are commonly given replies, both answers are wrong. Of course, only stargazers who observe far, far away from civilization, in those rarified neverlands of inky darkness, would know that from first-hand knowledge. For those privileged few, the correct answer is actually M81, the tight-armed spiral galaxy located 12 million light years away in Ursa Major.

Johann Elert Bode (1747–1826) bumped into this galaxy quite by chance on New Year's Eve 1774, while using his 7-foot (focal length) telescope. Today, spotting M81, or Bode's Galaxy as it has also come to be known, through 40-mm and smaller binoculars from suburban observing sites is considered by many to be a reasonably difficult task. But can its diffuse 7.9-magnitude glow actually be glimpsed without any optical aid? The answer is yes, but with a few important qualifications. Not only must the observing site be extraordinarily dark and completely absent of any atmospheric interferences, either natural or artificial, but the observer must have exceptionally keen vision.

M81 is positioned to the northwest of the Bowl of the Big Dipper. To find its general location, connect an imaginary line from Phecda [Gamma (γ) Ursae Majoris, the star marking the bottom left corner of the bowl] to Dubhe [Alpha (α) Ursae Majoris at the bowl's upper right corner, opposite the handle], and continue it an equal distance in the same northwesterly direction. After passing a lopsided diamond-shaped asterism formed by 32, 35, and 38 Ursae Majoris; SAO 15163, and SAO 15200, you will arrive at a pair of faint stars, 24 Ursae Majoris and SAO 14966, shining at magnitudes

4.6 and 5.7, respectively. If you can see them as well as a much fainter point, 7th-magnitude SAO 6918, to their north, then there is hope for spotting M81.

Look for a triangle of stars about 4° east-southeast of 24 Ursae Majoris and an equal distance north of the lopsided diamond. The stars in this second triangle – SAO 15147, SAO 15260, and SAO 15269 – shine at magnitudes 5.8, 5.8, and 5.0, respectively. M81 lies along a string of very faint stars that line up between these two triangles. From SAO 15147, look just to the west for 7.2-magnitude SAO 15100 and, a bit further along, the close-set pair of SAO 15046 and SAO 15054. These last stars, shining at magnitudes 7.9 and 8.0, are only 10 arc-minutes apart, and so are too close to be resolved individually by eye. Don't be fooled into thinking that their soft combined glow is M81. Instead, look just a bit farther west still for a very soft, exceedingly faint, not-quite-stellar point. That third step along the string is M81.

From his observing site in Virsbo, Sweden, Timo Karhula notes that "if I can't see SAO 15100 close by, then I know that I won't see the galaxy either." At his latitude (nearly 60°N), M81 passes almost directly overhead, far above any horizon-hugging haze and interference. "I have seen M81 most easily when I have been lying directly on the ground. When I have been concentrating and have identified all of the surrounding stars for at least 15 minutes, I feel confident of having positively viewed the galaxy by naked eye."

If you can duplicate Karhula's success, then you will join an elite group of stargazers who have seen the most distant naked-eye object.

2 Melotte 111

Target	Type	RA	Dec.	Constellation	Magnitude	Size	Chart
Melotte 111	Open cluster	12 25	+26	Coma Berenices	1.8	275′	2.2

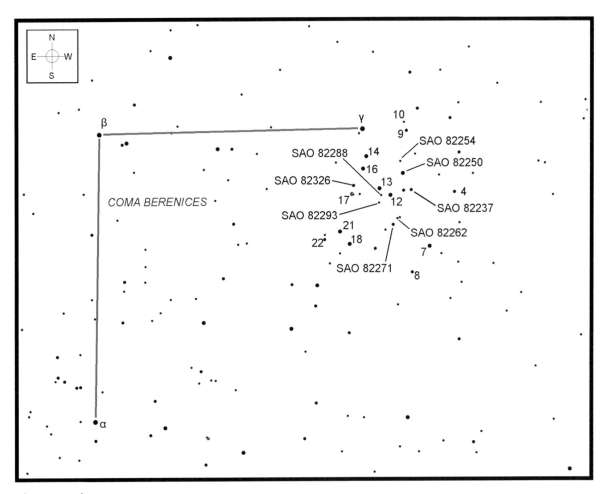

Chart 2.2 Melotte 111

If seeing M81 by eye alone was too tough a challenge, here is an easier target that is visible under less demanding conditions. The lovely little constellation of Coma Berenices lies north of Virgo and west of Boötes and appears almost directly overhead from central Europe and North America on spring evenings. The constellation's stars are much fainter than those in some of the more famous groups that surround it, so few casual stargazers pay it much attention.

Upon closer scrutiny, however, its dim stars appear "as if gossamers spangled with dew-drops were entangled there," as the early twentieth-century astronomy author and popularizer Garrett Serviss put it in his book *Astronomy with an Opera Glass* (D. Appleton and Company, 1888). He went on to describe the view as cobwebs that someone had failed to sweep out of this corner of the sky.

The constellation's name, which translates as Berenice's Hair, gives a hint of its romantic origin. In the third century BC, Egypt was ruled by Ptolemy III, the third generation of the Ptolemaic dynasty in Egypt. Ptolemy III came to power in 246 BC upon the death of his father, Ptolemy II. His queen, Berenice II, was well known for her long beautiful hair.

History remembers Ptolemy III primarily for his invasions of the northern kingdom of Syria. Legend has it that once, as her husband left on a particularly dangerous campaign, Queen Berenice vowed she would sacrifice her flowing locks to the gods if her husband returned to her safely. Upon his return, she kept her promise by cutting off her hair and placed it in the temple at Zephyrium. That same night, the tresses mysteriously disappeared. Ptolemy was furious, and Berenice wept inconsolably over the loss. There is no telling what might have happened to the guardians of the temple had it not been for Berenice's court astronomer, Conon of Samos. Thinking quickly, Conon led the king and queen outside into the night and pointed toward a gossamer patch of stardust not far from the star Arcturus. He assured them that the soft glow was her missing locks, which had been transfigured and placed among the stars. Since neither Ptolemy nor Berenice were knowledgeable in celestial lore, they believed Conon when he assured them that the silvery swarm had never been there before. Ever since, the world has recognized the constellation as Berenice's Hair. It's a good thing the king and queen were not stargazers and knew very little about the sky. If they had, then they would have recognized the queen's "hair" as the fluffy tip of Leo the Lion's tail, as earlier stargazers often portrayed it.

What Conon identified as his queen's heavenly locks is recognized today as an open cluster of stars. In catalogs, it is usually listed as either Melotte 111 or Collinder 256, but most amateurs prefer the more informal reference, the Coma Berenices Star Cluster. Call it what you will, this impressive collection spans 5° of sky and is both fun and challenging to study by eye. None of its 80[1] stars shines brighter than magnitude 4.4, placing them near the naked-eye limit for those of us suffering under suburban light pollution. The brightest half dozen form a distinctive triangular pattern reminiscent of the lower-case Greek letter lambda, λ.

---

[1] While most references cite the total number of stars in the Coma Berenices Star Cluster as 80, B. Archinal and S. Hynes count the number as high as 273 in their book *Star Clusters* (Willmann-Bell, 2003).

Table 2.1 *Selected naked-eye stars in the Coma Berenices Star Cluster*

Star	RA	Dec	Mag
Gamma (15)	12 26 56.27	+28 16 06.3	4.4
12 Comae	12 22 30.31	+25 50 46.1	4.8
7	12 16 20.54	+23 56 43.3	4.9
14	12 26 24.05	+27 16 05.4	4.9
16	12 26 59.32	+26 45 32.2	5.0
13	12 24 18.51	+26 05 55.2	5.2
17 (note 1)	12 28 54.70	+25 24 46.3	5.3
18	12 29 27.03	+24 06 31.9	5.5
21	12 31 00.59	+24 34 01.4	5.5
SAO 82250	12 20 19.65	+26 37 10.4	5.6
4	12 11 51.18	+25 52 12.9	5.7
SAO 82271	12 22 10.83	+24 46 25.4	6.2
8	12 19 19.21	+23 02 05.2	6.3
9	12 19 29.54	+28 09 24.8	6.3
SAO 82254	12 20 41.35	+27 03 16.0	6.3
22	12 33 34.23	+24 16 59.2	6.3
SAO 82293	12 24 26.77	+25 34 56.5	6.4
SAO 82237	12 19 02.01	+26 00 29.9	6.5
SAO 82326	12 28 38.16	+26 13 36.8	6.5
10	12 19 50.60	+28 27 51.6	6.7
SAO 82288	12 24 03.43	+25 51 04.3	6.7
SAO 82262	12 21 26.71	+24 59 48.8	7.4

Note 1. 17 Comae is a double star comprised of a 5.4-magnitude primary accompanied by a 6.6-magnitude secondary sun located 145 arc-seconds to its southeast. Although not possible to see with the naked eye, the companion does contribute light to the system, effectively raising the combined magnitude to 5.3.

By eye, the Coma Star Cluster appears to be engulfed by nebulosity, like the Pleiades (naked-eye Challenge 13). But, in reality, the group is completely cloud-free, which means that if you see any hint of softness it is probably due to the individual stars' proximity to one another or, perhaps, your own vision. Can you identify any of those individual stars by eye alone? If so, how many can you count?

Chart 2.2 plots all of the stars in the Coma Star Cluster to magnitude 7.5, while Table 2.1 lists selected representatives in order of increasing brightness. In all, 22 stars in and immediately surrounding Melotte 111 break this barrier.

3 Alcor and Mizar

Target	Type	RA	Dec.	Constellation	Magnitude	Size	Chart
Alcor and Mizar	Double star	13 23.9	+54 55.5	Ursa Major	2.2, 4.0	11.8′	2.3

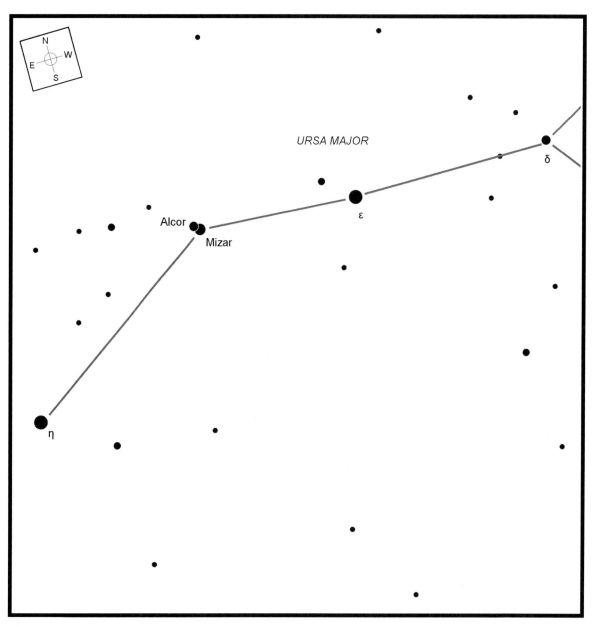

Chart 2.3 *Alcor and Mizar*

Is there any constellation in the sky more universally known than Ursa Major, the Great Bear? Most of us learned of it as a child, perhaps from a relative or friend, or possibly as a Scout working our way toward a merit badge in astronomy. The seven brightest stars in the group, known in North America as the Big Dipper and in the UK as the Plough, always draw our attention, especially in the spring when they ride highest in our sky.

Many cultures incorporated the stars of Ursa Major into their myths and legends. Ancient Hindu stargazers, for instance, saw the stars in the Big Dipper asterism as *Rishi*, or "seven scholars." In Scandinavia, they were known as Thor's Wagon, and also the Wagon of Odin, Thor's father. In ancient Egypt the Dipper's seven bright stars were seen as a bull's thigh, while in China they were symbolic of the government. Some Arabs and Hebrews saw the Dipper stars as a bier, or coffin; early Christians called it the Bier of Lazarus.

Who the first person was to see the seven Dipper stars, as well as the fainter surrounding points, as forming a huge bear is long lost to history, but it is interesting to note that this interpretation bridged many cultures that presumably didn't know the others even existed. We find references to a great bear in such diverse sources as ancient Greek and Roman texts and among Native American tribes. Was this strictly happenstance? Or is the common meaning indicative of a shared origin, perhaps with roots in ancient Asia?

There are many legends surrounding our skybound bear. Ancient Greek legend had it that Ursa Major was the heavenly incarnation of Callisto, a beautiful young woman with whom Zeus himself fell in love. Fearing the jealous rage of his goddess wife Hera, Zeus transfigured Callisto into a bear and placed her in the sky for safety. He also placed their son, Arcas (after the Greek word *arktos*, or "bear") near her side. Today, we know them as Ursa Major and Ursa Minor, respectively.

Half a world away and before European influence could be felt, many Native American peoples had myths of celestial bears portrayed among the same stars. My favorite comes to us from the Iroquois and Algonquin tribes of eastern North America. They each tell a story of how, every spring, a vicious bear left its den to wreak havoc on their tribes by killing many people and eating their food stores. Finally, in a desperate attempt to rid them of this menace, the tribal leaders sent their three mightiest Indian warriors out after the bear. In this story, the stars in the Dipper's handle represent those warriors rather than the bear's unusually long tail so often depicted in drawings. The Indians continue to chase the bear across the sky throughout the summer. Then, in the autumn, as the bear grows weary, one of the Indians injures it with an arrow. Blood drips from the wound and colors the leaves of the forest in pastel shades of red, orange, and yellow. What finally happens to the bear? Apparently the wound is not fatal, as both it and the relentless hunters return to the sky every spring.

If you look carefully, you can even see that the middle star in the Dipper's handle is joined by a nearby companion star. The Algonquins saw this faint cohort as the large pot that one of the hunters was carrying on his back to cook the bear in once they captured it. Other tribes knew the two close-set stars as a horse and rider, and often used them as a vision test for their own hunters, just as Greek warriors did.

Today, we know the horse and rider by their given Arabic names, Mizar and Alcor. Mizar, the brighter of the two, shines at magnitude 2.3, while Alcor is magnitude 4.0, or about five times fainter. To ancient Chinese stargazers, this was *Foo Sing*, the Supporting Star, while in Latin, it was called *Eques Stellula*, or Little Starry Horseman.

Although it is not the tell-all check that ancient starwatchers thought, seeing the Little Starry Horseman by eye remains a fun test to try from just about any spot in the northern hemisphere. The pair is separated by 11.8 arc-minutes, well within the resolution limit of most peoples' eyes. Both should be visible even through only marginally dark suburban skies. Look for Alcor to the east-northeast of Mizar.

That raises an interesting, and wildly disputed, question: what is the angular resolution of the human eye? Depending on the study, numbers can vary greatly. Part of the problem has to do with the conditions under which the tests are conducted. Recall from Chapter 1 that the human eye has two basic types of photo-receptors: cones, which respond to bright-light conditions, and rods, which respond to low-intensity light. It turns out that the resolving ability of rods is approximately 7 arc-minutes, only a fraction of the maximum resolution of the cones.

Of course, your personal results may vary depending on your own visual acuity. With sharp vision, a person may easily exceed this number, while a myopic observer may not come close. To judge your own eyes' angular acuity, try this simple test. Draw two black dots

Table 2.2 *Angular resolution test*

Distance to card (feet)	Angular resolution (arc-minutes)
1	27
2	13.5
3	9
4	7
5	5.5
6	5
7	4
8	3.5
9	3
10	2.5

separated by 0.0625 inch (2 mm) on a white index card, and tape the card to a wall in a properly lit room. Close one eye and back away from the card slowly. Stop when the two black dots can no longer be resolved, but instead merge into one. Measure your distance to the card to the nearest foot. Plug that distance into Table 2.2 to find your eye's angular resolution value.

Don't discount the resolving ability of the human eye. Today we rely on telescopes and attached instruments to determine the positions of sky objects to a small fraction of an arc-second, but early astronomers used only their eyes to make some startling discoveries. Perhaps the most amazing of all was Johann Kepler's (1571–1630) three laws of planetary motion, which he based largely on naked-eye observations of planetary positions by his mentor, Tycho Brahe (1546–1601). Brahe created a clever device that coupled a transit instrument with a giant protractor of sorts, and placed it in an observatory that he called Uraniborg. Brahe's sole mission was to record the times and angles when each of the naked-eye planets appeared in a narrow reference window in his observatory. He did so meticulously for more than two decades. After Brahe's death, Kepler tried to use these observations to confirm the Copernican model theory that placed Earth and the other planets in circular orbits around the Sun. Try as he might, he could not get the numbers to agree with the observations. Finally, Kepler was forced to conclude that, while the planets did indeed orbit the Sun, they did so in elliptical paths. Millennia of misconception had been corrected all because of the precision of the human eye.

4 M3

★
★★
★

Target	Type	RA	Dec.	Constellation	Magnitude	Size	Chart
M3	Globular cluster	13 42.2	+28 22.5	Canes Venatici	6.3	18′	2.4

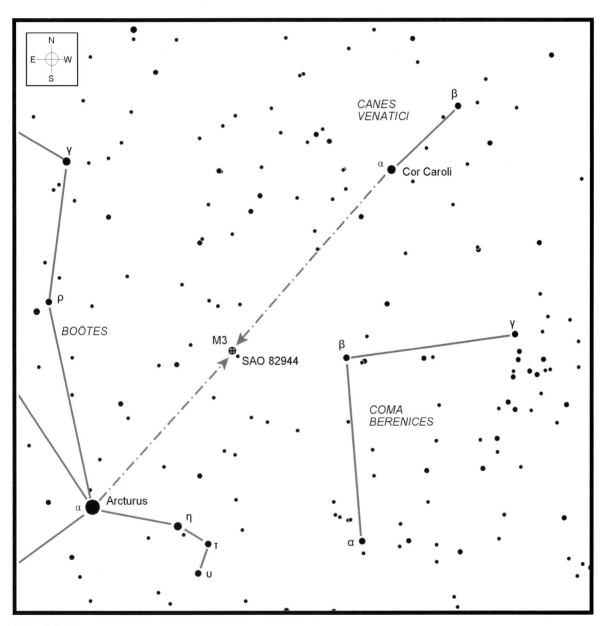

Chart 2.4 M3

Just south of the Great Bear, Ursa Major, lies one of the smallest, dimmest constellations in the northern spring sky. Canes Venatici, the Hunting Dogs, first appeared in *Firmamentum Sobiescianum*, the 1690 star atlas created by Johannes Hevelius (1611–1687). Mythologically, the two hunting dogs are said to belong to Boötes the Herdsman, who resides just to their southeast. Like winter's Canis Minor, the Little Dog, Canes Venatici is drawn primarily from two naked-eye stars, 3rd-magnitude Cor Caroli (α CVn) and 4th-magnitude Chara (β CVn). Technically, however, these two only represent the southern dog of the pair, which was originally named "Chara." The northern dog, called Asterion in mythology, is formed by a tiny triangular clump of 5th- and 6th-magnitude stars to the northeast of Cor Caroli. Spotting Asterion is itself a worthy challenge from suburban observing sites.

Its home constellation may be one of the sky's dimmest, but M3 is one of the sky's brightest globular clusters. In fact, it is so bright, coming in at magnitude 6.2, that M3 can be glimpsed with the naked eye from moderately dark observing sites. Can you see it? To find M3, imagine a line starting at Arcturus, in Boötes, and running 25° northwest to Cor Caroli. The cluster lies just short of the line's halfway point, 12° from Arcturus. To help you zero in, use 4th-magnitude Beta Comae (β Com), which is just 6° due west of M3.

We have a problem, however. If you have clear enough skies to see M3 by eye alone, then you should see not one, but two faint points here. The second object, cataloged as SAO 82944, is a golden K3 giant star shining at magnitude 6.2 and lying less than half a degree to the southwest of M3. Can you spot them both, only one, or do they merge together into a single blur? To help identify which you are seeing – if either, or both – use a pair of binoculars. The extra aperture and magnification of even the smallest pair will immediately separate the star from the globular. Once you have that clear in your mind, lower the binoculars and try again with your unaided eye.

5 Stars in Ursa Minor

Target	Type	RA	Dec.	Constellation	Magnitude	Size	Chart
Stars in Ursa Minor	Asterism	15 40	+80	Ursa Minor	2 to 6	–	2.5

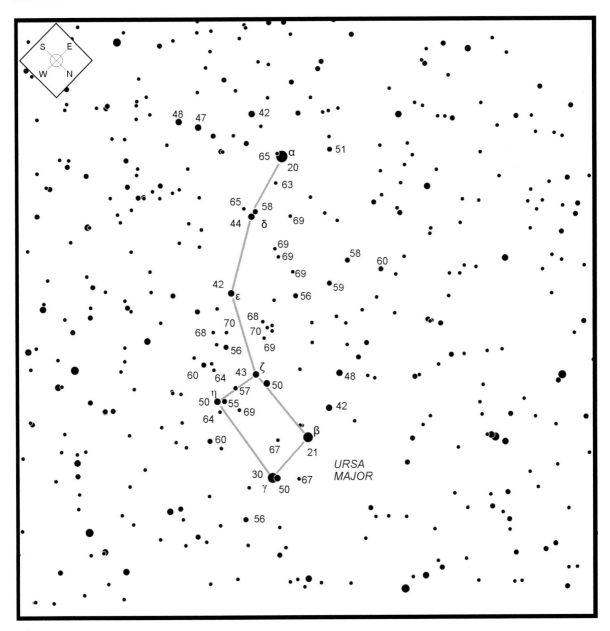

Chart 2.5 Ursa Minor

It's a sad testament to the modern world that seeing all seven stars in the Little Dipper asterism constitutes a challenge, but, in fact, for many suburbanites living in mid-northern latitudes, it does. The fact that the Little Dipper is circumpolar for much of the hemisphere, plus the fact that its stars span the magnitude gamut from 2nd magnitude to beyond naked-eye limit makes this a great tool for judging just how dark an observing site is . . . or is not.

As mentioned in naked-eye Challenge 3, the Little Dipper's home constellation of Ursa Minor traces its origins back to the story of the beautiful Callisto, and her son, the young hunter, Arcas. Callisto was transformed into the Great Bear, Ursa Major, while her teenage son, Arcas, Ursa Minor.

Chart 2.5 plots all of the naked-eye stars in the Little Dipper and its surroundings down to magnitude 7.0. How many of the Dipper's seven stars can you see? Polaris comes in at an even magnitude 2.0 (shown as "20" on the chart, the decimal omitted to avoid confusing it for another star), while Kochab [Beta (β) Ursae Minoris] is nearly its twin at magnitude 2.1 ("21" on the chart, etc.). The stars then take a greater-than-one-magnitude jump to 4.2-magnitude Epsilon (ε) Ursae Minoris in the handle. Coming in just

a tenth of a magnitude dimmer are Delta (δ) also in the handle and Zeta (ζ) Ursae Minoris in the Bowl. Extending westward from Kochab are the stars 5 and 4 Ursae Minoris, at magnitudes of 4.3 and 4.8, respectively. Finally, 5.0-magnitude Eta (η) Ursae Minoris completes the Dipper's figure. If you can see all seven stars by eye alone, then your suburban observing site is actually fairly dark.

You may be able to do even better. How many of the fainter stars in and around the Little Dipper can you see? Can you see the 5.6-magnitude star just "above" the bowl? How about the rest of the stars there, which appear to be flying in formation? If so, great; your limiting magnitude is 6.7, which is outstanding.

It is best to attempt these observations when Ursa Minor is highest above the horizon. The constellation culminates (that is, it crosses the meridian) at midnight on May 13. Remember that the time of culmination advances two hours for every month. Therefore, it reaches culmination at 10 p.m. local time on June 13, and so on. Of course, six months earlier or later finds the stars at their lowest points in the sky, where light pollution, haze, and clouds could well diminish limiting magnitude.

GLOBE at night

Just how bad is light pollution? That's the question being asked by GLOBE, short for Global Learning and Observations to Benefit the Environment. GLOBE is an interagency program funded by the National Aeronautics and Space Administration (NASA) and the National Science Foundation (NSF), supported by the U.S. Department of State, and implemented through a cooperative agreement between NASA and the University Corporation for Atmospheric Research (UCAR) in Boulder, Colorado.

Since 2006, GLOBE has conducted a light-pollution experiment called GLOBE at Night. The purpose of GLOBE at Night is to ascertain just how pervasive light pollution has become. Each winter around the February or March

New Moon, GLOBE at Night solicits educators, students, scientists, and amateur astronomers to look skyward at the constellation of Orion. Using GLOBE's star charts and data recording form, observers find the faintest stars they can detect within the Hunter, and then check off which star chart corresponds most closely with what they are seeing. Each GLOBE star chart corresponds to a whole magnitude difference, beginning at limiting magnitude 2 and going all the way down to magnitude 7. Reports are then filed through the GLOBE website for compilation and analysis.

To join the next GLOBE-at-Night challenge, visit www.globe.gov/gan.

6 M13

Target	Type	RA	Dec.	Constellation	Magnitude	Size	Chart
M13	Globular cluster	16 41.7	+36 27.6	Hercules	5.8	20′	2.6

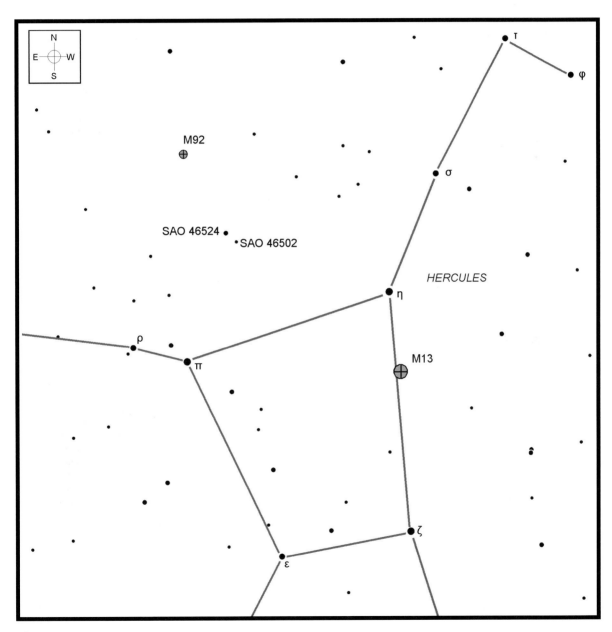

Chart 2.6 M13

Like M3 in springtime's Canes Venatici, spotting M13, the Great Hercules Globular Cluster, without optical aid is a good test of a suburban sky's darkness and transparency. Shining at magnitude 5.8, it lies along the western edge of the Hercules Keystone asterism.

The mythical Greek hero Heracles, son of Zeus, later named Hercules by the Romans, was the giant whose marvelous strength was celebrated so often in Greek legends. The most famous of his exploits was the twelve labors, which he performed at the bidding of Eurystheus. The constellations of Leo, Draco, Hydra, and Scorpius are all connected with the stories of these exploits.

The Hercules of yore is often depicted as wearing a lion skin and holding a club, an image adapted in artists' renditions of the constellation. Unfortunately for him, Hercules is also portrayed upside down in the sky, as if balancing on his head.

Four of the constellation's most prominent stars, often shown as framing a portion of his torso, are nicknamed the Keystone. The stars – Eta (η), Zeta (ζ), Epsilon (ε), and Pi (π) Herculis – all shine between magnitudes 2.8 and 3.9, and may themselves present a challenge to suburban stargazers. To check that you are in the right location, the Keystone lies two-thirds of the way between the orange giant star Arcturus in Boötes and the blue giant Vega in Lyra.

Once you spot the Keystone, can you see any faint stars inside? Eight shine between magnitudes 5 and 6.5, and make a good test of sky darkness. The brightest of the bunch are centered along the Keystone's southern edge.

The challenge presented here, however, is not how many interior Keystone stars you can spot, but rather whether or not you can catch a glimpse of globular cluster M13 along its western edge. This swarm of more than 100,000 stars lies one-third of the way southward from Eta to Zeta, and should be within naked-eye grasp from sites with a naked-eye limiting magnitude of 5.5 or better. If you can see some or all of the stars shown inside the Keystone in Chart 2.6, then the odds are good that you will also be able to spot M13 by eye, as a slightly nebulous "star."

Given M13's prominent location, it's a little surprising that pre-telescopic observers never recorded it. History records that the first person to lay eyes on it

was Edmond Halley (1656–1743) in 1714 when he accidentally hit upon it through his telescope. Two years later, *Philosophical Transactions* [Vol. 29, No. 347 (1716), pp. 390–2] published a paper by Halley in which he lists six "nebulae" that he could not resolve through his telescope. He opened the discussion by describing "certain luminous Spots or Patches, which discover themselves only by the Telescope, and appear to the naked Eye like small Fixt Stars; but in reality are nothing else but Light coming from an extraordinary great Space in the Ether." Today, the six objects in Halley's list include, in order of his listing: M42, the Great Nebula in Orion; M31, the Andromeda Galaxy; globular cluster M22 in Sagittarius; the grandest globular cluster of all, Omega Centauri in Centaurus; open cluster M11 in Scutum; and M13 in Hercules.

Halley was also the first to note M13's naked-eye visibility when he recorded "This is but a little Patch, but it shews it self to the naked eye, when the Sky is serene and the Moon absent." Later, in 1908, Garrett Serviss wrote in his book *Astronomy with the Naked Eye* that M13 appeared as "a faint speck, barely visible to a good eye . . . that indicates the location of one of the supreme wonders of the universe." Sixteen years hence, Mary Proctor recalled M13 as "the finest of all the clusters in the northern heavens and is just visible to the unaided eye on a dark night."

Trying to spot M13 by eye is one of my standard litmus tests for checking the darkness of a summer sky. I can spot it easily from the Stellafane amateur telescope makers' convention in Springfield, Vermont, where the naked-eye limiting magnitude averages between 6 and 6.5. From a semi-rural site on eastern Long Island not far from where I live (naked-eye limiting magnitude of 5.5), M13 is also often seen without optical aid. Even on some particularly transparent nights from my decidedly suburban backyard observing site, I have seen M13 by eye.

If seeing M13 by eye is no challenge at all, then try your luck with Hercules's second Messier globular, M92. M92 shines at magnitude 6.5, which makes it about 1.7 times fainter than M13. Look for its faint smudge north of the Hercules Keystone, about 3° beyond a close-set pair of dim stars, SAO 46502 and 46524. Together, those two suns mark the right angle of a triangle formed with Pi (π) and Eta (η) Herculis.

7 Great Dark Horse Nebula

Target	Type	RA	Dec.	Constellation	Magnitude	Size	Chart
Great Dark Horse Nebula	Dark nebula	17 10	−27	Sagittarius	–	~480′	2.7

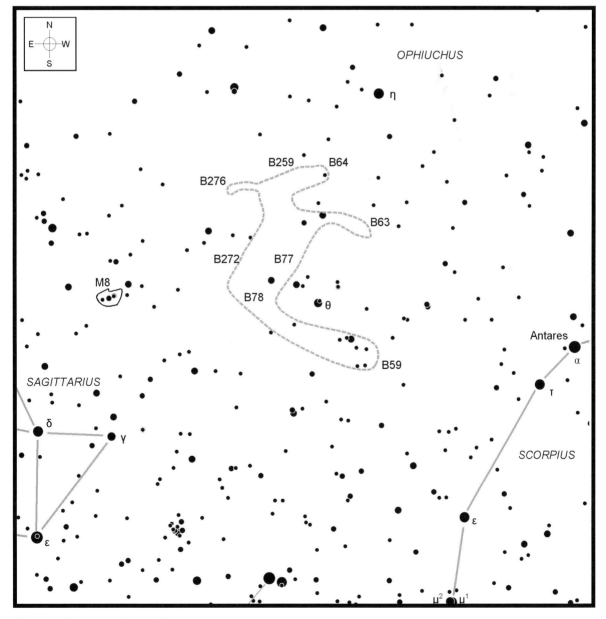

Chart 2.7 Great Dark Horse Nebula

One of the dead giveaways that the Milky Way is a spiral galaxy is the preponderance of dark nebulae that litter the plane of the galaxy. The summer Milky Way, stretching from Cygnus southward to Sagittarius, is especially polluted with clouds of opaque cosmic dust. One of the most obvious patches of dark nebulosity, the Great Rift, begins just south of Deneb in Cygnus and flows southward along the lane of our galaxy, through Aquila, to end about 120° later in Scutum. Photographs display the ghostly outlines of many dark nebulae prominently against the bright backdrop of more distant stars. But, as is often the case, when we look their way by eye, whether aided by telescope, binoculars, or eye alone, they are nowhere to be seen.

It is this subtlety that probably explains why so many classic observers from the golden age of visual astronomy made little mention of dark nebulae. After all, why look at something that can't be seen when there is so much out there that *can* be seen?

It wasn't until long after the marriage between photography and astronomy that the study of dark nebulae came into its own. Most of our early knowledge of these black clouds came about through the diligence of the American astronomer Edward Emerson Barnard (1857–1923), who was the first to photograph and study the lane of the Milky Way in detail. Four years after Barnard's death, his collection of photographs was published as *A Photographic Atlas of Selected Regions of the Milky Way* after his research had been completed by Edwin Frost, then director of Yerkes Observatory, and Mary Calvert. In addition to the extraordinary photos, Barnard also included regional maps that assigned a catalog number to each individual opaque cloud.

Many of Barnard's dark nebulae remain a supreme test for visual observers. To spot them, however, we have to change our way of thinking. In these cases, it's what you *don't* see that counts.

Besides the Great Rift, one of the largest and darkest dark nebulae in the summer sky lies midway between the Small Sagittarius Star Cloud (M24) and brilliant Antares in Scorpius. The nebula is so large that Barnard assigned it five separate entries in his catalog. By eye, we see them collectively as the Pipe Nebula. The Pipe Nebula extends over 7° in southern Ophiuchus. From a dark-sky site, its smoking-pipe shape stands out prominently against the star-studded backdrop. The "bowl" of the pipe, designated Barnard 78, looks roughly rectangular, while its stem, formed from

Figure 2.1 Great Dark Horse Nebula

Barnard 59, 65, 66, and 67, extends more than a degree to the west.

That's only part of this challenge. The Pipe Nebula is actually quite obvious when viewed under naked-eye limiting magnitude 6.0 or better skies, but by adding other, more subtle Barnard patches that float to the Pipe's northwest, the area transforms into the profile of a horse. In fact, the horse even seems to be strutting or prancing, as you might see in a circus act.

The first person to publicize the idea of the Dark Horse was author Richard Berry, former editor of *Astronomy* magazine. Recounting his "discovery," to me, Berry recalls:

After I graduated from college, I got into astrophotography, and became especially fascinated with the Milky Way. I started taking wide-angle pictures, eventually with large-format cameras. About the same time, I found a copy of Barnard's atlas at a university library, and it really inspired me. You can't shoot wide, deep images of the Sagittarius-Scorpius-Ophiuchus region without picking up the dark nebulae. Going back to my photos, sitting right there in plain sight was a huge dark horse against the star clouds! Barnard called part of it the Pipe Nebula because he was shooting relatively narrow angle views. Some years later, when I ran readers' images of the same region in *Astronomy*, I started mentioning the 'Great Dark Horse Nebula' in a few of the captions. The name (or at least the Horse) seems to have caught on.

The Pipe is unmistakable given a good sky, but the Horse takes a good eye and an even better imagination. Can you spot it? Figure 2.1 shows the horse standing on end, while Chart 2.7 identifies which clouds are which. Barnard 259 make up the horse's nose, Barnard 268 forms its mane, while a thin vein of darkness extending toward Barnard 276 completes the horse's head. Crescent-shaped Barnard 63 marks its bent front leg, while the remainder of the horse's torso is created by Barnard 67a, 72, 75, 261, 262, 266, 269, and 396. The horse's hindquarters are formed from the Pipe Nebula, with the bowl of the pipe as the horse's hip and the stem serving as its rear leg.

★★
★★

8 North America Nebula

Target	Type	RA	Dec.	Constellation	Magnitude	Size	Chart
NGC 7000	Emission nebula	20 58.0	+44 20.0	Cygnus	–	120′	2.8

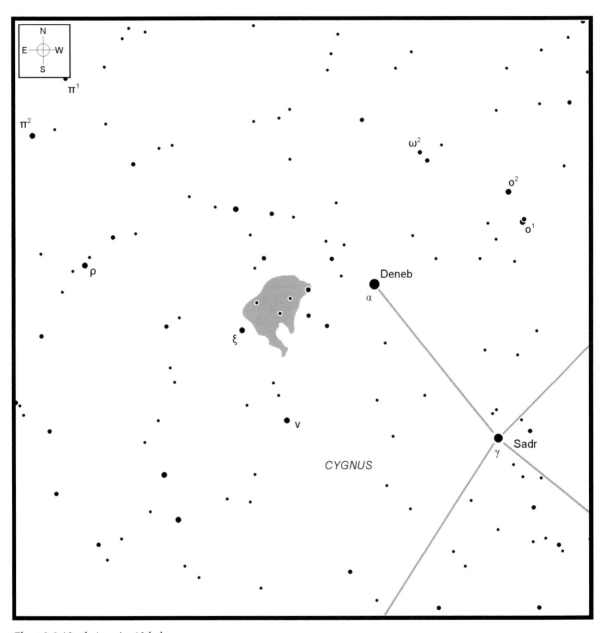

Chart 2.8 North America Nebula

The North America Nebula (NGC 7000) is a large
expanse of glowing hydrogen gas mixed with opaque
clouds of cosmic dust just 3° east of Deneb [Alpha (α)
Cygni] and 1° to the west of 4th-magnitude Xi (ξ)
Cygni. Famous as one of the most luminous blue
supergiants visible in the night sky, Deneb marks the
tail of Cygnus the Swan, or, if you prefer, the top of the
Northern Cross asterism.

The North America Nebula epitomizes how
observational astronomy has evolved over the years.
When he discovered it on October 24, 1786, Sir
William Herschel (1738–1822) described the view
through his 18.7-inch reflecting telescope as "very large
diffused nebulosity, brighter in the middle." Honestly,
I am surprised he could see it at all because of his
instrument's very narrow field of view. That's one of the
biggest challenges to seeing the North America Nebula
through a telescope – it spans an area nearly 2° in
diameter. From the sounds of Herschel's notes, the
thought of trying to spot it in anything less never
crossed his mind.

In 1890, the German astronomer Max Wolf became
the first to photograph the full span of NGC 7000.
Upon seeing his results, he christened it the North
America Nebula because of its eerie resemblance to that
continent. Since then, images of this vast emission
nebula have appeared in nearly every introductory
astronomy textbook and coffee-table astrophoto
album.

Before the 1970s, conventional wisdom had it that,
although this huge celestial continent was prominent in
photographs, it was nearly invisible to the human eye
because it was too large and too red. Part of that
mindset has to be attributed to the observing guides
that were in print at the time. Nearly all of them
concentrated solely on telescopic observing. They
were so fixated on what could be seen through
conventional long-focal-length instruments using the
comparatively narrow field eyepieces of the day that
they ignored other options. And since NGC 7000
could not squeeze into a single field of view it was
nearly impossible to isolate its clouds from their
surroundings.

One of the first authors to mention the North
America Nebula, though not by name, was the
Reverend Thomas W. Webb. In his classic book *Celestial
Objects for Common Telescopes*, Webb described the
nebula as having a "sharply defined south [edge], and
containing a dark opening like a cross; visible as a glow

Figure 2.2 North America Nebula

in a field glass, but brightest part scarcely visible in
$17\frac{1}{4}$-in reflector."

Still, most considered seeing the North America
Nebula by eye to be futile. The winds of change began
to blow, however, with the landmark work *Burnham's
Celestial Handbook*, first published in 1966. Author
Robert Burnham, Jr., advised "Binoculars show an
irregular glow more than $1\frac{1}{2}$ in diameter with the North
American shape becoming unmistakable on a clear
night. Perhaps the best view of the unusual outline is
obtained with a 3 or 4-inch rich-field telescope and
wide-angle eyepiece."

Another pioneer who urged amateurs to look for the
North America Nebula was Walter Scott Houston
(1912–1993). Throughout his nearly half century of
penning the Deep-Sky Wonders column in *Sky &
Telescope* magazine, Houston often mused about seeing
this difficult object. His evolving views captured the

changing attitudes over its visibility. For instance, in his September 1948 column, he wrote, "the North America nebula near Deneb cannot be observed readily without photography." But then, fast forward several decades to find Scotty advising "if observing conditions are very good, and you know what size and shape to expect, the North America Nebula can be made out easily with the naked eye."

Part of that change in attitude was undoubtedly due to the advent of narrowband nebula filters in the 1970s. As previously discussed, these filters block light at all but a few select visible wavelengths. Narrowband filters pass only the hydrogen-alpha (656 nm), hydrogen-beta (486 nm), and oxygen-III (496 nm and 501 nm) regions, where emission nebulae are strongest. The net result is increased contrast between the object under observation and the filter-darkened background. As a result, the North America Nebula is now a regular target at summertime star parties.

That still doesn't address the size issue, however. For that, we turn back to Burnham and Houston, who advise using binoculars or small, rich-field telescopes. Let's take it one step further. Can we spot the North America Nebula without any optical aid at all, save for perhaps a narrowband filter?

To try this yourself, first, can you see the Cygnus Milky Way clearly (Figure 2.2)? A reasonably dark suburban sky should be sufficient, but wait until Cygnus is nearest the zenith in order to isolate it from any errant clouds or light pollution. Zero in on the area between Deneb and Xi (ξ) Cygni, which lies at the edge of the Northern Coalsack, a large expanse of dark nebulosity at the northern terminus of the Milky Way's Great Rift. The Great Rift slices the galactic plane in half lengthwise, extending southward from Cygnus through Vulpecula, Sagitta, and Aquila on its way toward Sagittarius. The brightest part of the North America Nebula, marking "Mexico" and "Florida," juts into the Northern Coalsack in much the same way as their earthly counterparts mark the Gulf of Mexico's shoreline. The nebula's hook-shaped glow appears slightly brighter than the Milky Way immediately surrounding it, but be forewarned that, by eye alone, it is deceptively small. If you cannot see it by eye alone, try using a narrowband filter. It might be best to try the filter test with a friend who owns the same filter as you, since viewing through a pair of filters is preferred. Otherwise, you will lose the two-eyed advantage discussed in Chapter 1 by squinting through only one eye.

9 M2

★
★

Target	Type	RA	Dec.	Constellation	Magnitude	Size	Chart
M2	Globular cluster	21 33.5	−00 49.2	Aquarius	6.6	16′	2.9

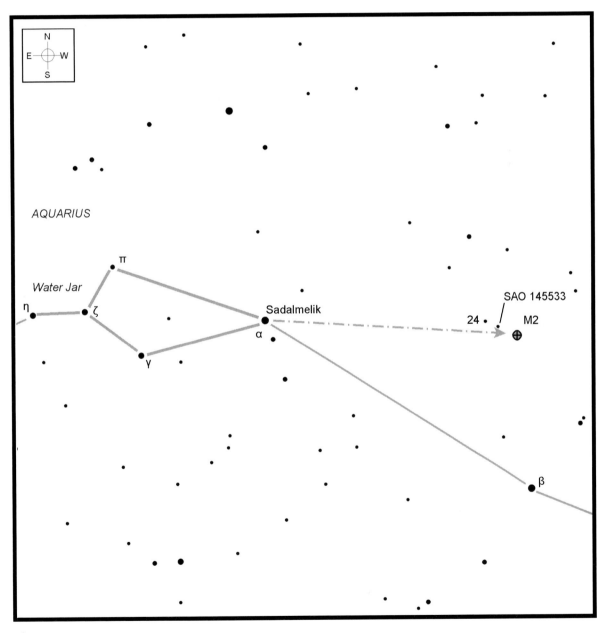

Chart 2.9 M2

Just as the springtime globular cluster M3 and summer's M13 challenge naked-eye observers, autumn brings with it the challenge of M2 in Aquarius to test our prowess. Admittedly, seeing the full outline of Aquarius is itself a challenge owing to its many faint stars. Fortunately, M2 lies near the constellation's northern border, just south of the Great Square of Pegasus (naked-eye Challenge 10). To make matters even easier, the stars of Aquarius provide an arrow that points directly at our quarry.

To begin, find Aquarius' famous Water Jar asterism, a distinctive triangular pattern formed by the 4th-magnitude stars Eta (η), Pi (π), and Gamma (γ) Aquarii, and punctuated by Zeta (ζ) Aquarii located centrally inside. By mixing-in the star Sadalmelik [Alpha (α) Aquarii] to the west of the Water Jar, an arrowhead is formed that points right to M2. As a visual cue for locating M2, the globular lies as far west of Sadalmelik as Eta lies to the star's east.

Be careful not to confuse the 6.2-magnitude star 24 Aquarii or 6.3-magnitude SAO 145533 with the globular. Although all three shine at nearly the same magnitude, M2's 16 arc-minute diameter is large enough to make it look nonstellar even to the unaided eye. Modern studies show that this globular cluster spans about 150 light years and claims 150,000 stars, making it one of the richer globulars in the Messier catalog. The oval shape is undoubtedly the result of the cluster's comparatively rapid rotation, which causes the perimeter to flatten at the poles, like Jupiter.

10   The Great Square of Pegasus

★

Target	Type	RA	Dec.	Constellation	Magnitude	Size	Chart
Stars in the Great Square	Asterism	23 40	+21 30	Pegasus	–	–	2.10a–f

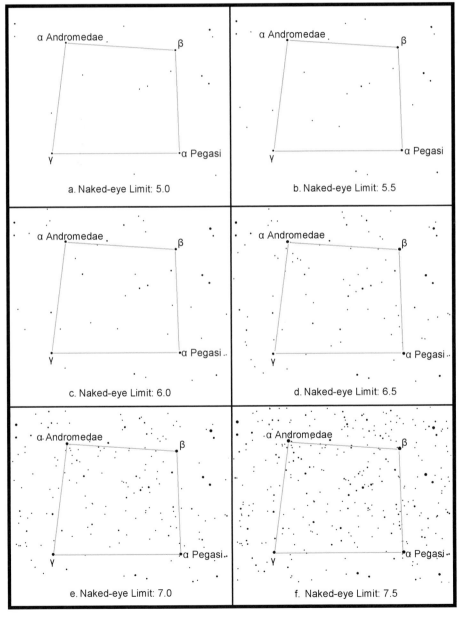

a. Naked-eye Limit: 5.0

b. Naked-eye Limit: 5.5

c. Naked-eye Limit: 6.0

d. Naked-eye Limit: 6.5

e. Naked-eye Limit: 7.0

f. Naked-eye Limit: 7.5

Chart 2.10 The Great Square of Pegasus

Pegasus is one of the best-known autumn constellations. Depicting the winged horse used by Perseus to rescue Princess Andromeda from the clutches of Cetus the Sea Monster, Pegasus flies high in our southern sky during October and November evenings.

Pegasus is usually drawn upside down in our sky, with four stars – Alpheratz,[2] Scheat, Markab, and Algenib – marking the corners of a great square that frames the horse's torso. A line of three stars – Homam, Baham, and Enif – hooking to the southwest of Markab is usually shown as the horse's neck and head, with Enif marking its nose. Unfortunately, this means that poor Pegasus is flying upside down. To make matters worse, we only see the front half of the horse; its hindquarters are nowhere to be found (there is a political joke in here somewhere, but I'll leave that to you).

Rather than strain to see a horse flying across the sky, planetarium lecturers often tell their audiences that the Great Square marks a baseball diamond, a perfect allusion for the World Series. In our cosmic baseball diamond, Scheat and Alpheratz, both 2nd magnitude, mark home plate and first base, respectively. The remaining two stars, Algenib and Markab, both 3rd-magnitude stars, are second and third base, respectively.

Let's fill in the rest of the team players out in the field. Pitching today's game is number 71: 5.3-magnitude 71 Pegasi, that is. It looks like he's about to be joined on the mound by 4.4-magnitude catcher Upsilon (υ) Pegasi and team manager Tau (τ) Pegasi, at magnitude 4.6. It would also appear that the home plate umpire, 4.8-magnitude 56 Pegasi, is heading out to see what's going on. With a time-out called, the batter, 3.5-magnitude Mu (μ) Pegasi, has headed back to the on-deck circle to talk to Lambda (λ) Pegasi, at magnitude 4.0, who is next at bat.

_____
[2] Although always shown as marking the northeast corner of the Great Square of Pegasus, Alpheratz is technically assigned to the neighboring constellation Andromeda. In fact, Alpheratz is Alpha (α) Andromeda.

Table 2.3 *Stars within the Great Square of Pegasus*

Naked-eye limiting magnitude	Number of stars	Chart 2.10 panel
5.0	4	a
5.5	7	b
6.0	9	c
6.5	24	d
7.0	37	e
7.5	76	f

Meanwhile, at first base, we have Psi (ψ) Pegasi, a 4.7-magnitude sun that appears to be playing in as if looking for the batter to bunt. Second baseman Chi (χ) Pegasi, at magnitude 4.8, is playing in the hole between first and second. Shortstop Phi (φ) Pegasi, at magnitude 5.1, and third baseman 70 Pegasi (magnitude 4.6) round out the infield, while HD 216489 is the 5.6-magnitude third base coach.

Depending on the darkness and transparency of your observing site, you may be able to see the full complement of players on our cosmic ball field, or possibly just the four bases. From dark, rural sites, however, the infield looks to be overrun by fans racing in from the stands, as if this must have been the final game of the World Series and the game has just ended. How many stars can you count inside the Great Square? Table 2.3 lists the number of stars visible in half-step increments beginning at magnitude 5.0, while the corresponding panels in Chart 2.10 plot the locations of those stars.

As has been advised before, wait for the Great Square to reach culmination, its greatest altitude above the horizon, before attempting a star count. That occurs at the stroke of midnight on September 1. Remember that stars rise two hours earlier every month, so Pegasus reaches culmination at 10 p.m. on October 1, 8 p.m. on November 1, and so on.

11 How big is M31?

Target	Type	RA	Dec.	Constellation	Magnitude	Size	Chart
M31	Galaxy	00 42.7	+41 16.1	Andromeda	4.4b	192.4′×62.2′	2.11

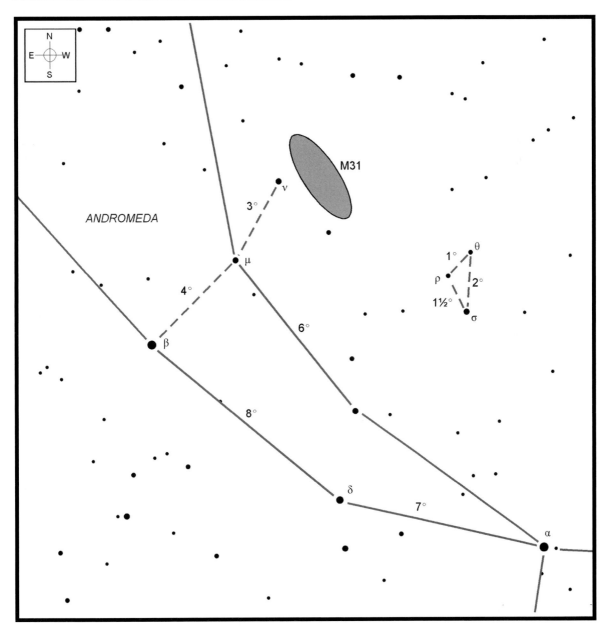

Chart 2.11 M31

How big is M31? In other words, how large an area of sky does the Andromeda Galaxy cover? Today, when we speak of the apparent size of a deep-sky object, we usually quote the full extent that has been measured very precisely on long-exposure photographic images. This is a perfectly correct way of doing it, but the listed numbers are often far larger than a visual observer can hope to see. This is especially true with galaxies, whether we are looking at them through a monstrous telescope, a pair of binoculars, or in the case we have here, the eye alone. Normally, visual observers can only see the brighter, central core of a galaxy and perhaps some indication of the surrounding halo of stars. The outermost edges of the galaxy, however, will almost always go unnoticed simply because of their comparative dimness.

The answer you get depends on whom you ask. For instance, the venerable *Burnham's Celestial Handbook* says that M31 spans 160′×40′, while *Sky Catalogue 2000.0* (Sky Publishing, 1985) lists it as 178′×63′. *The Deep-Sky Field Guide* (Willmann-Bell, 1993), companion to the *Uranometria 2000.0* atlas, has it spanning 185′×75′. Stephen O'Meara quotes 3°×1° (180′×60′) in his book *The Messier Objects* (Cambridge University Press, 1998), while Ronald Stoyan expands it 3.5°×1° (210′×60′) in his *Atlas of the Messier Objects* (Cambridge University Press, 2008).

How about you? How large does M31 appear to you? Figure 2.3 shows a wide-field view of this magnificent target. Although very precise measurement tools are used when sizing up celestial objects, to judge the size of M31 for yourself, all you need to do is lend yourself a hand. That's because the ratio of the size of the human hand to the length of the arm is the same proportion for just about everyone, regardless of age or gender. When you extend your arm fully, the various hand and finger configurations shown in Figure 2.4 can be used to make surprisingly accurate estimates of angular sky distances.

It's well known for instance that, at arm's length, your fist covers 10°. But did you know that the width of your extended pinky finger covers 1° of sky? Try it against the Full Moon one night and you'll see that it covers an area twice the Moon's $\frac{1}{2}$° span. The width of your extended thumb is twice as broad, corresponding to 2°. Extending your middle three fingers as in a scout salute, at arm's length will cover 5° of sky. The span between your pinky finger and forefinger equals 15°, while the span between your thumb and pinky equals

Figure 2.3 M31 and M33

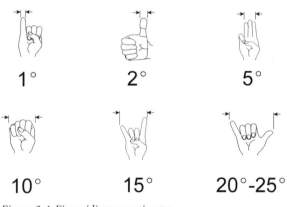

1°   2°   5°

10°   15°   20°-25°

Figure 2.4 Finger/distance estimates

25°. It should be noted that some people can stretch their hands more than others, which may throw this last measurement off a bit. To find out your hand span, hold it up against the Big Dipper. If your finger and thumb can cover its length fully, your span is 25°; a little less and it is probably closer to 18° to 20°.

From your observing site, how large does M31 look? Can you completely block it with just your pinky? If so, you're only seeing the galaxy's central nucleus. If you can cover the galaxy with your thumb, then you are making out some of the extensions associated with the galaxy's spiral-arm disk. From the darkest sites, perhaps in the desert southwest of the United States or on mountain tops in the Rockies, M31 may actually extend the better part of a 5° Scout salute.

By trying this test from many different sites, both dark and not-so-dark, you will see first hand the effect that light pollution, atmospheric haze, and other interferences can have on the visibility of extended objects. And as you're performing these tests, repeat them with both eyes as well as one eye to prove to yourself the benefit of binocular vision over monocular viewing.

12   M33

Target	Type	RA	Dec.	Constellation	Magnitude	Size	Chart
M33	Galaxy	01 33.8	+30 39.6	Triangulum	6.3b	65.6'×38.0'	2.12

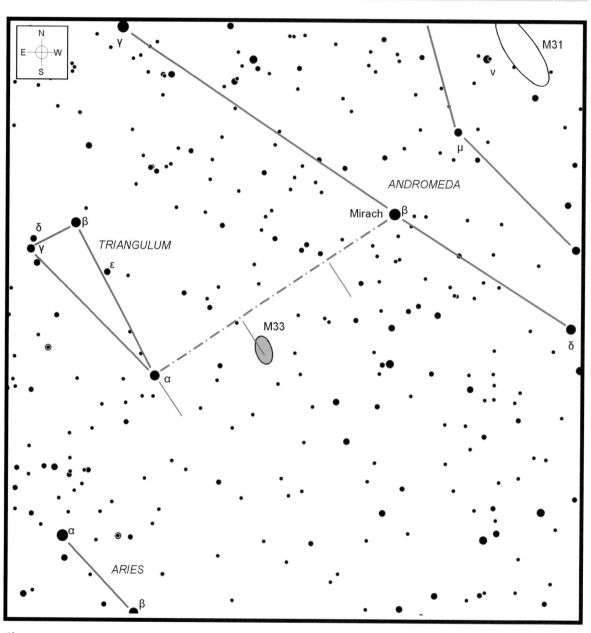

Chart 2.12 M33

M33, the Great Triangulum Spiral Galaxy, is considered by many to be one of the ultimate tests of an autumn night's darkness and transparency. Infamous for its low surface brightness, many an amateur has spent long hours searching for it through telescopes to no avail. It's a wonderfully frustrating contradiction. With its brightness usually listed at magnitude 6.3, many new observers believe that it should be visible easily through their telescopes. But it is not. That's because, despite its integrated magnitude value, M33 has a very low surface brightness; that is, its brightness per unit of area is so close to that of the surrounding sky that the dim glow of the galaxy blends seamlessly into the background. It is very easy to pass right over M33 without even noticing it.

Spotting M33 by eye alone is a fun sport for late evenings during the summer star party season. Can you spot its dim glow about a third of the way between the stars Alpha (α) Trianguli and Mirach [Beta (β) Andromedae]? Look for a very dim glow about a Full Moon in diameter.

I fondly recall my first naked-eye encounter during the 1974 Stellafane convention in Springfield, Vermont.

There in the early morning hours, a group of us had informally gathered on the summit of Breezy Hill to enjoy the beauty of the seasonal sky. It was an exceptionally clear night, with the full span of M31 visually accounted for. A few of us thought we also saw M33 nearby; others were convinced we were crazy. As the debate began to heat up, who should meander by but Walter Scott Houston? He must have been listening in to our debate. With a certain knowing tone to his voice, he announced that "M33 is quite obvious tonight" and wondered if any of us had spotted it. The Doubting Thomases in the crowd were quickly silenced! But unless the sky is absolutely, perfectly pristine, seeing M33 with the eye alone is an impossible task.

Classified as an Sc spiral galaxy, M33 is tilted almost face-on to our view, which is one reason why its surface brightness is so low. Its spiral arms can be suspected faintly through large amateur telescopes, pinwheeling away from the galaxy's central hub. Several knots of stars and interstellar matter dot the arms. We will meet them in later challenges.

13 Pleiades

Target	Type	RA	Dec.	Constellation	Magnitude	Size	Chart
Pleiades	Open cluster	03 47.5	+24 06.3	Taurus	1.2	110′	2.13

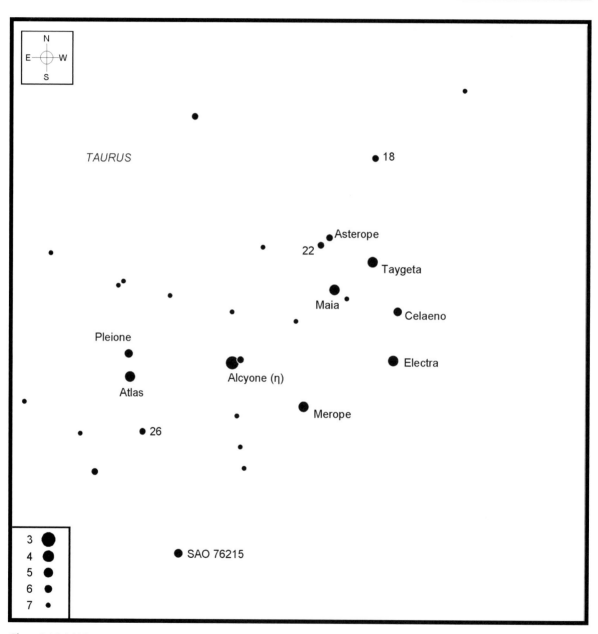

Chart 2.13 M45

It sounds like a silly question on the surface, but how many stars can you see in the Seven Sisters? A neophyte would undoubtedly answer "seven, naturally." But of course, anyone who has ever looked at the Pleiades with the unaided eye knows that seeing seven stars is not always doable. Can you? Or is the star count limited to only four, or five, or six?

Sometimes, the greatest challenge facing urban stargazers is simply trying to see *any* hint of the Pleiades at all. The brightest member of the pack, Alcyone, shines at third magnitude. If you are observing from an inner city, then spotting Alcyone might constitute the start and end of your personal Pleiades challenge.

From typical suburban skies, where the naked-eye limiting magnitude is about 4.5, we can expect to see five additional stars, or Pleiads as they are known. These stars – Atlas, Merope, Electra, Maia, and Taygeta – all shine between magnitudes 3.6 and 4.2. Together with Alcyone, they create the Pleiades' familiar tiny dipper shape that new recruits often misidentify as the Little Dipper (see naked-eye Challenge 5).

As light pollution fades, the family of Pleiades grows. A naked-eye limiting magnitude of 5.5 increases the total number of naked-eye Pleiads to 9. Celaeno shines meekly between Taygeta and Electra, while a dim point cataloged as SAO 76215 lies to the south-southeast of Atlas, Alcyone, and Merope. The toughest star joining the horde is Pleione, a faint speck just north of Atlas. Not only does its closeness to Atlas hide its existence, Pleione, also cataloged as BU Tauri, varies erratically between magnitudes 4.8 and 5.5. (By the way, for students of mythology, Pleione is not one of the seven sisters, but rather, their mother. Atlas was their father. Therefore, even on a good night, we are really only seeing five of the sisters.)

Moving farther away from civilization to skies of naked-eye limiting magnitude 6.0 skies brings the Pleiad tally to 11, while the darkest skies (naked-eye limiting magnitude 7.5) raise the count to 22. Not only does that account for all seven sisters, but it would appear that we are catching them in the middle of a big family reunion! Of course, how many of those stars you see depends not only on sky darkness, but also on your eye's visual acuity. I know that, personally, my vision has certainly taken a turn for the worse as I've gotten older. As a teenager, I used to see seven or eight stars routinely from my suburban backyard. Today, under similar sky conditions and even with eyeglasses to correct for my nearsightedness, I can only make out five or six. That's aging for you . . . although it still beats the alternative (which is to die young!).

The problem is actually two-fold. First, the stars are packed fairly tightly together, which can make resolving some close-set pairs difficult for some. If you have 20/20 vision, however, you should be able to see stars separated by only 1 or 2 arc-minutes. The tightest naked-eye Pleiads, 6th-magnitude 21 (Asterope) and 22 Tauri, are separated by 2.4 arc-minutes and should be faintly resolvable under dark skies.

Compounding the problem with the Pleiades is the fact that the cluster is immersed in a cloud of cosmic dust. Although seeing the nebulosity that engulfs the Pleiades cluster is a challenge left to giant binoculars and small telescopes (small-scope Challenge 69), the clouds can wash out the faintest naked-eye Pleiads under very dark skies. The problem is similar to how the Ring Nebula's glow makes it difficult to see its central star (monster-scope Challenge 170).

The first intensive study of the Pleiades was conducted three decades before the invention of the telescope. In 1579, the German astronomer and mathematician Michael Moestlin (1550–1631; sometimes spelled "Maestlin") inventoried the stars that he saw in a two-coordinate list of relative positions. His results were masterful, matching the correct positions of the stars as we know them today with uncanny accuracy. His catalog included 11 stars, although there is some additional evidence that he may have seen as many as 14.

Much more recently, the famous variable-star observer Leslie Peltier (1900–1980) regularly noted from 12 to 14 stars on moonless nights from his home in Delphos, Ohio. That's an impressive number, although to my knowledge the record for the most Pleiads seen by eye is still held by the late Walter Scott Houston. In one of his Deep-Sky Wonders columns in *Sky & Telescope* magazine, Scotty recalled that he counted 18 cluster members while stargazing near Tucson, Arizona, back in 1935. (In that same article, he noted that four decades later, he could only make out the five brightest from the same location because of light pollution.)

How about you? Take the Pleiades challenge on the next clear, moonless night and see how many you can count. But don't just go out and casually look up. Make a concerted effort to count the stars. If you are a purist, you will want to do your star count before consulting a chart, in much the same way as Moestlin had to, to avoid the power of suggestion. Or if you'd like a hand instead, bring a copy of Chart 2.13 with you. In either case, sit in a comfortable lounge chair, bundle up to match the temperature, and see how many stars you can count.

| 14 | Barnard's Loop | | | | | | ★★★★★ |

Target	Type	RA	Dec.	Constellation	Magnitude	Size	Chart
Barnard's Loop	Supernova remnant	05 54	−01 00	Orion	–	420′×60′	2.14

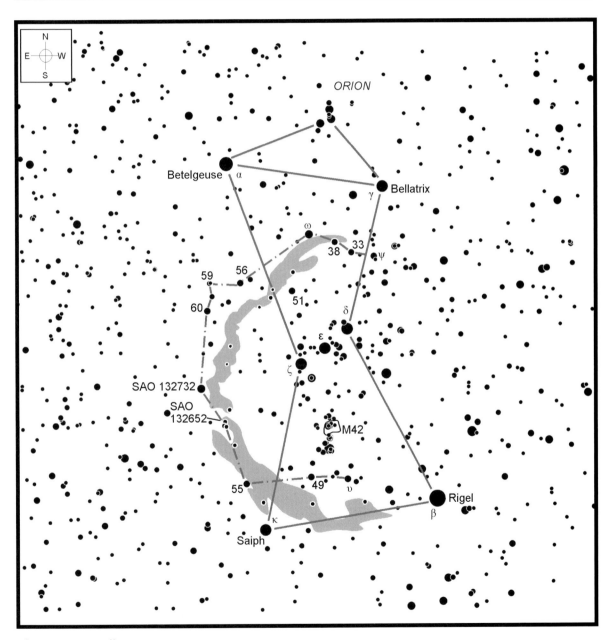

Chart 2.14 Barnard's Loop

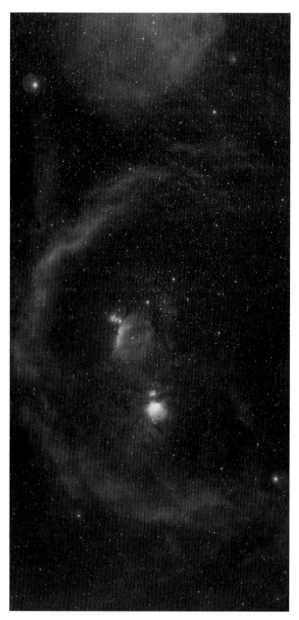

Figure 2.5 Barnard's Loop

One of the greatest naked-eye challenges goading amateur astronomers around the world is trying to spot the elusive arc of nebulosity known as Barnard's Loop (Figure 2.5). Cataloged officially as Sharpless 2-276, Barnard's Loop is a ghostly, 10°-wide semicircular bow of nebulosity that wraps around the eastern side of Orion the Hunter. In long-exposure photographs, it bears the unmistakable resemblance to portions of the

Veil Nebula supernova remnant in Cygnus (binocular Challenge 32). Spotting it by eye stands as a monumental test for observers.

Although Barnard's Loop was named for the renowned American astronomer Edward Emerson Barnard (1857–1923), who described the scene he captured on photographs made in October 1894 as a "great nebula extending in a curved form over the entire body of Orion," Barnard was not the first person to glimpse the Loop. Records show that Barnard's Loop was discovered visually by Sir William Herschel. Herschel published observations of 52 broad regions of the sky that he thought contained traces of nebulosity. The region around Barnard's Loop is listed as Area #27 and is centered at right ascension (RA) 05h 48.3m, Declination +01° 09.9'. Practicing an economy of words, Herschel simply described his 27th entry as "affected with milky nebulosity."

Few confirming observations were made of Herschel's 52 nebulous regions, igniting a debate over their existence that raged in certain astronomical circles for more than a century. While some of Herschel's 52 regions have subsequently been proven false, Barnard's images of Area #27 left little doubt about its existence.

Debate over Herschel's Area #27 continues to this day, but now it revolves around seeing the Loop by eye. Many amateurs have noted sections of the Loop through surprisingly small apertures, ranging in size from 50-mm binoculars to 3- to 5-inch (76- to 127-mm) telescopes. But can Barnard's Loop be seen by eye alone? It certainly is large enough, spanning the height of Orion's star-studded torso. Is it too faint, or more correctly, too red for the human eye to detect? The answer is "no;" it can and has been glimpsed without optical aid. But there are a few caveats.

I have read many accounts of observers claiming to have seen Barnard's Loop by eye, but I suspect many of them are false. That's not to say the observers are falsifying what they saw. I don't doubt their honesty in the least. But from their descriptions, I suspect that they did not see the *real* Loop, but rather a string of faint stars that follow very nearly the same path through Orion. The False Loop is formed by 10 stars that shine between magnitude 4.5 and 5. The illusion begins north of the Belt stars at Psi (Ψ) Orionis, and then hooks counterclockwise around the Belt, connecting the stars 33, 38, Omega (ω), 56, and 60 Orionis. The False Loop then winds to the southwest, linking the faint stars SAO 132732, and 55, 49, and Upsilon (υ)

Orionis as it curves between Orion's Sword and the stars Saiph and Rigel. Although these stars are widely separated, as evidenced in Chart 2.14, the brain tends to play tricks on us when we're not careful. Rather than interpret the False Loop as a series of faint stars, our eye–brain system tends to fill in empty gaps to create a single image, especially at low light levels. This optical illusion is caused by our psychological tendency to connect indistinct features into some sort of comprehensible whole, and is exactly why Percival Lowell saw straight canals crisscrossing Mars.

In order to see the real Barnard's Loop, several factors have to come together. First, a clear, dark night free of any trace of moonlight, haze, and clouds is an absolute must. Light pollution, especially in the direction of Orion, is also a no-no. It is best to wait for Orion to be highest in the sky, to further remove any terrestrial interference. You, the observer, should be seated or lying down; standing will only cause eye strain and interference. The best solution would be to lie on a chaise longue tilted so that you are looking at Orion more or less straight on. You also need to know the point where your eyes' peripheral vision is most sensitive. Review the discussion in Chapter 1 for further thoughts.

If you have them available, try narrowband and hydrogen-beta filters to improve image contrast. If possible, hold identical filters in front of both eyes simultaneously to take advantage of binocular vision. Some observers report good results with these, but oxygen-III filters seem to offer little benefit.

Start with the Loop's brightest segment, which lies just south of 56 Orionis and ends just west of SAO 132732. If you spot that segment successfully, see if you can extend it toward the region in between 56 and 51 Orionis. Try staring toward the Belt stars, while focusing your peripheral vision toward SAO 132732.

Okay, take a breath. Unfortunately, the southern half of Barnard's Loop is quite a bit fainter than the northern half. To spot the segment lying between Saiph [Kappa (κ) Orionis] and Rigel [Beta (β) Orionis] try blocking both stars with your fingers using the "V for Victory" sign. Doing so just might make you victorious.

15   M35

Target	Type	RA	Dec.	Constellation	Magnitude	Size	Chart
M35	Open cluster	06 09.0	+24 21.0	Gemini	5.1	28.0′	2.15

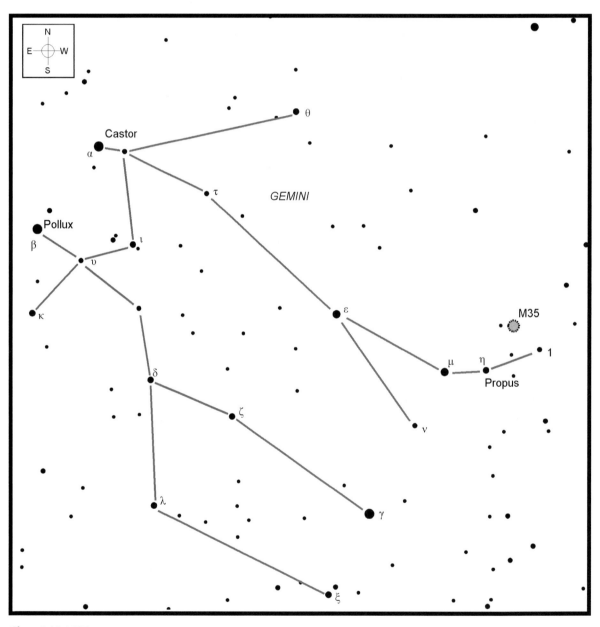

Chart 2.15 M35

A battle is portrayed among the constellations of our winter sky, as Orion the Hunter tries his best to rescue the Seven Sisters (naked-eye Challenge 13) from the clutches of Taurus the Bull. If we enlarge our view, it's almost as if this is a street brawl, with many onlookers watching the action, but doing nothing to help either side. We have Auriga the Charioteer to the north of Taurus, and the Gemini brothers to the Bull's northeast. Were this an actual fight, however, the odds are good that the Gemini twins, Castor and Pollux, would have been the first to jump into the fray. In ancient Greek mythology, Castor was a famous horse trainer and soldier, while Pollux (the Latinized version of his original Greek name, Polydeuces), is remembered as Sparta's leading boxer. The circumstances of their birth are nothing short of soap opera material. Though considered twins in that their mother, Leda, Queen of Sparta, gave birth to both at the same time, Castor's father was Leda's husband, King Tyndareus, while Pollux was fathered by Zeus.

Castor and Pollux were also members of the Argonauts, the legendary crew of the ship *Argo Navis*. Tales recount how they saved the crew of the *Argo* on more than one occasion, including once when Pollux used his boxing skill to defeat Amycus, a son of Poseidon, Greek god of the sea.

The heads of the Gemini twins are marked by the bright stars Castor [Alpha (α) Geminorum] and Pollux [Beta (β) Geminorum]. Unlike the twins that they represent, the stars are not related. Castor, a spectral type A system, lies some 50 light years away, while Pollux, a type K orange giant, is 34 light years from Earth. Each twin's stick-figure-like body extends

southwestward from its head star. The brothers are portrayed as arm-in-arm, appearing as if standing on the hazy band of our wintertime Milky Way galaxy, which flows along the constellation's southern border.

If you trace Castor's stick-figure body southwestward, you will eventually come to the star Propus, which literally means "forward foot." Under dark skies, Castor's foot and ankle are formed from a three-star arc consisting of Propus, Mu (μ), and 1 Geminorum. Look carefully and you might notice a soft, 5th-magnitude blur of starlight just north of his big toe, 1 Gem. That's the open cluster M35. I like to call it the "Soccer Ball Cluster," since it looks like Castor is about to score a goal!

Spotting M35 by eye is a good test for darker suburban skies. In his March 1989 Deep-Sky Wonders column,[3] Walter Scott Houston noted that he could see M35 naked eye with the help of a nebula filter. Although he didn't elaborate further, I would imagine that he was using a so-called broadband "light-pollution" filter. Narrowband filters would likely block too much light to be of much use in this application.

In that same column, Houston mused as to whether or not anyone ever glimpsed any of the cluster's individual stars without optical aid. Estimates place more than 400[4] stars within M35, although none shines brighter than 8th magnitude.

---

[3] Walter Scott Houston, "Deep-Sky Wonders," *Sky & Telescope*, March 1989, pp. 338–40.
[4] Although many popular references state that M35 contains more than 200 stars, Archinal and Hynes *Star Clusters* identifies 434 stars within the cluster.

16   M41

Target	Type	RA	Dec.	Constellation	Magnitude	Size	Chart
M41	Open cluster	06 46.0	−20 45.3	Canis Major	4.5	38.0′	2.16

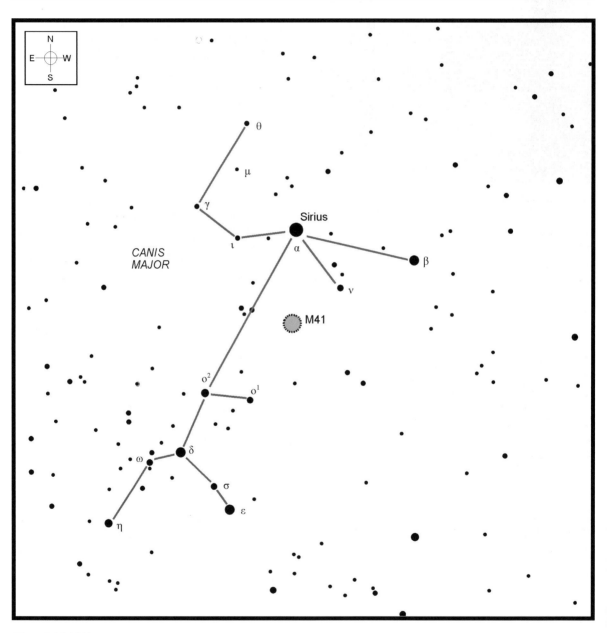

Chart 2.16 M41

Figure 2.6 M41

mentions a star cluster *"iuxta Syrium ad ortum,"* which translates to "near Sirius, to the east." Despite Hodierna's error placing the mysterious object to the east of Sirius instead of to its south, two maps found among his manuscripts clearly show M41 in its proper location.

Hodierna's observations were not widely known at the time, however, so he never received due credit for his discoveries by his contemporaries. Instead, John Flamsteed's independent discovery of M41 in 1702 became the first widespread notice of the cluster's existence. Messier added it to his catalog as entry number 41 in 1765.

All of these early observations of M41, apart from Aristotle's possible sighting, were made through telescopes. Yet the cluster today is listed as having an integrated magnitude of 4.5. Did any of these observers, or anyone else for that matter, ever notice it by eye alone? Perhaps, but few observing guides mentioned its visibility. The only popular works to mention its naked-eye detectability were *Celestial Objects for Common Telescopes* by the Rev. T. W. Webb (first published 1859), and William Tyler Olcott's *In Starland with a Three-Inch Telescope* (G. P. Putnam's & Sons, 1909).

To find M41 for yourself, look 4° due south of Sirius [Alpha (α) Canis Majoris] for a soft glow that is clearly not a starlike point. My own observing notes from years ago recall my first naked-eye encounter with M41. In the late winter of 1986, during the peak of "Halley Fever," many people headed south to see the famous comet at its predicted peak brightness. From my vantage point in Flamingo, Florida, at the southernmost tip of Everglades National Park, Halley was rather disappointing, but the view of the far-southern sky was spectacular. M41 was not just suspected by eye; it was easily visible as a blurry smudge.

Two important factors undoubtedly contributed to M41's obviousness back then. Without question, Everglades Park is very dark. Beyond that, the Everglades' southerly latitude, 25° north, allows M41 to culminate 45° above the southern horizon, far above any extinguishing haze or fog.

As is so often the case, however, once a challenging observation is made under pristine conditions, it is often repeatable under less than ideal circumstances. Back home on Long Island at latitude 40° north, I can see M41 frequently from an observing site that typically only garners a naked-eye limiting magnitude of 5.5.

When we think of classical deep-sky observers, names that come to mind include Charles Messier; William, John, and Carolyn Herschel; Pierre Méchain; and others from the 18th and 19th centuries. As great as they all were, however, they were really johnnies-come-lately. Who was history's first deep-sky observer? Technically, that distinction probably falls to an unsung cave dweller who first looked up and saw the Pleiades star cluster (naked-eye Challenge 13). One of the first stargazers in historic times to record the appearance of a deep-sky object, however, was the classic Greek philosopher and scientist Aristotle (384–322 BC). In 325 BC, he noted in his treatise *Meteorologies* a mysterious "cloudy spot" to the south of Sirius (Figure 2.6). Was Aristotle's observation the first sighting of open cluster M41? Some astronomical historians believe so. Others, however, are not convinced. Instead, they believe Aristotle saw a knot in the winter Milky Way and not the cluster itself. Whatever that glowing patch was, Aristotle could tell that it was not a single star.

We need to travel nearly two millennia through history until we find reference to the open cluster we know today as M41. In 1645, Giovanni Batista Hodierna (1597–1660), priest and court astronomer for the Duke of Montechiaro on Sicily, published a list of 40 sky objects that he discovered with a simple 20-power Galilean refractor. One of his entries

That raises an interesting question. How far north can M41 still be successfully spotted without optical aid? Try your luck on the next cold, clear, crisp winter's night, after a burst of dry arctic air sweeps away accumulated haze. While you're at it, be sure to enjoy M41 through binoculars or a telescope. Even the smallest instruments will resolve many of its 80 stars. Study the colors of those stars, as many show subtle tints of yellow and orange. An orange sun just to the south of the cluster's center, cataloged as SAO 172290, stands out especially well.

17 Young Moon/ old Moon

Target	Type
Young Moon/old Moon	Solar System

One of the most beautiful naked-eye sights in the sky is the thin crescent Moon (Figure 2.7) hanging low in the western sky during evening twilight, or low in the east as dawn approaches. Whether positioned over a tranquil beach, a distant mountain, or a bustling cityscape, this monthly scene is sure to attract the attention of dyed-in-the-wool stargazers and give pause to passersby who just happen to look skyward as they go about their earthly affairs.

There is more to see than just the slender crescent, however. Look carefully and you should notice the entire outline of the lunar disk, including the greater portion not yet illuminated directly by sunlight. The nighttime portion of the lunar disk is being illuminated softly by sunlight reflecting off the Earth. If we were on the Moon, we would see Earth going through phases just as we see the Moon doing throughout the month from Earth. The Earth's phase at any given moment, however, is the opposite, or more correctly, the *complement*, of the Moon's phase. In other words, when we see a thin crescent Moon here on Earth, someone on the Moon would see a fat gibbous Earth. And that's exactly why we can see the "full" lunar disk during the waxing (and waning) crescent. Sunlight reflecting off the Earth bounces back into space and onto the Moon, an effect known as *earthshine*.[5]

Just how thin a crescent Moon can you see? One of the greatest challenges posed to naked-eye observers is trying to spot the crescent Moon less than 24 hours after the instant of New Moon. By definition, New

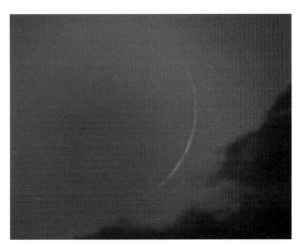

Figure 2.7 Young Moon

Moon occurs when the position of our satellite along the *ecliptic* (that is, its celestial longitude) is precisely the same as the Sun's. If timing works out so that the exact moment of New Moon occurs in the early predawn hours from your location, then a very thin crescent may just be visible that same evening immediately after sunset.

The Moon's celestial latitude also plays a big role. If the Moon was at the same celestial latitude as the Sun, referred to as an *orbital node*, we would have a total solar eclipse. The lunar orbit, being tilted 5° with respect to Earth's solar orbit, usually carries the Moon to the north or to the south of the Sun at New phase. The farther away the Moon appears from the Sun at New, the better your chance of spotting the young crescent.

Besides good timing, what does it take to spot the Moon just hours after New (or before that point, in the early morning)? First and foremost, just as in real estate, the three Ls come into play: location, location, location! A perfectly clear view in the correct direction is a must.

---

[5] During the days of prohibition in the 1920s, when consuming alcohol was banned, as well as in the years of self-imposed temperance leading up to the passage of the 18th Amendment, stills were set up far and wide to manufacture illegal alcohol. The mood of the nation, at least outwardly, was such that these clandestine operations were conducted mostly at night, by the light of the Moon. From this, the liquor became known as *moonshine*.

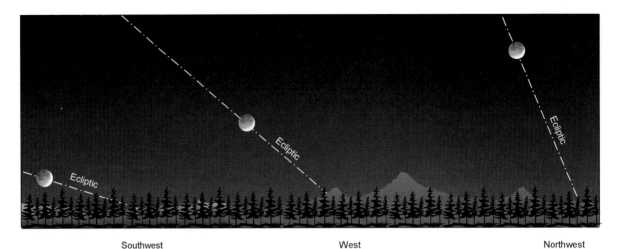

Figure 2.8 The angle of the ecliptic to the evening western horizon during the northern hemisphere's autumn (left), summer/winter (center), and spring (right)

Not only must the horizon be free of earthly obstructions, there also must be no hint of any obscuring clouds, haze, smog, or fog. As discussed in Chapter 1, the clearest skies occur when nautical and aviation forecasts quote horizontal visibilities between 60 and 100 miles (96 and 160 km).

Elevation above sea level is also a vital consideration. Not only will the chances of an unobstructed horizon improve, but the air is likely to be less hazy and darker than from a beach or a valley.

Ideally, the Moon should be positioned above the horizon at the same azimuth as the setting Sun is below the horizon. Owing to the Earth's rotational axis being tilted $23\frac{1}{2}°$ with respect to its orbit of the Sun (the *ecliptic*), this only happens twice a year, around the equinoxes. As Figure 2.8 shows, the best months for spotting a very young Moon from the northern hemisphere are March and April evenings. If you want to find the Moon less than a day before New, then your best chance from north of the equator is during the predawn hours of September and October. The ecliptic is nearly perpendicular to the horizon at these times, which places the Moon as high in the sky as it can be. (This issue is addressed again in naked-eye Challenge 18, Spotting Mercury.)

The Moon should also be at or near the perigee point in its orbit (closest distance to Earth) at New Moon, since its closer distance will result in faster

motion away from the Sun in our sky than when it's at or near apogee (farthest from Earth).

Clearly, spotting a very young Moon takes a lot of planning! Begin your own attempt by confirming when both the Sun and the Moon will set from your observing site. You can find that out, as well as the altitude and azimuth of each, through the US Naval Observatory Astronomical Applications Department's Data Service's website, http://aa.usno.navy.mil/data. Select what you are interested in finding, such as the altitude and azimuth of the Moon and Sun on a given date, enter your location (or longitude and latitude, if not in the location database), and press "compute."

Head out at sunset and immediately scan the area around the point where the Sun slipped below the horizon. Slowly scan along the horizon. Although this is a naked-eye challenge, binoculars will prove immensely helpful, since the very young Moon will lie deep in the bright twilight sky.

To date, the record[6] for seeing the youngest Moon visually with optical aid is held by Mohsen G. Mirsaeed

---

[6] Using a refractor outfitted with special baffling, visual and infrared filters, and an optimized digital camera, German astronomer Martin Elsässer was able to capture images of the crescent Moon just 10 minutes before New Moon in May 2008. There's a record that is likely to stand for quite a while! Unless, of course, you count images of total solar eclipses that have been overexposed to show the darkened lunar maria on the silhouetted lunar disk. If so, then many have captured images of the 0-hour-old Moon.

of Tehran, Iran. Using 150-mm binoculars, Mirsaeed spotted the Moon on September 7, 2002, when it was just 11 hours 40 minutes after New and only 7.3° from the Sun. Wow! Mirsaeed's feat is very impressive, but it also brings up a good point. Never look for the Moon when it is so close to the Sun that you could accidentally aim toward the latter by mistake. Unfiltered sunlight can damage your vision beyond repair, possibly even resulting in permanent blindness.

The record for spying the youngest crescent Moon without any optical aid is held by *Astronomy* magazine contributing editor Stephen James O'Meara. He spotted the young crescent with the unaided eye just 15 hours 32 minutes after New Moon on May 24, 1990. Records are meant to be broken, however. Can you do better?

If you meet this challenge and catch a glimpse of an extremely young (or extremely old) crescent, there is always the additional test to spot so-called opposing crescents. These are extremely rare sightings of the waning crescent within 24 hours of New, followed the next day with a sighting of the waxing crescent less than 24 hours after New. New York comet expert John Bortle was the first to report success catching sight of opposing crescents in April 1990. Bortle is quick to point out, however, that his success came only after "two decades of attempts; most were thwarted by weather."

★

| 18 | Spotting Mercury |

Target	Type
Mercury	Solar System

A popular "urban legend" that has been around for years claims that Nicolas Copernicus, who conceived the heliocentric Solar System, never saw the planet Mercury, not even once, during his lifetime. Whether or not that is true remains open to debate, but one thing is for certain. Despite its shining as brightly at magnitude −2 at times, Mercury is the most difficult naked-eye planet to observe.

That's because Mercury is a victim of circumstance. Part of the dilemma is that, being so close to the Sun, Mercury never gets more than 28° away from the Sun in our sky, even under the best of conditions. Like spotting the young Moon in the last challenge, seeing Mercury requires good timing. That is not to say that Mercury will be 28° from the Sun *in our sky*, but rather that it is never more than that distance away from the Sun *along the ecliptic*. These instants are referred to as points of greatest elongation. Mercury is said to be at greatest eastern elongation when it lies at maximum distance to the east of the Sun. This places it low in the western evening twilight immediately after sunset. Greatest western elongation occurs when Mercury is at its farthest distance west of the Sun, making it visible low in the eastern sky just before sunrise. As a result, the best chance for spotting Mercury in the evening is when it is at or near greatest eastern elongation. For early risers, Mercury is best seen at or near greatest western elongation.

Like the young-Moon challenge, the angle that the ecliptic makes with the observer's horizon is critical. Simply put, steeper is better. For those of us living in the northern hemisphere, the ecliptic is most steeply

pitched immediately after sunset around the Vernal Equinox. Six months later, the narrow angle that the evening ecliptic makes with the horizon is poorly placed for spotting Mercury. In the morning sky, however, it is nearly perpendicular to the eastern horizon, making the dates around the Autumnal Equinox the best for spotting the innermost planet before dawn.

Actually, there's more to it than this. Compounding the problems facing observers looking for the innermost planet is the fact that Mercury's orbit is very eccentric. At *perihelion* (the point in a planet's orbit where it is closest to the Sun), Mercury is 28.5 million miles (46 million km) from the Sun. Half an orbit later at *aphelion* (the orbital point farthest from the Sun), Mercury is 43.3 million miles (69.8 million km) away. This means that Mercury is about 1.5 times more distant at aphelion than at perihelion. Clearly, the best time to spot Mercury is when it is at or near greatest elongation and at aphelion, since both together will maximize the planet's distance from the Sun in our sky.

Like seeing the young Moon, seeing Mercury requires a clear horizon that is free of earthly obstructions and atmospheric haze and clouds. That doesn't necessarily mean a mountain peak thousands of feet above sea level. In fact, one of my most memorable sightings of Mercury occurred one evening years ago as I was riding the Staten Island Ferry across New York Harbor. I can still close my eyes and see Mercury pass behind the Statue of Liberty as we sailed across the placid harbor − an image that I still carry with me in the photo album of my mind.

19 Spotting Uranus

★
★★
★

Target	Type	Chart
Uranus	Solar System	2.17

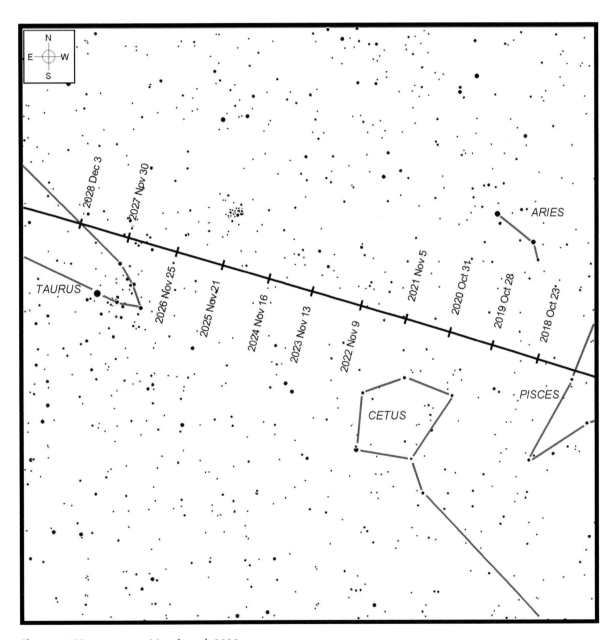

Chart 2.17 Uranus at opposition through 2020

On March 12, 1781, the Solar System was a simple, very-well-behaved place that was best summed up with the phrase "what you see is what you get." There was the Sun, the Moon, and the five planets: Mercury, Venus, Mars, Jupiter, and Saturn. Apart from a handful of moons orbiting some of the planets that required a telescope to be seen and the occasional faint comet, the entire contents of the Solar System were naked-eye territory.

All that changed on the very next night. While looking through his homemade 6.2-inch (157-mm) *f*/13.6 reflecting telescope, William Herschel stumbled onto something unexpected. Herschel, a German-born amateur astronomer living in England, had found a greenish star that was not shown on any of his star charts. "What a peculiar looking star," he must have thought. In fact, if he looked carefully, he could see it was not a star at all, but rather a fuzzy disk.

Returning to the same spot in the sky over the next several nights, Herschel found that his greenish discovery had moved just slightly against the background of stationary stars. The fact that the Solar System had five planets was well established, leading him to believe initially that he had discovered a comet. Only after many additional observations were compiled and examined was a definitive orbit calculated. Whatever Herschel had discovered was located far beyond Saturn, the most distant planet known at the time. Further, it was following a roughly circular orbit, quite a contrast to the highly elliptical orbits of comets.

It eventually became clear that Herschel had discovered a new member of the Sun's planetary family. Herschel referred to his new find as Geogium Sidus in honor of King George III, the English monarch at the time. His suggestion didn't stick, however. Instead, the name Uranus, proposed by Johann Elert Bode, was soon adopted by the rest of the astronomical world.

The fact that Uranus hovers near the naked-eye magnitude limit immediately raised the question in Bode's mind whether Uranus had ever been seen by others, either with or without optical aid, before Herschel's discovery. Looking at western records only, it appears that Uranus had been seen more than 20 times prior to that fateful night in 1781, yet no observer ever recognized it as more than a faint star. That's likely due to the planet's slow solar revolution. Uranus takes just

over 84 earth years to complete one voyage around the Sun.

The first recorded sighting of Uranus was in 1690 by John Flamsteed, the first Astronomer Royal at the Royal Greenwich Observatory. Flamsteed is most remembered for compiling his famous star catalog in which he assigned numbers to stars within each constellation in order of increasing right ascension. Although his catalog was initially created to help British navigators determine their longitude when sailing the high seas, astronomers still use Flamsteed's catalog to this day. While surveying Taurus on December 23, 1690, Flamsteed noted several faint stars between the Pleiades and the Hyades. He assigned one star in particular the designation 34 Tauri and moved on. As did "34 Tauri!" In fact, 22 years later, Flamsteed met "34 Tauri" again after both had progressed to Leo the Lion. Flamsteed recorded Uranus in 1712 near the star Rho (ρ) Leonis. He also recorded it four more times in 1715, when Uranus was south of Sigma (σ) Leonis.

No fewer than three other astronomers also saw Uranus before Herschel, but did not recognize it as more than a faint star. The third Astronomer Royal, James Bradley, glimpsed it three times, first as a dim point in Capricornus in 1748, and again in Aquarius in 1750 and 1753. Tobias Mayer, professor of economics and mathematics at Georg-August University of Göttingen, Germany, made a single observation in 1756 when Uranus was also within Aquarius. Finally, there is the hapless tale of Pierre Charles Le Monnier, a French astronomer and physicist. Remembered best for his temper and squabbling nature, Le Monnier apparently saw Uranus as many as ten times between the years 1764 and 1771. Six of those observations came within a four-week window during January 1769, when Uranus had wound its way into Aries.

Uranus has completed almost three orbits of the Sun since Le Monnier's time. Over the decade 2010 to 2020, the seventh planet will span barren Pisces on its way to crossing into Aries in 2018. The stark backdrops of these ecliptic constellations means that the chances are good that you will be able to see Uranus without any optical aid provided your sky is dark and light-pollution free.

To better the odds, look for Uranus when it is near opposition and its distance away is minimal. Chart 2.17 plots the planet's location among the stars on each of the dates of opposition until the year 2020.

Unfortunately, Uranus passed aphelion (its farthest point from the Sun – 1.9 billion miles or 3 billion kilometers) in February 2009. As a result, it is near its dimmest magnitude for the next several years. Things will slowly improve over the next half century, as Uranus slowly progresses toward perihelion (its closest point to the Sun, 2.73 billion kilometers) in 2050. Uranus's brightness will change from magnitude 5.7, where it is now, to a maximum of 5.3 near perihelion.

20 Glimpsing Vesta

Target	Type	Chart
Vesta	Solar System	2.18

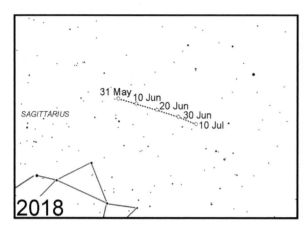

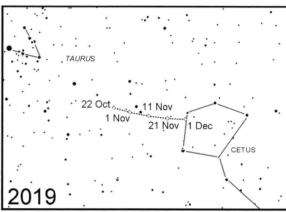

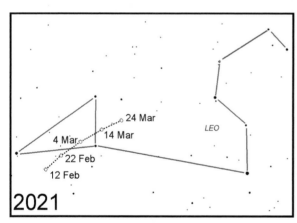

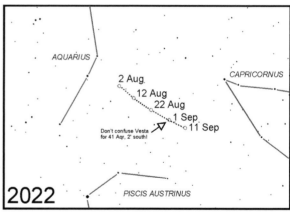

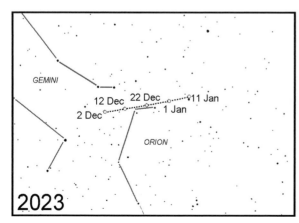

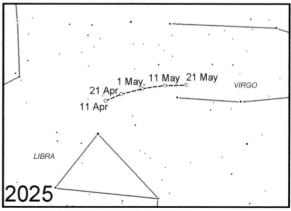

Chart 2.18 Vesta at opposition

On January 1, 1801, the Solar System became a little more crowded still. That night, the Sicilian astronomer Giuseppe Piazzi stumbled upon the first asteroid, which he named Ceres after the Roman goddess of grain. Now reclassified as a dwarf planet by the International Astronomical Union, Ceres was the first object discovered to be orbiting the Sun between Mars and Jupiter.

As is so often the case in astronomy, once the first of something is discovered, the flood gates open. That certainly has been the case with asteroids. The second asteroid, Pallas, was found in 1802, while the third, Juno, was discovered in 1804. By the end of the nineteenth century, several hundred were known.

Surprisingly, even though Ceres is the largest member of the clan at about 600 miles (960 km), it is not the brightest. That honor goes to Vesta, the fourth asteroid discovered. Vesta was found on March 29, 1807, by the German physician Heinrich Olbers. Olbers had also discovered Pallas two years earlier.

Although oblong, Vesta only has an average diameter of about 320 miles (520 km), it can outshine Ceres by more than a full magnitude at times. While Ceres reflects about 10% of the sunlight striking it, Vesta's bright surface reflects more than 30%. Why that is exactly remains a mystery, although studies show that Vesta's surface is apparently covered with basalt, a byproduct of volcanism. This suggests that, at some point in the distant past, Vesta was volcanically active, or perhaps was part of a larger body that was.

These mysteries, and more, are part of the reason why NASA launched the Dawn spacecraft in September 2007. Dawn's mission is to explore both Vesta and Ceres. As it stands now, Dawn will arrive at Vesta in

Table 2.4 *Dates of Vesta oppositions*

Year	Date of opposition	Constellation	Magnitude	Chart
2018	20 Jun	Sagittarius	5.3	2.18a
2019	11 Nov	Cetus	6.5	2.18b
2021	4 Mar	Leo	6.0	2.18c
2022	22 Aug	Aquarius	5.8	2.18d
2023	22 Dec	Orion	6.4	2.18e
2025	1 May	Libra	5.6	2.18f

August 2011 and spend the next nine months in orbit. It is due to leave Vesta in May 2012 and move on to Ceres, arriving there in February 2015.

Meanwhile, Vesta has the distinction of being the only asteroid to crack the naked-eye barrier. For one to three weeks either side of its reaching opposition, Vesta can be seen without any optical aid provided the sky is clear and dark, and you know exactly where to look. Chart 2.19 plots the paths that Vesta will follow in the sky during the next six oppositions, while Table 2.4 lists the dates of each. Vesta's magnitude, omitting the decimal, on each date of opposition is shown in parentheses. Each chart plots stars down to approximately 7th magnitude.

Of these, Vesta will be magnitude 5.3, about as bright as it ever gets, around its 2018 opposition. Unfortunately, that same opposition has Vesta passing through the crowded constellation Sagittarius, where many faint stars will give it plenty of competition. It is nearly as bright during the 2011 and 2014 oppositions. These passages carry Vesta through the constellations Capricornus and Virgo, respectively, both of which have far fewer faint stars to masquerade as the asteroid.

21   Seeing the zodiacal light and the Gegenshein

Target	Type
Zodiacal light and Gegenshein	Solar System

We all know that our Solar System consists of the Sun, eight major planets, three dwarf planets, hundreds of satellites, tens of thousands of asteroids and probably millions of comets and Kuiper Belt Objects. We are also usually told that in between those bodies are huge gaps of nothingness. In reality, interplanetary space is anything but empty. In fact, it's littered with debris and dust left over from the formation of the Solar System itself, as well as scattered rubble from collisions between asteroids. Each grain of this so-called interplanetary dust is like a tiny microplanet orbiting the Sun.

Interplanetary dust particles typically range between 10 and 100 micrometers across, which is smaller than the width of a human hair, although some may be a meter or more in size. Seeing them from Earth is out of the question. Or is it? While we can't hope to see a single particle, we can enjoy their collective presence in the night sky immediately after sunset or right before dawn. When sunlight strikes these uncountable grains of dust, we can see their combined glow in the sky. Since most of this material lies along the ecliptic, just as the planets do, we call this phenomenon the *zodiacal light* (Figure 2.9).

With proper conditions, the zodiacal light can be seen as a very dim, wedge-shaped cone of light rising above the western horizon about 60 to 90 minutes after sunset and for about the same length of time above the eastern horizon before sunrise. The glow, however, is much dimmer than the Milky Way's hazy band, so dark conditions are a must. Any hint of moonlight, light pollution, clouds, or haze will obscure it completely.

As with spotting the young Moon and the planet Mercury, your best chance of seeing the elusive zodiacal

Figure 2.9 Zodiacal light

light is when the ecliptic is angled steeply to the horizon. Again, for evening observations from the northern hemisphere, that is, from mid-March to mid-April, the weeks immediately around the Vernal Equinox. Early morning observations are most successful around the Autumnal Equinox in September.

The window of opportunity is even smaller, however, because of the zodiacal light's tenuous nature. Not only must the ecliptic be angled steeply against the horizon, any foreign light, whether from our Moon or an errant streetlight, must also be absent from the sky. Any interference whatsoever will quench the dim glow of the zodiacal light.

My first encounter with the zodiacal light was in April 1986 when I was in Florida's Everglades National Park to view Halley's Comet. As I enjoyed the darkening evening sky, scanning with my binoculars what for me are far-southern skies, I couldn't help but notice the western sky staying brighter long after twilight should have ended. Only later did it dawn on me, if you'll pardon the pun, that this wasn't a protracted sunset at all; this was the zodiacal light.

That same confusion, incidentally, led the twelfth-century Persian astronomer Omar Khayyam to refer to it as a "false dawn" when he saw it in the morning sky. It also struck me as an extended sunset when I next saw it years later at the Winter Star Party. Again, the southerly latitude of the star party, held annually in the Florida Keys, tilted the ecliptic nearly perpendicular to the western horizon. The image of this ghostly cone of light rising up from the western horizon and passing through the dim stars in Pisces and Aries before ending near the gentle Pleiades star cluster, is another one of those mental "Kodak moments" that will last a lifetime.

Although the cone of the zodiacal light appeared to end near the Pleiades that night, it actually continues to extend along the ecliptic and completely encircle the sky. The dim, all-encompassing extension of the zodiacal light is called the *zodiacal glow*. The zodiacal glow runs right along the ecliptic, averaging between 5° and 10° in width. Seeing it takes a very special sky indeed.

Related to both the zodiacal light and the zodiacal band is the *Gegenshein*. The Gegenshein, also known as the counterglow, is a brightening in the zodiacal band at a point in the sky exactly 180° away from the Sun. At this point, all of the particles of interplanetary dust are effectively at full phase, just as the Moon is when opposite the Sun in our sky. Because of this, the amount of reflected sunlight is peaked, resulting in the brighter appearance.

Seeing the Gegenshein is one of the toughest challenges in this book. Not only do you need perfectly dark skies and perfectly dark-adapted eyes, you also have to know when and where to look. In general, look upward at midnight to the point where the ecliptic passes through the meridian, the imaginary line that divides the sky in half, separating the eastern half from the western.

This point must also be far from the glow of the Milky Way, since that will be more than enough to hide the Gegenshein. The best time to see the Gegenshein is during late autumn, when it will appear against the star-poor constellations of Pisces and Aries, and again in late winter, when it is located in Cancer and Leo. Just be careful not to confuse the glow of the Beehive Cluster, M44, in Cancer for it. The Gegenshein is much fainter.

3
Binocular challenges

Two eyes are better than one. That's been my astronomical gospel ever since I got hooked by this hobby because of a homework assignment to view a total lunar eclipse. I watched the eclipse unfold through a pair of 7×35 binoculars and was transfixed by what I saw.

Those 7×35s helped introduce me to observational astronomy. Although those binoculars are long gone, my enjoyment for binocular astronomy has never waned. The binocular universe is a very personal one to me. I never head out to view through any telescope without a pair of binoculars flung over my shoulder.

What about you? Do you bring a pair of binoculars with you when you head outside for a night under the stars? If not, you really should! You might be surprised how much of the universe they will show you.

22 M82

Target	Type	RA	Dec.	Constellation	Magnitude	Size	Chart
M82	Galaxy	09 55.9	+69 41.0	Ursa Major	9.3b	11.3′×4.2′	3.1

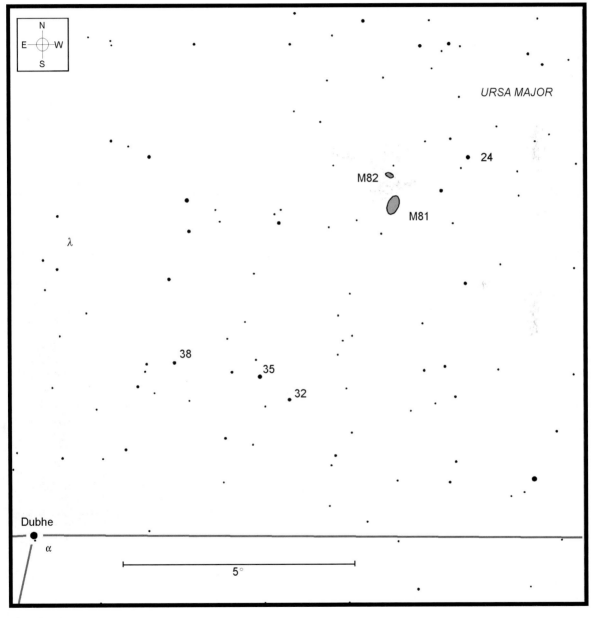

Chart 3.1 M82

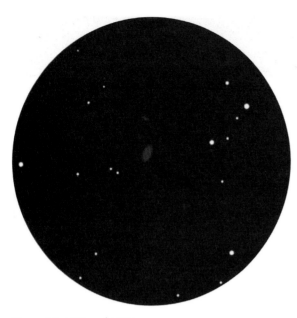

Figure 3.1 M81 and M82

As it dutifully circles the north celestial pole, the Great Bear, Ursa Major, offers a picture window into the depths of the universe. Thousands of galaxies lie within its borders, although nearly all are beyond the range of most binoculars. Except for two, that is.

M81 and M82 (Figure 3.1) form one of the most famous pairs of galaxies found anywhere in the sky. The former has already been described in the previous chapter, where it is listed as naked-eye Challenge 1. We return to M81 again here, as seeing it along with companion galaxy M82 is a good test for suburban binocularists.

Like its brighter companion, M82 was discovered by Johann Bode on the final day of the final month of 1774 when he also noticed a small, fainter splinter of light less than 1° to the larger galaxy's north. His observation was forgotten and both galaxies went unobserved for another five years until they were

independently rediscovered by Pierre Méchain. Charles Messier incorporated Méchain's find into his burgeoning catalog some 19 months later.

Zeroing in on M81 and M82 is easy through wide-field binoculars by following the same instructions given in naked-eye Challenge 1 in the last chapter. Now, some readers may be wondering why I would include M82 as a binocular challenge. To some, seeing M82 through common-size binoculars offers no challenge at all. Others, however, find it almost insurmountable. Even two observers standing side-by-side may have two totally different experiences seeing M82. It seems that spotting M82 through binoculars depends as much on a binocular's exit pupil as it does on sky darkness. In fact, I will go out on a limb and say that using the exit pupil is even more important than sky darkness when it comes to seeing M82.

To test this theory, I conducted a simple test using three different pairs of 50-millimeter binoculars from my suburban backyard. Since all three had the same aperture, their light-gathering prowess was the same. Each was mounted on a tripod to minimize shaking. I also waited until both galaxies were high in the sky to eliminate as much sky glow as possible. First, I raised my old 7×50s (7.1-mm exit pupil) their way. M81 was clearly visible, but I could do no better than suspect M82's existence with averted vision. Putting the 7×50s aside, I switched to my 10×50 binoculars (5-mm exit pupil). Sure enough, M81 was again clearly visible, as was M82 with averted vision. Finally, I tried a pair of 16×50 binoculars (3.1-mm exit pupil). The field background was noticeably darker but, sure enough, M82 was also more obvious.

The reason for the steady improvement is two-fold. First, by shrinking the exit pupil, the brightness of the background sky also decreased, resulting in improved image contrast. At the same time, the higher magnification increased image size, which also helps to bring out low-contrast targets.

If you can, try this experiment yourself and see if your results mirror mine.

23 M84 and M86

	Target	Type	RA	Dec.	Constellation	Magnitude	Size	Chart
23a	M84	Galaxy	12 25.1	+12 53.2	Virgo	10.1b	6.4′×5.5′	3.2
23b	M86	Galaxy	12 26.2	+12 56.8	Virgo	9.8b	8.9′×5.7′	3.2

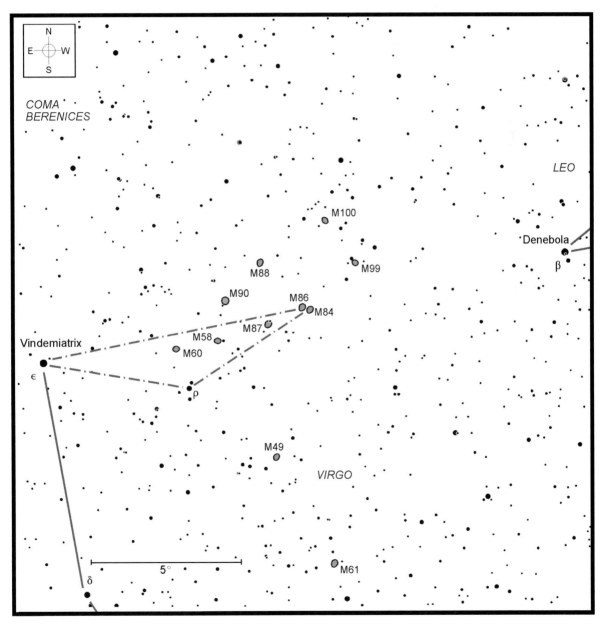

Chart 3.2 *M84 and M86*

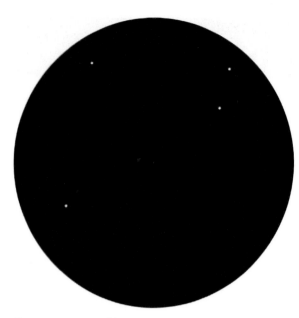

Figure 3.2 M84 and M86

Look at any deep-sky map of the spring sky and it's pretty clear that, when it comes to galaxy hunting, the constellations of Coma Berenices and Virgo are the places to be. The Virgo Cluster plays host to an estimated 1,300 individual galaxies and forms the heart of the Local Supercluster of which our Milky Way's Local Group of galaxies is thought to be an outlying member.

Of those 1,300 galaxies, two of the brightest are M84 and M86, found not far from the Virgo Cluster's geometric center. Both shine at about 9th magnitude, bringing them within the grasp of 40-mm binoculars under dark skies. Figure 3.2 shows them through 10×50 binoculars. They are separated in our sky by a scant 20 arc-minutes, making them look more like a

faint, fuzzy double star than a pair of extragalactic targets. Even after careful study, it's hard to tell their true identity through binoculars.

Studies show, however, that their twin appearance does not necessarily imply twin galaxies. While many sources continue to classify each as an elliptical galaxy, the most recent evidence indicates that M84 is actually one of those strange missing links, a lenticular galaxy. Also known as type S0 galaxies, lenticular galaxies share some of the traits of traditional spirals and other qualities found in elliptical systems. Lenticular galaxies contain very little interstellar matter and are made up of mostly old stars. Any dust is generally found only near the galactic core, not in outlying regions. Their spiral arms are difficult to make out even in deep photographs, which is why many refer to lenticular galaxies as "armless spirals."

Finding M84 and M86 through common binoculars takes a little extra effort because there are no bright stars in their immediate area. Center your binocular view on Vindemiatrix [Epsilon (ε) Virginis] and then shift 5°, or about about one field of view, west-southwest to a pair of suns formed by 5th-magnitude Rho (ρ) Virginis and a fainter 7th-magnitude star. M84 and M86 lie another 5° to the northwest of Rho. If you can imagine a large, flattened triangle with the corners formed by Vindemiatrix, Rho, and M84/M86, you should be able to zero in on the galaxies' location fairly easily. Actually seeing them, however, may require using some or all of the tricks mentioned in Chapter 1, including averted vision, mounting your binoculars to a steady support, and lightly tapping the side of the barrel to impart the slightest vibration to the field.

If you can find M84 and M86, then try your luck with the other Messier objects in the area. All shown on Chart 3.2 should be visible to patient observers using 50-mm or larger binoculars.

| 24 | **M104** | | | | | | | |

Target	Type	RA	Dec.	Constellation	Magnitude	Size	Chart
M104	Galaxy	12 40.0	−11 37.6	Virgo	9.0b	8.8′ × 3.5′	3.3

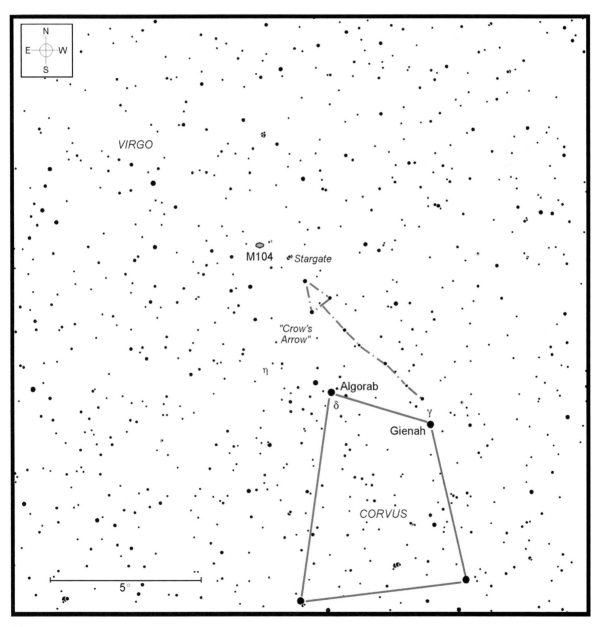

Chart 3.3 M104

M104, the famous Sombrero Galaxy in Virgo is one of
the most photogenic galaxies in the entire sky. The
"Sombrero" nickname comes from the broad lane of
dark nebulosity that slices along the rim of this edge-on
spiral. The appearance, strongly reminiscent of a
broad-brimmed Mexican hat, makes this a truly unique
sight among the tens of thousands of galaxies that
populate spring's night sky.

Pierre Méchain discovered M104 on May 11, 1781.
Although Méchain quickly notified Charles Messier of
his find, it was too late to be included in the published
list of 103 nebulous objects, which had been sent to the
printer's not long before. Once the printed version of
the catalog was released, Messier penciled in a
handwritten note about M104 in his personal copy.
Whether he ever intended to include it in a supplement
or perhaps in a second edition, we will never know, but
neither ever materialized.

Many years later, the French astronomer Camille
Flammarion (1842–1925) stumbled upon Messier's
personal copy of his catalog while browsing through a
bookstore. After perusing his incredible find,
Flammarion thought it fitting to add M104 to the list in
1921, the first of seven posthumous additions to the
Messier catalog. (An historical aside: Flammarion also
identified the galaxy NGC 5866 in Draco as the missing
M102. This latter decision, however, is still debated
among astronomical historians, as most now believe
that M102 was a mistaken duplicate observation of
M101 in Ursa Major.)

Today, we know that M104 (Figure 3.3) is probably
an outlying member of the Virgo Cluster, which also
includes M84 and M86 (Challenge 23). At a projected
distance of 65 million light years, however, M104 is on
the far side of the cluster from our vantage point. The
center of the Virgo Cluster lies some 20° to the north of
M104, and is approximately 10 million light years
closer.

Even at this vast distance, M104 still shines at 8th
magnitude, which is within reach of 7×35 binoculars.
You just have to know where to look. There are several
ways of finding M104 but, for me, the easiest is simply
to follow the arrow. Which arrow is that? Aim your
binoculars toward the star Gienah [Gamma (γ) Corvi]
marking the northwestern corner of the distinctive

Figure 3.3 M104

trapezoidal body of neighboring Corvus the Crow. Just
to its north is a 6th-magnitude star (HD 106819) that is
the first in a line of faint stars extending to the northeast.
Add to them three more stars set in a triangle at the
opposite end of the line, and you have the "Crow's
Arrow," as I call it. The tip of the Crow's Arrow points
right at a tiny knot of even fainter stars. With 15× or
higher power binoculars, the tiny knot can be resolved
into six stars that form a strange little asterism that
looks like a triangle within a triangle. Several years ago,
Texas amateur astronomer John Wagoner christened it
the "Stargate," which seems like a great name since it
looks like some sort of strange space portal.

M104 lies just to the northeast of the Stargate in the
same field of view. Most binoculars show it as a tiny
oval glow extending east–west. Owing to its southerly
declination, however, trees, clouds, and light pollution
may hamper seeing it. If that's the case, wait until it
culminates (that is, is due south in your sky and highest
above the horizon) and give it another go. Giant
binoculars (greater than 10× and 70-mm aperture)
may also add just a hint of the Sombrero's brim of dark
nebulosity.

25 NGC 5128

Target	Type	RA	Dec.	Constellation	Magnitude	Size	Chart
NGC 5128	Galaxy	13 25.5	−43 01.0	Centaurus	7.8b	25.8′×20.0′	3.4

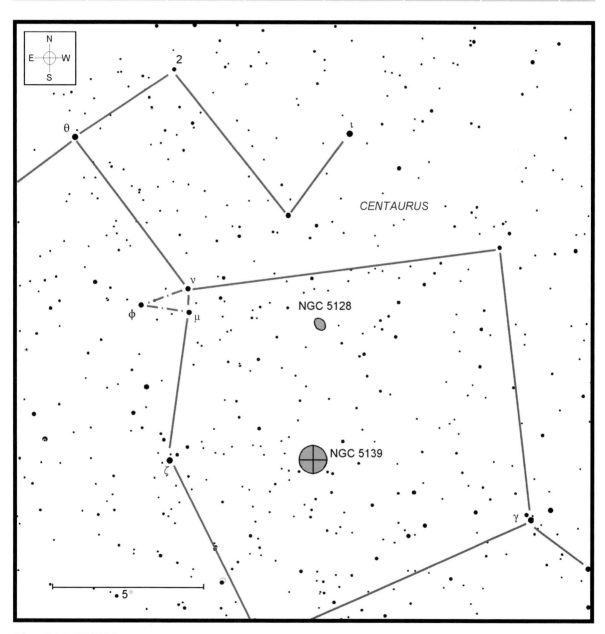

Chart 3.4 NGC 5128

Figure 3.4 NGC 5128

On August 4, 1826, Scottish astronomer James Dunlop discovered what was to prove to be a puzzle for the ages. Even his notes sound more like a riddle than scientific evidence. While observing from Parramatta, New South Wales, Dunlop found "a very singular double nebula." He went on to describe how "these two nebulae are completely distinct from each other, and no connection of the nebulous matters between them." Decades later, Dunlop's find had been assigned the designation NGC 5128 in the *New General Catalogue*.

What had Dunlop found? That question remained unanswered for many more years to come. Most astronomers, even the father of galactic astronomy, Edwin Hubble, believed that it was a nebula inside the Milky Way.

The first authoritative source to suggest NGC 5128's extragalactic nature was the *Harvard Survey of the External Galaxies Brighter than 13th Magnitude* published in 1932 by Harlow Shapley and Adelaide Ames, and better remembered as the *Shapley–Ames Catalog*. Shapley later described NGC 5128 as "pathological . . . a large 'problem' galaxy" in his book *Galaxies*.

Less than 20 years later, the burgeoning field of radio astronomy had uncovered that NGC 5128 was the source of strong radio emissions. Now also known as the radio source Centaurus A, the radio emissions from this pathological galaxy are more than 1,000 times as intense as those from our own Milky Way. Strangely,

Table 3.1 *Greatest altitude of NGC 5128 from various cities*

City	Latitude (north) (°)	Greatest altitude of NGC 5128 (°)
Honolulu, Hawaii	21	26
Miami, Florida	26	21
Houston, Texas	30	17
San Diego, California	33	14
Atlanta, Georgia	34	13
Los Angeles, California	34	13
Charlotte, North Carolina	35	12
Las Vegas, Nevada	36	11
San Francisco, California	38	9
St. Louis, Missouri	39	8
Washington, DC	39	8
Kansas City, Missouri	39	8
Denver, Colorado	40	7
Philadelphia, Pennsylvania	40	7
Salt Lake City, Utah	41	6
New York, New York	41	6
Lincoln, Nebraska	41	6
Chicago, Illinois	42	5
Rome, Italy	42	5
Albany, New York	43	4
Milwaukee, Wisconsin	43	4
Toronto, Ontario	44	3
Minneapolis, Minnesota	45	2
Montreal, Quebec	46	1
Seattle, Washington	48	not visible
Paris, France	49	not visible
London, England	52	not visible
Edinburgh, Scotland	56	not visible

the strongest radio emissions come not from the visible galaxy, but rather from two large, albeit optically invisible, lobes located to either side.

The most recent studies seem to confirm what some had suggested decades ago. When we look at NGC 5128, we are looking at a case of galactic cannibalism. Inside the heart of the beast lies a massive black hole that is consuming a smaller spiral galaxy. The two are believed to have collided between 160 and 500 million years ago. Part of the evidence supporting this claim is the large number of globular clusters containing young stars that are associated with NGC 5128. Most globulars are made up of the oldest stars known. Here, however, we see massive waves of starburst, where new stars spring up in far greater numbers than in normal, noncolliding galaxies. Some of the stars in these new globular clusters are undoubtedly the result of this violent process.

Your binoculars certainly will not detect any of the internal strife that NGC 5128 is enduring, but they can show you this intergalactic enigma that continues to fascinate astronomers. That is, if you can see the galaxy at all. NGC 5128 lies 43° south of the celestial equator, always confining it to the far southern sky for most of the northern hemisphere. It never rises more than 7° above my own horizon at latitude 40° north. Go farther north, and its maximum altitude shrinks. Travel southward and NGC 5128 will get correspondingly higher in the sky.

I first met NGC 5128 in April 1986, when my family and I had traveled to the Florida Everglades to see Halley's Comet. Although the nearly tailless comet was disappointing, the galaxy was anything but. Indeed, with averted vision, I could even make out the distinctive dust lane that slices the galaxy in half through my tripod-secured 10×50 binoculars.

Later, after returning home to Long Island, I was amazed to find that, from a beach overlooking the Atlantic Ocean along the island's south shore, I could still weakly see NGC 5128 through my 10×50s (Figure 3.4). It was faint, no doubt about it, but by waiting until it culminated, it was unmistakable. In fact, the view was quite dramatic, as both the galaxy and the ocean's far-distant horizon just barely squeezed into the same field.

The best time to look for NGC 5128 is when it is on the meridian and highest in the sky. As luck would have it, the bright star Spica in Virgo has nearly the same right ascension as the galaxy, so use it as a guide. Starting at Spica, scan straight southward for about 12°, or maybe two fields of view, to 3rd-magnitude Gamma (γ) Hydrae. Gamma is a good benchmark to show that you're on the right track. Continue on and in another 13° you will come to 3rd-magnitude Iota (ι) Centauri. It won't be long now. Look for the 4th-magnitude star HD 117440 to Iota's southeast, and a distinctive triangle of stars formed from Nu (ν), Mu (ν), and Phi (φ) Centauri farther southeast still. Those last three suns all lie at the same declination as the galaxy, so if you can see them, then the galaxy is waiting for you. Place HD 117440 toward the top of your binocular field, and Nu and Mu toward the eastern edge. Look carefully along the western side of the field and you should see the galaxy's faint, fuzzy disk.

Table 3.1 lists the maximum altitude that NGC 5128 will reach above the southern horizon from various cities. For my fellow amateur astronomers in the southern tier of the United States, it's a simple catch. Even for those at the latitude of San Francisco, it still gets 10° above the horizon, which should be far enough above any haze on dry spring nights. North of that, however, seeing this target depends very much on low humidity and a nearly perfect horizon.

26 Leo Trio

	Target	Type	RA	Dec.	Constellation	Magnitude	Size	Chart
26a	M65	Galaxy	11 18.9	+13 05.6	Leo	10.3b	9.8′×2.8′	3.5
26b	M66	Galaxy	11 20.3	+12 59.0	Leo	9.7b	9.1′×4.1′	3.5
26c	NGC 3628	Galaxy	11 20.3	+13 35.4	Leo	10.3b	14.8′×2.9′	3.5

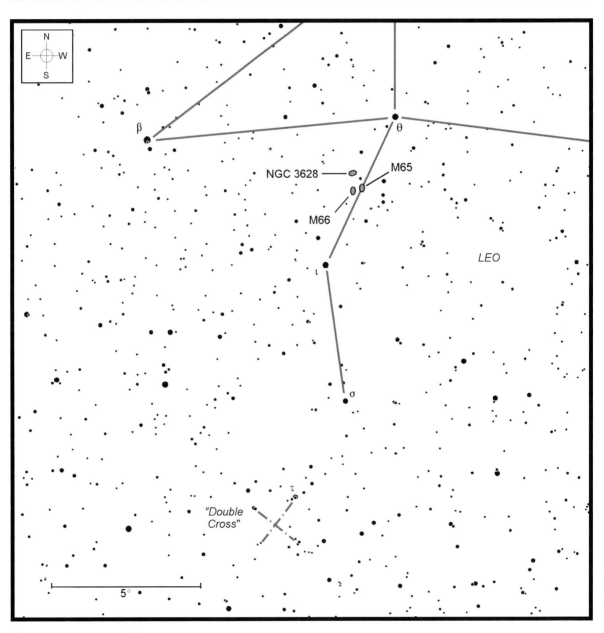

Chart 3.5 M65, M66, NGC 3628

Being far from the plane of the Milky Way, Leo can't boast beautiful star clusters, rich star fields, or glowing clouds of nebulosity. But it can flaunt five galaxies from the Messier catalog. The two brightest, M65 and M66, are found southeast of the star Chertan [Theta (θ) Leonis], which marks the right angle in the triangle of stars forming Leo's back and tail. They are joined by a third galaxy, NGC 3628, which is found just to the others' north. Pierre Méchain missed this last target when he discovered M65 and M66 in 1780, as did Messier when he first observed Méchain's galactic pair in the same year.

NGC 3628 is actually the largest of the three at close to 80,000 light years across. M65 and M66 each measure about 50,000 light years across, roughly half the diameter of the Milky Way. They lie a scant 125,000 light years from one another, closer than the Magellanic Clouds are from us. Imagine the dazzling show each must put on if viewed from a hypothetical planet orbiting a star within the other galaxy. NGC 3628 is more distant, some 250,000 light years away from the others.

M65 shows very tightly wound spiral arms in photographs, leading to it being classified as a type Sa spiral. Neighboring M66 is slightly looser in overall structure, and is therefore cataloged as a type Sb spiral. M66's northern spiral arm appears quite distended, probably due to the game of gravitational tug-o'-war it is playing with its two neighbors. There also appears to be a large amount of cosmic dust and reddish nebulae scattered throughout M66's spiral halo, both signs of extensive star-formation regions triggered by intergalactic tidal forces.

NGC 3628, with the loosest spiral structure of the three, is an Sc spiral. From our vantage point, we see this distant island universe almost exactly edge-on. Telescopes show a conspicuous band of dark dust clouds girdling the galaxy's edge, which effectively obscures the bright central region and hides most of the young stars in its spiral arms.

The Leo Trio (Figure 3.5) is found halfway between Chertan [Theta (θ) Leonis] and Iota (ι) Leonis. To locate the group, drop about half a binocular field ($2\frac{1}{2}°$) south of Chertan to a line of three 7th-magnitude stars. Look very carefully about half a field to their east for a very dim smudge of light lying below a faint star. That will be M66.

Figure 3.5 M65, M66, NGC 3628

M65 is just west of M66 but, at half a magnitude fainter, is more challenging. I can routinely see M66 through my 10×50s from my suburban backyard, but M65 takes a better-than-average night. It also helps to brace the binoculars.

Finally, NGC 3628 appears as a faint smudge of grayish light just to the north of the other two. Although its full disk spans 15′×3′, binoculars will probably only uncover its central, starlike core with perhaps a hint of fuzziness extending to the east and west.

While we are in the area, let's follow a zigzag line south of Chertan, past the stars Iota (ι) and Sigma (σ) Leonis to 4th-magnitude Tau (τ) Leonis. Tau teams with a fainter point to its southeast to create a pretty double star through binoculars. Place Tau toward the upper right (northwest) corner of your field and take a careful look around. You should spot not one, not two, but three more faint doubles. Add them together and you get an asterism that I like to call the "Double Cross," since all four pairs collectively form an X pattern. None of the four pairs is a true physical binary, since all eight stars are located at very different distances away. But that doesn't detract from their unique appearance through our binoculars.

27 M101

Target	Type	RA	Dec.	Constellation	Magnitude	Size	Chart
M101	Galaxy	14 03.2	+54 20.9	Ursa Major	8.3b	28.9′×26.9′	3.6

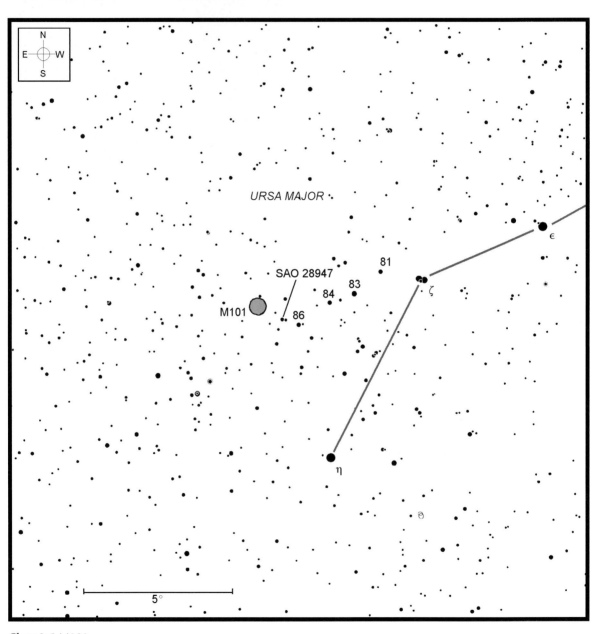

Chart 3.6 M101

The Pinwheel Galaxy, M101, is one of those objects that presents different challenges depending on the instrument being used. Here we discuss finding the galaxy through binoculars, but later, in Chapter 6, large-scope Challenge 137 will detail seeing structural detail within the galaxy's spiral arms.

I still recall my frustration at trying to find M101 on my first time through the Messier catalog. I was using a 4-inch reflector at the time. Try as I might, it just wasn't there. "What's wrong with this galaxy?" I thought.

Of course, nothing was "wrong" with the galaxy. What was wrong were my expectations. I was only looking at numbers. At a rated magnitude of 7.9, M101 sounds as though it should be reasonably bright. But it is not. Remember that when we look at the magnitude ratings for deep-sky objects, those values indicate the equivalent magnitude of the object if it could somehow be squeezed down to a stellar point. But a galaxy like M101 does not look like a star. Its full disk spans nearly half a degree. This causes its surface brightness, or brightness per unit area, to be very low. And there's the rub. Only the galactic core and innermost spiral halo are bright enough to be seen through binoculars and small telescopes. The full extent of the spiral arms is considerably fainter.

Pierre Méchain discovered M101 during the winter of 1781. When Messier finally spotted M101 on March 27 of the same year, he undoubtedly found his catalog's 101st entry to be a difficult target, as well. His notes recall a "nebula without a star, very dark and extremely large."

Surprisingly, M101 may actually be easier to see through conventional binoculars than conventional telescopes. Their low magnification means that M101's feeble light is dispersed less, keeping the galaxy more concentrated and "starlike" in appearance.

To zero in on M101, it's easiest to starhop off the famous visual double star, Alcor and Mizar (naked-eye

Figure 3.6 M101

Challenge 3), located at the crook in the Big Dipper's handle. Follow a trail formed by four dim stars (81, 83, 84, and 86 Ursae Majoris) to the east-southeast. At 86 Ursae, hop a little to the northeast, past the faint field star SAO 28947 to M101.

Through my 10×50 binoculars, M101 (Figure 3.6) looks more like a specter than a galaxy floating in a very attractive star field. When the field is high in the sky, M101 can be glimpsed directly as a softly glowing sphere that diffuses evenly into the background; otherwise, averted vision is needed to pull it out from the surrounding sky. The light of the silvery Moon, however, will be enough to extinguish the galaxy's dim light, which has lost much of its luster after traveling across some 27 million light years to reach us.

★

28 Draconian doubles

	Target	Type	RA	Dec.	Constellation	Magnitude	Separation	Chart
28a	16+17 Dra	Binary star	16 36.2	+52 54.0	Draco	5.5, 5.6	84″	3.7
28b	Nu (ν) Dra	Binary star	17 32.2	+55 11.1	Draco	4.9, 4.9	60″	3.7
28c	Psi (ψ) Dra	Binary star	17 41.9	+72 08.9	Draco	4.6, 5.8	30″	3.7

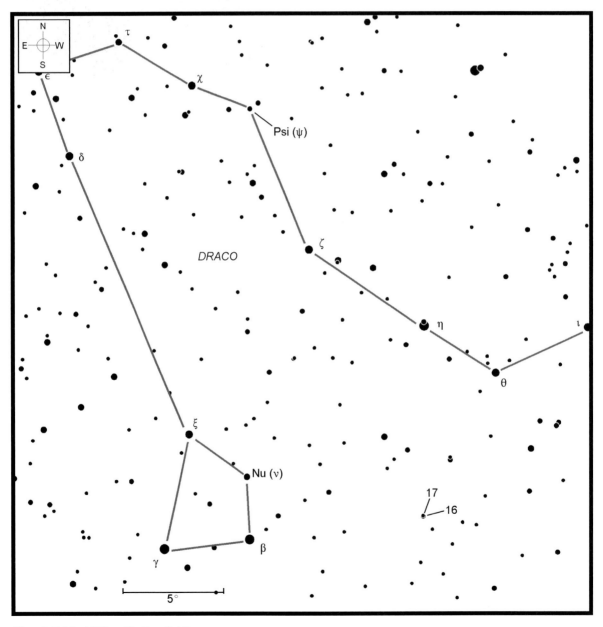

Chart 3.7 16+17 Dra, Nu Dra, Psi Dra

Recall from the discussion in Chapter 1 that the resolving power of a telescope depends on its aperture, but that the resolving power of a binocular is more a function of magnification. Table 1.4 listed the expected resolving ability for many different magnification values. It's time to put a few of those to the test.

The northern constellation of Draco the Dragon winds its way through our northern summer sky, and, while most of its stars are fainter than 3rd magnitude, it holds some fun resolution tests for binoculars. The first, which should be fairly easy to resolve even with modest 6× pocket binoculars, is the attractive pairing of 16 and 17 Draconis. These 5th-magnitude suns are separated by 84″ in our sky, and put on a fine show even through light-polluted skies. To find them, set your sights on the four-sided head of Draco. From the star Nu (ν) Draconis, the faintest of the four stars, shift one binocular field to the west, where you'll encounter 6th-magnitude Mu (μ) Draconis. Another binocular field west and you'll meet up with 16 and 17 nestled closely together.

As you set off from Draco's head, look carefully and you'll see that Nu is actually a very pretty pair of twin white suns known to many observers as the "Cat's Eyes." Both are separated by 60″, which brings them just within range of 6× binoculars. Seeing them both with 10× binoculars should be straightforward, provided the binoculars are supported steadily.

Figure 3.7 16+17 Dra

How about a tougher test? Try your luck with Psi (ψ) Draconis, lying partway along the dragon's sinuous body. What looks like a single 4th-magnitude point to the eye is actually a pair of 5th- and 6th-magnitude stars separated by 30 arc-seconds. That's right at the estimated limit for 8× binoculars, but again, they will need to be steadily supported for any hope of resolving this pair.

29   M20

Target	Type	RA	Dec.	Constellation	Magnitude	Size	Chart
M20	Reflection nebula	18 02.4	−22 59.0	Sagittarius	9.0	17′ × 12′	3.8

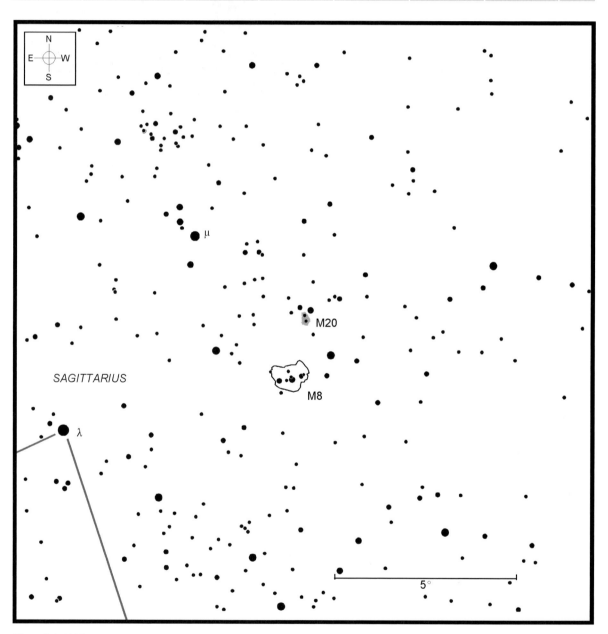

Chart 3.8 M20

If there is a better way to spend a clear, moonless summer's night than reclining in a chaise longue and scanning the Milky Way with a pair of binoculars, I don't know what it is. I've spent more than my fair share of binocular time focused squarely on M8, the Lagoon Nebula in Sagittarius, and simply letting my eyes wander around. The $6\frac{1}{2}°$ field of my 10×50 binoculars is wide enough to envelop the nearby open cluster M21 as well as the enigmatic Trifid Nebula, M20.

Many observers erroneously believe they have seen the gossamer clouds of the Trifid through binoculars, or even with the unaided eye. But, as we know by now, looks can be deceiving. Like neighboring M8, M20 is actually a combination of an open star cluster surrounded by faint tendrils of emission and reflection nebulae. The open cluster portion of M20, cataloged separately as NGC 6514, is a tight collection of 70 stars shining as brightly as 6th magnitude and spanning half a degree. It's their collective glow that many interpret as the Trifid *Nebula*, when in reality, they are really only seeing the Trifid *Cluster*. Messier's initial impression also seems to reflect this latter interpretation, when he described his 20th catalog entry as a "star cluster...between the bow of Sagittarius and the right foot of Ophiuchus." It's interesting to note that his subsequent description of M21 notes that "the stars in these two clusters [M20 and M21]...are surrounded by nebulosity." While most interpret this as confirming that Messier saw the Trifid's nebulosity as well as the cluster, I've always wondered if that were so. After all, Messier also said that he saw nebulosity in M21, which we know today to be cloud-free. Perhaps what he actually saw in both cases were artifacts resulting from his telescope's poor optical quality.

Figure 3.8 M20

The true test of whether or not you are seeing the Trifid Nebula (Figure 3.8) or simply the scattered glow of NGC 6514 is to swing your attention back and forth between it and M21. If, like Messier, you see "nebulosity" in M21, then suspect that your binoculars are suffering from either dirty optics or lens fogging. Both will scatter light to make it look as though most stars are surrounded by clouds. If, on the other hand, M21 looks clean, then you are probably seeing the true Trifid Nebula. Personally, I have never seen the intertwining lanes of dark nebulosity that give rise to the "Trifid" nickname through my 10×50 binoculars. They are, however, fairly evident through 16×70 binoculars from under dark, rural skies. What is the smallest binocular that will show them from your location?

30 "Houston's Triangle"

	Target	Type	RA	Dec.	Constellation	Magnitude	Size	Chart
30a	M57	Planetary nebula	18 53.6	+33 01.7	Lyra	9.7	1.8′×1.4′	3.9
30b	M11	Open cluster	18 51.1	−06 16.0	Scutum	5.8	13′	3.9
30c	M27	Planetary nebula	19 59.6	+22 43.2	Vulpecula	7.6	6.7′	3.9

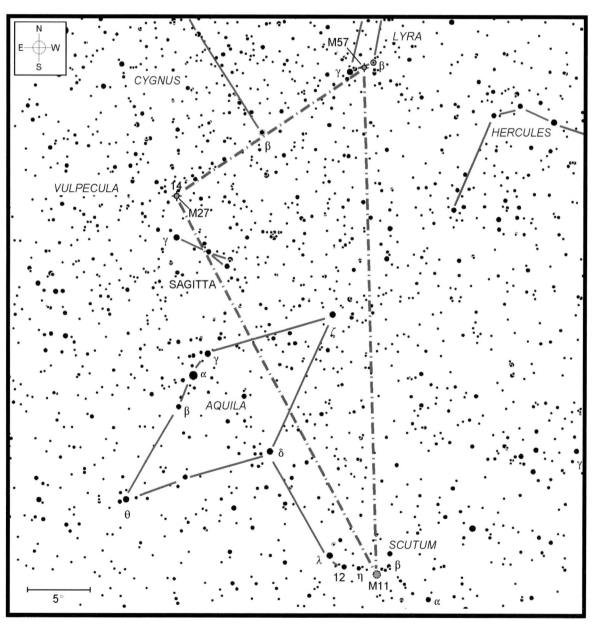

Chart 3.9 M57, M11, and M27

The face of amateur astronomy changed forever in September 1946. That month, a seed was planted that would blossom into a decades-long infatuation with deep-sky observing among amateurs everywhere. Without any fanfare at the time, a man named Walter Scott Houston took over the reigns of a small column called Deep-Sky Wonders in *Sky & Telescope* magazine. Another prolific observer, Leland Copeland, originated the column and penned it for four years, but it was under Houston's watch that it evolved from a modest tabulation of three or four suggested objects into an engaging dialogue that would capture the imagination of readers for nearly half a century.

That first column written by Houston, a mere dozen lines in length that belied this amazing man's tale-spinning nature, listed three objects in the late summer sky – M57, M27, and M11 – that I like to think of as Houston's Triangle. All three are visible through common binoculars, although they present three different levels of difficulty.

Of the three, M11, the famous Wild Duck Cluster in Scutum, is the easiest to see through binoculars. To find it, trace Aquila the Eagle's diamond-shaped body from Altair to the curve of its tail-feather stars, Lambda (λ) and 12 Aquilae. Together with Eta (η) Scuti, Lambda and 12 Aquilae form a three-star arc that curves right toward M11. More than 1,000 stars belong to this open cluster, making it the densest of its type among the Messier objects and a real showpiece in telescopes. Through binoculars, M11 actually looks more like an unresolved globular cluster than an open cluster. Except for a solitary 8th-magnitude point buried within, all of M11's suns shine between 11th and 14th magnitude, too faint to be resolvable with common binoculars (it's very doable if you have keen eyes and 20× binoculars).

Next comes M27, nicknamed the Dumbbell Nebula for its telescopic resemblance to a weightlifter's barbell. Even the view through 10×50 binoculars (Figure 3.9) begins to reveal its unusual shape. M27 carries the distinction of being the first planetary nebula ever discovered, by Messier himself in 1764, and only one of four listed in his famous catalog. Located about 1,250 light years away, M27 is one of the biggest and brightest planetary nebulae visible in northern skies.

While it is easy to tell M27 apart from the surrounding stars through binoculars, locating it in the first place can be another matter. That's because its home constellation, Vulpecula the Fox, is nearly impossible to make out against the background Milky

Figure 3.9 M27

Way field. Rather than scratch your head in frustration, do as most others do and start your search in neighboring Sagitta the Arrow. Even though Sagitta is also very small, its four main stars create a distinctive pattern that stands out more prominently. Concentrate your aim on Gamma (γ) Sagittae, the star marking the Arrow's eastern tip. Look about half a field due north of Gamma for a triangle of stars formed from 14, 16, and 17 Vulpeculae. M27 will look as a smudge of grayish light less than half a degree southeast of 14 Vulpeculae, the easternmost star in the triangle. Even 7× binoculars will show the nebula's box-shaped glow floating in a field strewn with stardust. Catching a hint of its hourglass form, however, will take more magnification that most handheld binoculars can muster.

Lastly, we have M57, the Ring Nebula in Lyra. M57 lies twice the distance from us as M27, and as a result, appears much smaller. While M27 measures 8′ × 6′ across, the Ring is just 1.4′ × 1′ in size. Even though M57 at magnitude 8.8 is bright enough to be visible through most binoculars, it will look like nothing more than another faint star.

The challenge to binocularists is to figure out *which* faint star is actually the Ring Nebula, since its tiny donut-shaped disk needs at least 25× to be resolved. Although that task sounds daunting at first, as luck would have it, the field surrounding M57 actually simplifies the task. Center your aim along the bottom of

Lyra the Lyre's rectangular frame, halfway between the star Sulafat [Gamma (γ) Lyrae] marking the eastern corner and Sheliak [Beta (γ) Lyrae] at the western corner. M57 is found almost exactly halfway between the two. To confirm that you are looking in the right place, M57 is the last point in a diagonal line of five faint stars that starts just north of Sheliak (Beta). If you can see all four real stars in that line as well as a slightly fuzzy point just beyond, then you've hit M57.

We will return to the Ring Nebula again in Chapter 7, when it presents other challenges for observers viewing through the largest backyard telescopes.

31 Barnard's E (Barnard 142 and Barnard 143)

★★★★★

Target	Type	RA	Dec.	Constellation	Magnitude	Size	Chart
Barnard's E	Dark nebula	19 41.1	+10 54.3	Aquila	–	40′×23′	3.10

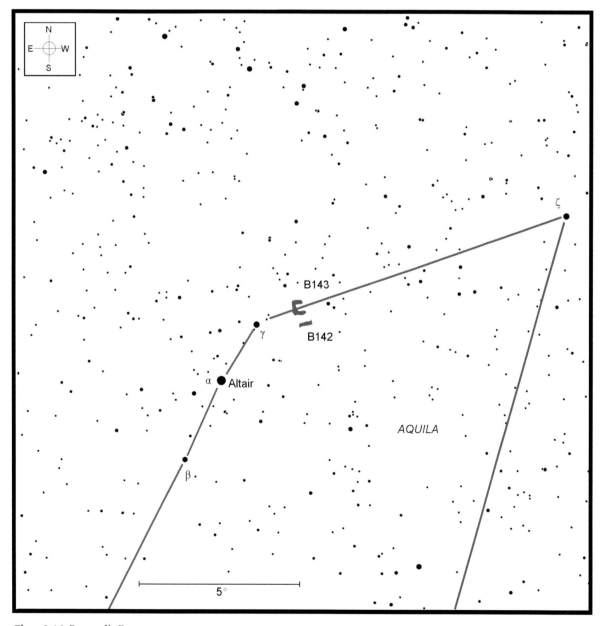

Chart 3.10 Barnard's E

Figure 3.10 Barnard's E

The Milky Way in and around the constellations of
Aquila and Scutum is littered with huge wafting clouds
of cosmic dust. These vast expanses of dark nebulosity
are clearly visible on long-exposure photographs of the
area as they strike a pose against the bright, starlit
backdrop. Trying to spot them visually, however,
without the benefit of being able to amass photons
from the distant starry milieu, can prove very difficult.
After all, when it comes to observing dark nebulae, it's
what you *don't* see that counts.

One of the more visually interesting dark nebulae
within Aquila carries the dual designation of Barnard
142 and Barnard 143, for its multiple entries in Edward
Barnard's *Catalogue of 349 Dark Objects in the Sky*. Figure
3.10 shows the view through a pair of 10×50
binoculars. Barnard 143, the northernmost of the two,
resembles a pickle fork with the prongs opening
toward the west. Barnard 142 lies just to the south
and is shaped like an irregular, starless rectangle.
Both stand out nicely through binoculars, provided
there is no moonlight or light pollution to spoil the
view.

The human mind always likes to create something
that is tangible out of the intangible. Though Barnard
142 and 143 are included as two separate items in
Barnard's listing, recent reckonings often lump both
together for their combined visual interest. Today,
Barnard 142 and 143 are better known as "Barnard's
E." That's because, from under dark skies, they
collectively form a very distinctive capital E against the
bright background. Barnard's E, also sometimes called
the Triple Cave Nebula, is easy to locate in the sky just
3° northwest of brilliant Altair and 1° due west of
Tarazed [Gamma (γ) Aquilae]. A few summers ago,
while observing from the Stellafane telescope makers
convention in Springfield, Vermont, the E struck me as
especially vivid when I pointed my 10×50 binoculars
its way. The effect, however, was lost through a pair of
16×70 binoculars, probably because of their narrower
field of view. The E spans an area approximately
1°×2°.

32 Veil Nebula

★
★★
★★
★

	Target	Type	RA	Dec.	Constellation	Magnitude	Size	Chart
32a	NGC 6960	Supernova remnant	20 45.9	+30 43.0	Cygnus	–	60.0′ × 9.0′	3.11
32b	NGC 6992	Supernova remnant	20 57.0	+31 30.0	Cygnus	–	80.0′ × 26.0	3.11

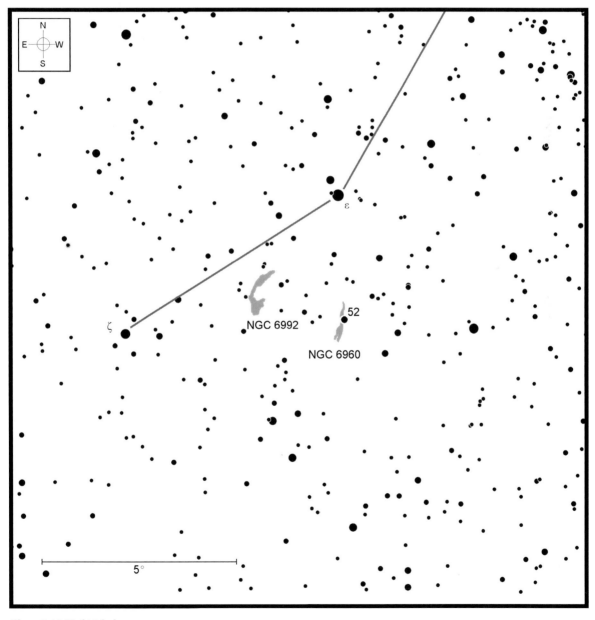

Chart 3.11 Veil Nebula

Floating among the myriad points of light in eastern
Cygnus is all that remains of a once mighty superstar.
When it exploded as an all-consuming supernova, it
was as bright as magnitude −8.

That was between 5,000 and 8,000 years ago. When
we swing our telescopes and binoculars its way today,
we find the ethereal echo of that once mighty
supergiant star. The star's dense core has never been
found, but its cloud of stellar shrapnel continues to
expand away from ground zero.

The remains of the ancient Cygnus supernova,
known today as the Veil Nebula Complex, float some 3°
south of Epsilon (ε) Cygni, the star at the eastern end of
the Northern Cross's crossbeam. Over the course of the
last 15,000 years, the remains of the supernova
explosion have slowly begun to disperse, and now
cover an area spanning nearly $3° \times 3\frac{1}{2}°$. Eventually, the
remains will disappear completely, scattering atoms of
hydrogen, helium, oxygen, iron, silver, gold, and other
elements into the cosmos to seed future generations of
stars, planets, and perhaps even intelligent beings.

The principal sections of the Veil were discovered by
William Herschel through his 18.7-inch reflector on
September 5, 1784, and carry several separate
designations in the *New General Catalogue*. NGC 6992
marks the bow-shaped, eastern branch of the Veil, while
a small, fragmented portion to the south is NGC 6995.
The western, fainter edge of the Veil is known as NGC
6960. While NGC 6992/6995 lie among a field of faint
stars, NGC 6960 appears to "touch" the 4th-magnitude
star 52 Cygni. Appearances notwithstanding, they are
actually nowhere near each other in space.

Trying to see NGC 6992/6995 and NGC 6960 can
confound even experienced amateurs viewing through
large telescopes. Yet, under dark skies, all three can be
seen through 40-mm binoculars. The trick is to know
what to look for.

First, try for the brightest section, NGC 6992. It lies
tucked under the Swan's eastern wing, just southwest of

Figure 3.11 Veil Nebula

a small triangle of 7th- and 8th-magnitude stars
between Epsilon (ε) and Zeta (ζ) Cygni. Averted vision
will probably be required at first, although one summer
in the New York State Catskill Mountains, I easily saw it
and NGC 6995 through my 10×50 binoculars (Figure
3.11). Both combined to resemble a faint wisp of
smoke wafting away from an invisible chimney.

NGC 6960 was also visible through the same
binoculars that night, but with much more difficulty.
Not only is it fainter, but the overpowering light from
52 Cygni really plays havoc. It's interesting to note that,
although just a coincidence, the portion of NGC 6960
to 52's north is easier to see than that to the south. In
fact, the southern half of NGC 6960 went completely
undetected through my 10×50s that night, even with a
pair of narrowband nebula filters jury-rigged over the
eyepieces.

33 Pelican Nebula

Target	Type	RA	Dec.	Constellation	Magnitude	Size	Chart
Pelican Nebula (IC 5067/5070)	Emission nebula	20 51.0	+44 00.0	Cygnus	–	60.0′ × 50.0′	3.12

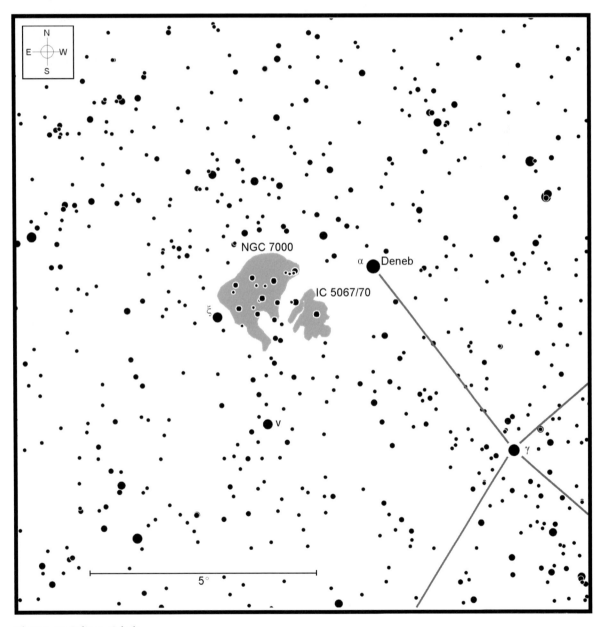

Chart 3.12 Pelican Nebula

Naked-eye Challenge 8 in the last chapter dared you to see the North America Nebula without optical aid. How did you do? If you passed that trial, then pick up your binoculars and see how you do with this next test. The North America Nebula is easy to see compared with spotting our next target, the Pelican Nebula.

In reality, the Pelican and the North America Nebulae are both part of the same huge complex of glowing hydrogen gas. An opaque cloud of interstellar dust that slices in front of the background emission nebula gives the illusion that we are looking at two different entities. That absorption cloud is cataloged separately as Lynds 935, or LDN 935, its listing in the catalog of dark nebulae compiled by the astronomer Beverly T. Lynds and published in the *Astrophysical Journal Supplement* (Vol. 7, p. 1) in 1962.

The Pelican carries two catalog designations – IC 5067 and IC 5070 – that point to two portions of the nebula. The eastern edge of Lynds 935 carves out the outline of the North America Nebula's (terrestrial) east coast, while its western edge forms the long beak and pointy head of the celestial pelican, listed as IC 5070. Photographs of the region reveal two small, circular dark dust clouds marking the bird's eyes, while a brighter tuft to their northwest, IC 5067, suggests the curved shape of its head and neck.

The Pelican's ionized hydrogen is easy to record in photographs, but seeing any hint of it by eye is usually frustrated by its deep-red emission. Conditions have to be nearly perfect to see even the slightest hint. A good rule of thumb is set by the North America Nebula itself. If it is *easily* visible by eye, then the sky might be clear enough to see the Pelican through binoculars.

The most prominent part of the Pelican is its east-facing silhouette against Lynds 935. Use the star 57 Cygni as a further guide, as it is positioned just east of the "bill." From here, the bill slices diagonally

Figure 3.12 *Pelican Nebula*

southeastward toward the fainter field star HD 199373. The back of the Pelican's head curves westward toward the star 56 Cygni.

Interestingly, the North America/Pelican complex lies an estimated 2,000 light years away, which is only a few hundred light years farther than the Orion Nebula. But, while the Orion Nebula is an easy target to spot even under less than ideal conditions, this similar stellar nursery is far more difficult to study even under superior skies. My best view of the Pelican through binoculars came a few summers ago while attending a star party in the Catskill Mountains. There, with a naked-eye limiting magnitude of better than 6.5, the Pelican's profile was distinct through my 10×50 binoculars (Figure 3.12).

34   NGC 7293

Target	Type	RA	Dec.	Constellation	Magnitude	Size	Chart
NGC 7293	Planetary nebula	22 29.6	−20 50.2	Aquarius	7.3	16′	3.13

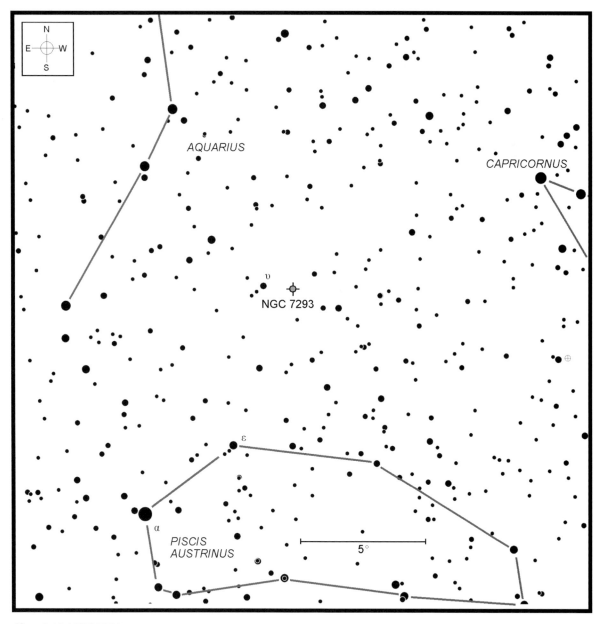

Chart 3.13 NGC 7293

Binoculars and planetary nebulae usually do not mix. Even through telescopes, many look like nothing more than fuzzy stars, only revealing their true identity after careful examination at high magnifications. Low-power binoculars are usually left in the dust. There is one planetary nebula in the autumn sky, however, that measures half the size of the Full Moon in our sky and shines at 7th magnitude. Sounds like a perfect combination for binoculars.

NGC 7293, the Helix Nebula, is found among the dim stars of Aquarius, the Water-Bearer. Unlike so many objects listed in the *New General Catalogue*, NGC 7293 was not discovered by the father–son team of William and John Herschel. Instead, discovery is credited to German astronomer Karl Ludwig Harding, who spotted it around the year 1824 using a 4-inch refractor at Göttingen Observatory.

Now, wait a minute! How could classic observers like Charles Messier, Pierre Méchain, and William and John Herschel have missed a 7th-magnitude object that measures 16′ in diameter? The answer is obvious to anyone who has ever tried to find the Helix through a long-focus telescope and narrow-field eyepieces, which is what these observers used for their discoveries. When you look for the Helix, it's just not there.

That's because simple data do not tell the whole story of the Helix Nebula. NGC 7293 is, in fact, the brightest and largest planetary in the sky. It is also the closest, at about 520 light years away. Why, then, is it so hard to see? Simply because it is so large. Imagine how dim a 7th-magnitude star would be if you defocused it until it measured the size of the Helix Nebula. It would pretty much fade into the background. The surface brightness, or brightness per square area, drops precipitously. Couple that fact with an object so large that it would more than fill the field of a narrow-field eyepiece and it's easy to understand how the Helix could be passed over and escape unseen. But aim toward the Helix through a pair of 50-mm binoculars on a dark autumn night, and its dim, ethereal disk should be visible without too much difficulty.

Figure 3.13 NGC 7293

To find the Helix, begin at autumn's brightest star, Fomalhaut in Piscis Austrinus. By placing Fomalhaut along the southeastern edge of the field, 4th-magnitude Epsilon (ε) Piscis Austrini should pop into view along the opposite side. Move 6°, or about another field of view, due north to find 5th-magnitude Upsilon (υ) Aquarii. Finally, slide Upsilon just slightly to the east; the dim glow of the Helix should be nearly centered in view.

Through my own 10×50s, the Helix looks slightly oval, its long axis tilted northwest–southeast. The outer edge is slightly irregular in texture, with brighter sections toward the northeast and southwest, coincident with the disk's minor axis. I can't say with any certainty that I can detect the nebula's central hole through my 10×50s (Figure 3.13), even though it is so obvious in photographs. My 11×80s and 16×70s turn the Helix into the familiar ring shape, but only with averted vision.

35 | M32 and M110

★
★

	Target	Type	RA	Dec.	Constellation	Magnitude	Size	Chart
35a	M110	Galaxy	00 40.4	+41 41.4	Andromeda	8.9b	21.9′×10.9′	3.14
35b	M32	Galaxy	00 42.7	+40 51.9	Andromeda	9.0b	8.7′×6.4′	3.14

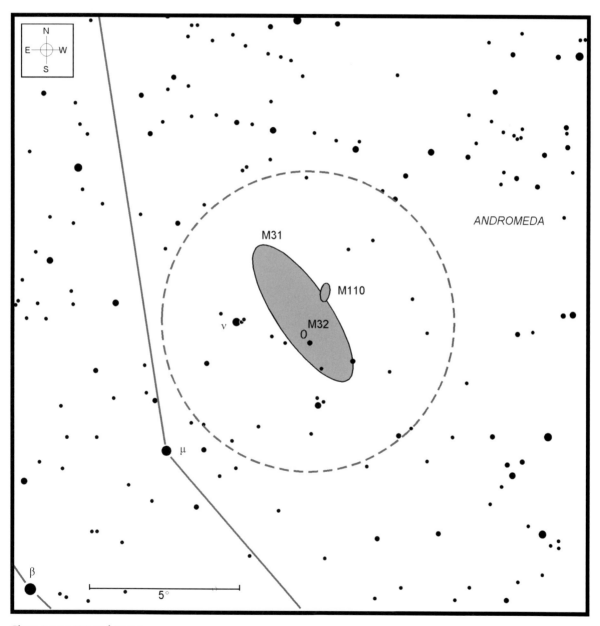

Chart 3.14 M32 and M110

Figure 3.14 M32 and M110 (with M31)

No other deep-sky object north of the celestial equator appears as large, as bright, and as complex as M31, the Andromeda Galaxy. M31 holds something for everyone, whether it is the naked-eye challenge of simply seeing the galaxy through suburban light pollution or looking for some of its many distant globular clusters through the largest backyard telescopes.

For binocularists, the challenge lies not in the galaxy itself, but rather in spotting its four companion galaxies. Two of them, M32 and M110, are discussed here. The other two lie about 7° farther north, across the border in Cassiopeia. This latter pair, NGC 147 and NGC 185, may be beyond the grasp of our handheld binoculars, but they will be captured through the giant binoculars included in the next chapter.

For now, let's concentrate on M32 and M110 (Figure 3.14). M32 is the brighter, but smaller, of the two, and the easier to spot through binoculars. Credit French observer Jean-Baptiste Le Gentil (1725–1792)

with its discovery on October 29, 1749. You'll find this dwarf cE2[1] galaxy to the south of M31's core, looking like a fuzzy "star" superimposed against the spiral arm halo. M32 shines at 8th magnitude, within reach of 35-mm binoculars.

The other companion, M110, is larger, fainter, and not nearly as conspicuous. Unlike M32, which shows a strong central concentration, M110's disk looks homogenous. Its dull, featureless appearance hid M110 for 24 more years until Messier himself spotted it on August 10, 1773. He marked its position precisely with respect to the other two on a drawing he made that evening but, for reasons that are lost to history, decided not to mention it in his catalog. Looking at this drawing nearly two centuries later, British amateur astronomer and Messier expert Kenneth Glyn Jones proposed adding Messier's find as M110.

M110, now categorized as either a dwarf E5 or E6 elliptical galaxy,[2] shines at magnitude 8.5, about half a magnitude fainter than M32, and spans $22' \times 11'$. Its diffuse nature and corresponding lower surface brightness means that M110 takes more effort to see than its compact neighbor. Still, with patience and perseverance it is visible through 35-mm binoculars. Look for it to the northwest of M31's core, about twice as far as M32 is to its south. I've been able to make out both from my light-polluted suburban backyard through 10×50 binoculars, although M110 required averted vision. In fact, M110 makes a good gauge of the night's darkness. I know that if I can see it through my 10×50s, then the night is a good one for hunting big game through larger instruments.

----

[1] The "c" in M32's morphological designation indicates the presence of a centralized supermassive object, probably a black hole.
[2] M110 is an odd example of an elliptical galaxy, as it contains some dark dust clouds, absent in generic ellipticals. As a result, some astronomers now refer to M110 as a spheroidal galaxy, like some of the companion galaxies to the Milky Way.

36 NGC 247 and NGC 253

	Target	Type	RA	Dec.	Constellation	Magnitude	Size	Chart
36a	NGC 247	Galaxy	00 47.1	−20 45.6	Sculptor	9.1	21.4′×6.0′	3.15
36b	NGC 253	Galaxy	00 47.5	−25 17.3	Sculptor	8.0b	27.7′×6.7′	3.15

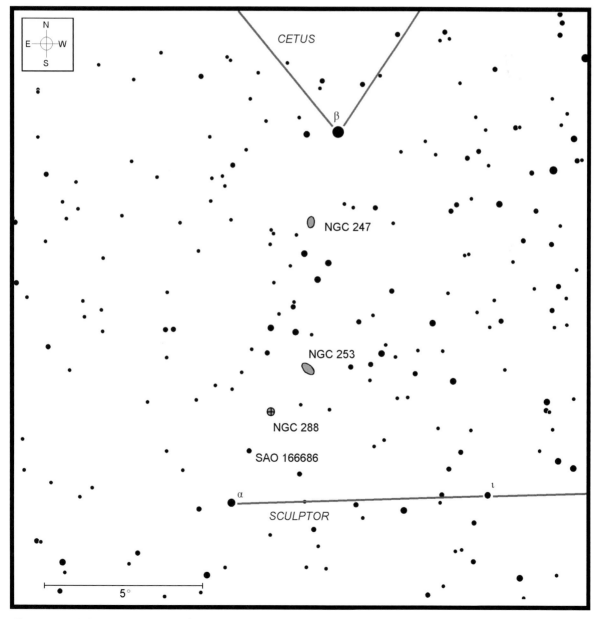

Chart 3.15 NGC 247, NGC 253, and NGC 288

NGC 253, the Silver Coin Galaxy, was discovered by
Caroline Herschel, the sister of William Herschel, in
1783 as she scanned the sky for comets. Observing
from the south of England, neither she nor her brother
could envision her discovery was anything more than
an extended, nebulous glow.

Under better conditions, however, NGC 253 proves
to be one of the finest edge-on spirals in the entire sky.
It is easy to see through binoculars from the southern
hemisphere. In fact, at magnitude 7.6, some have even
glimpsed it by eye alone from very dark sites, such as
the Australian Outback, where it passes nearly overhead
in November and December.

Unfortunately, from mid-northern latitudes NGC
253 is a much harder catch because of its southerly
declination. Even when it rides highest in our sky, light
pollution, haze, and clouds can easily obscure its
magnificence.

Fortunately, NGC 253 is easy to find, lying just 7°
south of 2nd-magnitude Deneb Kaitos [Beta (β) Ceti].
With the star on the northern edge of your binocular
field, look toward the south-southeast for a small
triangular asterism of 5th- and 6th-magnitude stars.
Place that triangle against the field's northern rim
where Deneb Kaitos was a moment ago and NGC 253
should be visible just south of center. If you continue
too far south, you'll come to 4th-magnitude Alpha (α)
Sculptoris, which tells you that you've overshot your
quarry.

From my backyard, where the light dome from a
distant shopping center rises some 25° above the
southern horizon, the galaxy is barely visible as a
cigar-shaped glow through 10×50 binoculars (Figure
3.15). Moving to better skies, the same binoculars
reveal a brighter central core surrounded by the faint
sheen of the galaxy's complex spiral arm halo. In many
ways, it looks like a tiny version of M31, the
Andromeda Galaxy (naked-eye Challenge 11).

That's where the resemblance ends, however. While
M31 is relatively placid as galaxies go, NGC 253 is
where the action is. Known as a "starburst galaxy," NGC
253 has recently undergone the pangs of rapid star birth

Figure 3.15 NGC 253 and NGC 288

within dense regions of dust around its core. During a
period of starburst, which could last for millions of
years, stars can form at tens, even hundreds, of times
the rate in normal galaxies. Since these newly formed
stars are very massive and very bright, starburst galaxies
are among the most luminous galaxies in the universe.

NGC 253 is the brightest and largest member of the
Sculptor Group, the closest collection of galaxies to our
own Local Group. The Sculptor Group includes a total
of 14 members.

If NGC 253 was just too easy for you, then try your
luck with another Sculptor Group member. NGC 247
appears nearly as large as its neighbor, but is about two
magnitudes fainter. Look for it just to the north of the
triangle's eastern tip. A 9th-magnitude star
superimposed on the southern tip of the galaxy is just
barely visible through binoculars, creating the illusion
of a small comet with a very faint tail. I've only been
able to spot it through my 10×50s under very dark,
moonless skies, and even then, only with averted vision
and with the binoculars securely braced.

37 NGC 288

★
★ ★
★

Target	Type	RA	Dec.	Constellation	Magnitude	Size	Chart
NGC 288	Globular cluster	00 52.8	−26 34.9	Sculptor	8.1	13′	3.15

Most globular clusters are gathered around the cores of galaxies, like moths surrounding the flickering warmth of a flame or porch light. But every now and then, we come upon a renegade, a straggler who bucks the system. That's what we have when we look at NGC 288. Lying less than a degree away from the south galactic pole, NGC 288 seems far out of place, a quarter of the sky away from most of its fellow globulars in Ophiuchus, Sagittarius, and Scorpius.

If you were able to spot NGC 253 in the previous challenge, then don't move a muscle! Both it and NGC 288 appear near enough to one another to fit into the same binocular field easily, creating an attractive odd couple of celestial proportions. Look for the globular just $1\frac{3}{4}°$ to the southeast of NGC 253, halfway between the galaxy and the reddish 6th-magnitude star SAO 166686. SAO 166686, in turn, is also exactly $1\frac{3}{4}°$ northwest of Alpha (α) Sculptoris, creating a nice, evenly spaced line for us to hop from one to the next, to the next. Of course, the fact that they just happen to lie along the same line of sight as seen from Earth is purely coincidental. They are nowhere near each other in space. NGC 253 is estimated to be 10 million light years away, while NGC 288 is right next door, so to speak, a mere 28,700 light years from us.

Shining at 8th magnitude, NGC 288 looks like a distant ball of cotton framed by several faint field stars, through binoculars. Its loose stellar structure, rated as class 10 on the 1-to-12 Shapley–Sawyer globular density scale,[3] causes its surface brightness to be lower than some other 8th-magnitude globulars. As a result, you will probably need to use averted vision and brace your binoculars against a sturdy support to spot it. Its southern declination also contributes to the challenge, since its disk, like that of NGC 253, is often smothered in atmospheric schmutz (that's a technical term).

By studying the spectra of representative stars, astronomers estimate the age of NGC 288 to be approximately 13 billion years, give or take a billion. That makes it one of the oldest globulars known. This conclusion was reached by Michelle Bellazzini of the Osservatorio Astronomico di Bologna (Astronomical Observatory of Bologna, Italy) and colleagues in a study published in 2001.[4] Interestingly, the study, which compared NGC 288 to another stray globular, NGC 362 in far southern Tucana, found that the latter is about 2 billion years younger. Although stars in globular clusters are among the oldest known, they came into existence across a broad range of time in the early universe.

---

[3] Published in *Harvard Observatory Bulletin*, No. 824 (1927), the Shapley–Sawyer scale rates globulars from 1 to 12 based on stellar concentration. The smaller the assigned number, the more highly concentrated the stars toward the center of the respective globular.
[4] Michele Bellazzini, Flavio Fusi Pecci, Francesco R. Ferraro, Silvia Galleti, Márcio Catelan, and Wayne B. Lansman, "Age as the Second Parameter in NGC 288/NGC 362? I. Turnoff Ages: A Purely Differential Comparison," *The Astronomical Journal*, Vol. 122 (2001), pp. 2569–86.

★

38 Autumn binary tests

	Target	Type	RA	Dec.	Constellation	Magnitude	Separation	Chart
38a	Psi-1 Psc	Double star	01 05.7	+21 28.4	Pisces	5.3, 5.6	29″	3.16
38b	Zeta Psc	Double star	01 13.7	+07 34.5	Pisces	4.9, 6.3	23″	3.16
38c	Lambda Ari	Double star	01 57.9	+23 35.8	Aries	4.8, 7.3	36″	3.16
38d	30 Ari	Double star	02 37.0	+24 38.8	Aries	6.5, 7.4	38″	3.16

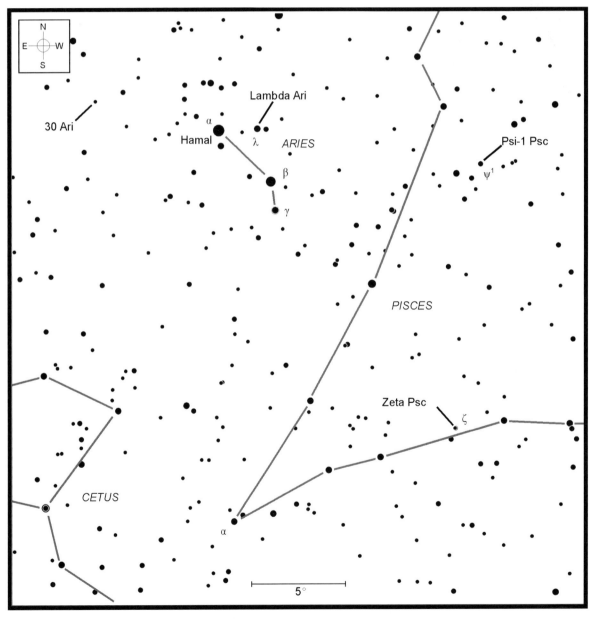

Chart 3.16 Psi-1 Psc, Zeta Psc, Lambda Ari, and 30 Ari

Figure 3.16 Psi-1 Psc

Regardless of your observing site, be it under dark rural skies or from a more cosmopolitan setting, hunting down challenging binary stars is a great way to confirm the quality of your optics, your eyesight, and your powers of observation. The four presented here pose exciting tests for stargazers using $9\times$ and $10\times$ binoculars.

Let's start with Lambda (λ) Arietis, a close-set pair just west of Hamal [Alpha (α) Arietis]. Here, you'll find a 4.8-magnitude primary sun teamed with a 7.3-magnitude companion set 36″ to the northeast. From Table 1.3 (binocular resolution limits), both may be resolvable in a good pair of $7\times$ binoculars, although

the disparate magnitudes can confound that effort. Seeing both should pose little problem for steadily mounted $10\times$ binoculars, however.

If Lambda was a little too tough, then 30 Arietis might prove easier. Although the pair is fainter, its yellowish, 6.5-magnitude primary is separated from the slightly bluish 7.4-magnitude companion by 38″. Look for 30 Ari about 7° to the northeast of Hamal and 3° to the north-northwest of 5th-magnitude Nu (ν) Arietis.

Crossing into Pisces, focus next on Psi-1 (ψ 1) Piscium. You'll find it 11° west of Beta (β) Arietis. As shown in Figure 3.16, Psi-1 is the northwesternmost of four stars in a pleasant little wedge-shaped asterism. Psi-1 consists of two nearly identical white stars, magnitude 5.3 and magnitude 5.6, separated by 29″. That places them right at the limit of steadily mounted $8\times$ binoculars. If you don't have a tripod to hold your binoculars, try propping them against a fence post or tree. Without something to brace against, Psi-1 will remain a single star even at $10\times$.

Finally, Zeta (ζ) Piscium will test your visual acuity even through tripod-mounted $10\times$ binoculars. It will also test your starhopping skills, since there are few naked-eye stars in its immediate area. The best way to locate it is to begin at Hamal and Beta (β) Arietis. Extend the line connecting them farther toward the southwest. In 8°, you'll come to Eta (η) Piscium. Continue straight along for another 9° to Zeta. Once you spot it, look for the 4.9-magnitude primary sun paired with a 6.3-magnitude companion just 23″ to its northeast. Some observers comment that Zeta looks yellowish through telescopes, but I have never noticed any coloring through binoculars, even after defocusing the image slightly to enhance color. Perhaps your eyes' cones are more sensitive than mine.

39   M74

Target	Type	RA	Dec.	Constellation	Magnitude	Size	Chart
M74	Galaxy	01 36.7	+15 47.0	Pisces	10.0b	10.5′×9.5′	3.17

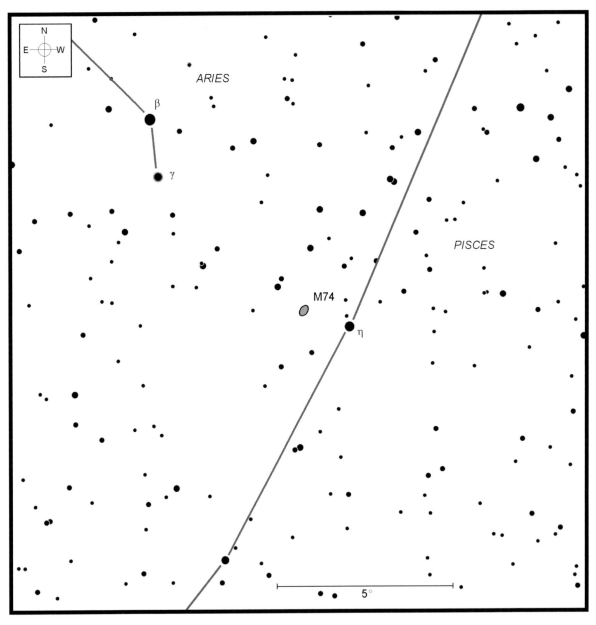

Chart 3.17 M74

Many judge M74 to be the most difficult of the Messier
objects to see, and with good reason. M74 is a face-on
spiral galaxy, which as a breed can be notoriously
difficult to detect visually because of their low surface
brightness. M74 is no exception. Although it is rated at
magnitude 10, the surface brightness of this
broad-armed Sc spiral is down around 14th magnitude,
barely above the surrounding sky under the darkest
conditions. Pierre Méchain hit the nail on the head
when he described it as "quite broad, very dim, and
extremely difficult to observe" after discovering it in
September 1780.

William Parsons, the third Earl of Rosse
(1800–1867), included M74 as one of 14 "spiral or
curvilinear nebulae" in a paper that he published in
1850. Parsons was on the right track but, ten years after
his publication, Friedrich Argelander mistakenly listed
M74 as a star designated BD +15°238 in his famous
Bonner Durchmusterung star survey. The purpose of the
survey was to determine accurate positional and visual
magnitude values for every star – all 324,188 of them –
visible with Argelander's 78-mm refractor in Bonn,
Germany. Through a small telescope like Argelander's,
M74's nearly stellar core is all that's visible; the
embracing spiral arms are too faint to be seen.
Apparently, Argelander was either unaware of Parsons'
observations or chose to ignore them.

If you are doing these challenges in order, then you
just passed M74 on the way to Zeta (ζ) Piscium in the
previous listing. Recall that, to find Zeta, it's best to start
at Hamal and Beta (β) Arietis, both in neighboring
Aries the Ram. Again, extend the line connecting them
to the southwest a little more than a binocular field of
view, to Eta (η) Piscium. Stop right there! You may not
know it, but M74 is now in view. Focus your attention
about $1\frac{1}{2}°$ east-northeast of Eta. Using every observing
trick in the book (supporting the binoculars on

Figure 3.17 M74

something, using averted vision, and so on), look for a
very subtle glow about halfway between Eta and a pair
of 7th-magnitude field stars.

Can M74 really be observed with binoculars? My
own notes, made several years ago through 10×50
binoculars (Figure 3.17), recall a "suspected sighting;
very dim glow seen fleetingly with averted vision,
subsequently confirmed with 16×70 binoculars." There
are confirmed observing reports by others of M74 being
spotted through 42-mm binoculars. But the most
impressive report I know of was made by super-eyed
Stephen O'Meara. He recalls in his book *The Messier
objects* that he once spotted M74 through 7×35
binoculars from his "sparsely populated corner of
Hawaii."

40   NGC 1499

Target	Type	RA	Dec.	Constellation	Magnitude	Size	Chart
NGC 1499	Emission nebula	04 00.5	+36 33.0	Perseus	–	160′×42′	3.18

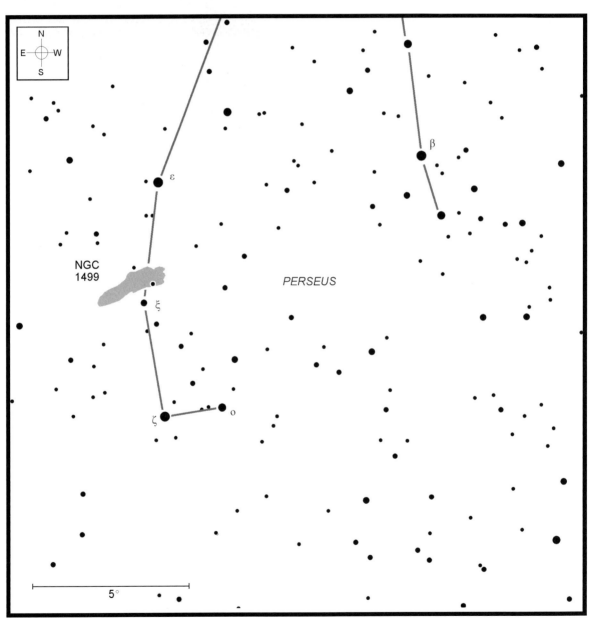

Chart 3.18 NGC 1499

Okay, now we're getting serious. NGC 1499, nicknamed the California Nebula, for its likeness in long-exposure photographs to the Golden State's outline, is one of those targets that puts even the most experienced observers to the test.

Most historians say that the first European explorer to discover California was Portuguese explorer Juan Rodriguez Cabrillo in 1542. It would take another 343 years before the state's celestial namesake was discovered by American astronomer Edward Emerson Barnard. On November 3, 1885, Barnard visually spotted its faint glow just northeast of Xi (ξ) Persei through the 6-inch Cooke refractor at Vanderbilt University in Nashville, Tennessee. This is arguably one of the greatest visual discoveries of all time because of the California Nebula's huge expanse. From tip to tip, NGC 1499 measures $2\frac{1}{2}°$ long and $\frac{3}{4}°$ wide, which is far larger than the narrow field of Barnard's refractor. Add to that the fact that, like so many clouds of glowing hydrogen, NGC 1499's surface brightness is incredibly low because its primary emissions are restricted to the red portion of the visible spectrum. How Barnard first spotted its feeble glow without the benefit of today's contrast-enhancing filters and sophisticated wide-field eyepieces only emphasizes further his incredible talents as an observer.

Even with those modern conveniences, seeing the California Nebula is a crowning achievement for twenty-first-century observers. The problem is trying to determine where the nebula starts and the surrounding sky ends. The narrower the field of the telescope, the tougher it is to tell. In fact, it is not uncommon for NGC 1499 to remain invisible through 12- to 14-inch telescopes, even under the darkest skies, and yet be visible through with binoculars or, yes, even the unaided eye.

First, let's talk strategy. The California Nebula is easy to pinpoint thanks to 4th-magnitude Xi (ξ) Persei, which is set just 1° to its south. But that star, which is the ionizing energy source for the nebula's glowing hydrogen clouds, is both a blessing and a curse. While it makes zeroing in on the nebula a snap, its light can easily overwhelm the nebula's weak glow. That is why knowing *how* to look for low-surface-brightness objects

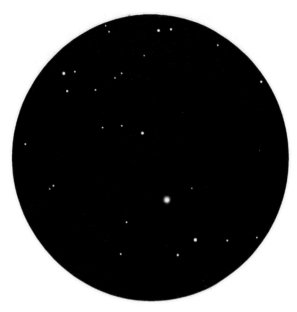

Figure 3.18 NGC 1499

is just as important as knowing *where* to look. By mounting your binoculars on a tripod or other support, the star can be moved out of view, extinguishing its interfering presence and improving the odds. The California Nebula is also one of the very few targets that is noticeably improved with hydrogen-beta line filters. If you have access to two, try affixing them to your binocular eyepieces.

Using averted vision, start with the slightly brighter northeastern edge of the cloud, opposite Xi, and then trace out the nebula's "Pacific coast." Its glow is very dim, but the long, slender profile is unmistakable once you realize the shape and size (Figure 3.18). Several superimposed stars remind me of lights from distant cities. The brightest, a 6th-magnitude sun, lies just offshore about where San Francisco would be, and marks a crook in the coastline.

When it comes to spotting the California Nebula, remember that a clear, dark sky is an absolute must. There is also no substitute for experience. Indeed, with both, some observers have actually reported seeing the California Nebula without any optical aid, save for looking through a pair of hydrogen-beta filters.

41 M1

Target	Type	RA	Dec.	Constellation	Magnitude	Size	Chart
M1	Supernova remnant	05 34.5	+22 01.0	Taurus	8.4	6.0′×4.0′	3.19

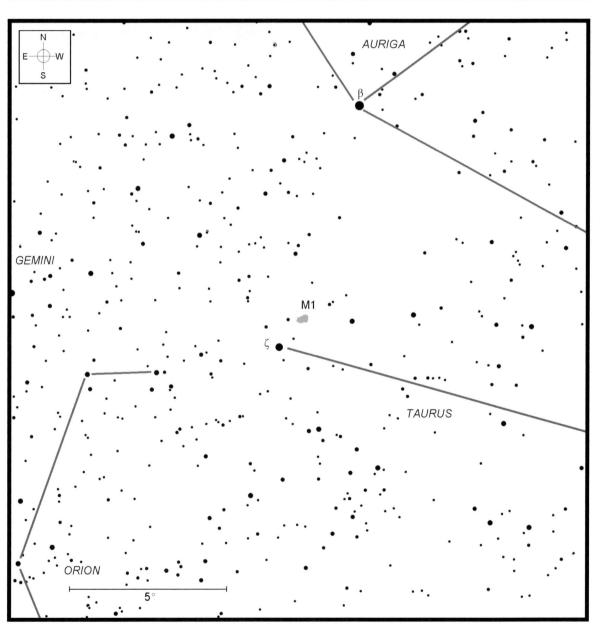

Chart 3.19 M1

The universe changed on August 28, 1758. Early that morning, as he scanned the predawn sky for Halley's Comet on its predicted return, the French comet hunter Charles Messier found something he had never seen before through his telescope. The mysterious object, lying just off the southern "horn" of Taurus, the Bull, looked just like a comet through his small refractor. Returning over the course of the next two weeks, Messier realized that what he had found could not be a comet, since there was no telltale indication of orbital motion. Instead, it remained anchored in place. Messier noted the position of his unexpected discovery,[5] and was later to include it as the first entry in a catalog of annoyances that might fool him and his fellow comet hunters into thinking they too had found a new comet. The rest, as they say, is history.

The story of M1, of how it is the expanding remnant of a massive star that detonated in a huge supernova explosion, is well known to most. That explosion was witnessed by ancient Chinese and Anasazi Native American stargazers in July AD 1054. Although half a world away from each other, both recorded the exploding star's sudden appearance. At its peak, the supernova shone as brightly as magnitude −6 and was visible in broad daylight for nearly a month. Today, when we look toward the site of the supernova explosion, all we see is the expanding cloud of debris we call M1.

On a much more personal level, the universe changed for me in January 1972 and, again, M1 was the reason. It was a cold winter night, with snow on the ground and crystal-clear skies overhead. Outside in my backyard, I had my new 8-inch reflector, my trusty *Skalnate Pleso* star atlas, and a pair of 7×35 binoculars. Wearing a heavy parka against the cold, I laid back against a snow drift, enjoying the sky with the binoculars as I waited for my telescope's optics to acclimate to the cold temperature. As I made my way past many winter favorites, I happened to scan along the southern horn of Taurus. The blue 3rd-magnitude

[5] Actually, Messier was not the first person to see M1. That distinction belongs to London physician and amateur astronomer John Bevis, who discovered it 27 years earlier.

Figure 3.19 M1

Zeta (ζ) Tauri marks the tip of the horn. Through the binoculars, I could also see nearby 6th- and 7th-magnitude stars that, with Zeta, formed a small trapezoid that I had used in the past to find M1 through my reflector. Looking more closely at the field, I thought I could see the dimmest smudge of light nearby. Sure enough, I was seeing, ever so faintly, M1 through the binoculars. That one observation hooked me on binoculars and began my lifelong pursuit of pushing them to their limits, which I enjoy to this day.

Through 10× and lower-power binoculars, M1 looks like a rounded rectangle of grayish light (Figure 3.19). That's an unusual shape in a universe of round and oval nebulae. Like many smaller deep-sky objects, M1 benefits from magnification and smaller exit pupils. Under dark skies, it appears much more obvious through a pair of quality 16×50 binoculars than through similar 10×50s, and even more so than through 7×50 binoculars. The secret to seeing it is to brace the binoculars against something, whether that something is a tripod, a fence post, or, yes, even lying in a pile of snow.

| 42 | **M78** | | | | | | |

Target	Type	RA	Dec.	Constellation	Magnitude	Size	Chart
M78	Reflection nebula	05 46.8	+00 03.5	Orion	8.3	8.4′×7.8′	3.20

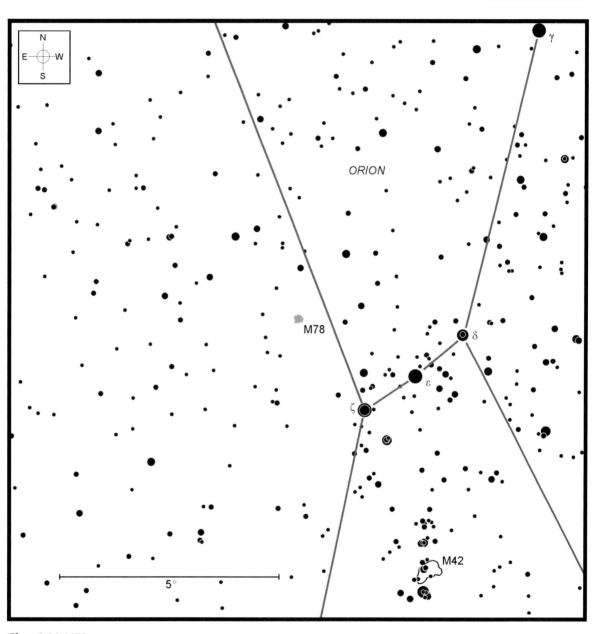

Chart 3.20 M78

Figure 3.20 M78

While M42 hogs the Orion deep-sky spotlight, a lesser-known Messier object also lies within the Hunter's broad borders. M78, a small tuft of reflection nebulosity, was discovered by Pierre Méchain in 1780 and added to Messier's catalog later that same year. It lies about $3\frac{1}{2}°$, or probably a little better than half a binocular field, due east of Mintaka [Delta (δ) Orionis] and $2\frac{1}{2}°$ northeast of Alnitak [Zeta (ζ) Orionis].

Like M42, M78 belongs to a vast expanse of hydrogen gas called the Orion Molecular Cloud drifting across the constellation. Both objects are denser clumps within that cloud. While the hydrogen gas in M42 is being ionized into fluorescence by ultraviolet energy from the many stars embedded within, M78 is visible only by reflecting the light from nearby stars. The starlight reflects off the grains of interstellar dust scattered throughout the cloud, creating a reflection nebula. Reflection nebulae are always easy to identify in photographs by their deep-blue color, caused by the dust grains scattering blue light. The effect is similar to the scattering process that makes our daytime sky blue.

Nebula filters offer little or no help improving the visibility of reflection nebulae. Therefore, to see objects like M78, we need a reasonably dark, transparent sky. Actually, I find M78 to be an excellent test of a winter sky's quality. Under a sky with a naked-eye limiting magnitude of 4.5, M78 is barely visible through my 10×50 binoculars as a small, very faint blur (Figure 3.20). Improving conditions to a naked-eye limiting magnitude of 5.5, M78 becomes much more obvious when the same binoculars are aimed its way. The oval cloud appears punctuated by a brighter, slightly offset core. Although the overall appearance reminds many of a faint comet, the brighter "nucleus" is actually a pair of close-set 10th-magnitude stars that are embedded within the cloud. Despite the fact that the ultraviolet radiation emissions from these stars is not enough to turn the cloud into an emission nebula, their visible light is sufficient to light up the multitude of dust grains within.

43   NGC 2158

Target	Type	RA	Dec.	Constellation	Magnitude	Size	Chart
NGC 2158	Open cluster	06 07.4	+24 05.8	Gemini	8.6	5.0′	3.21

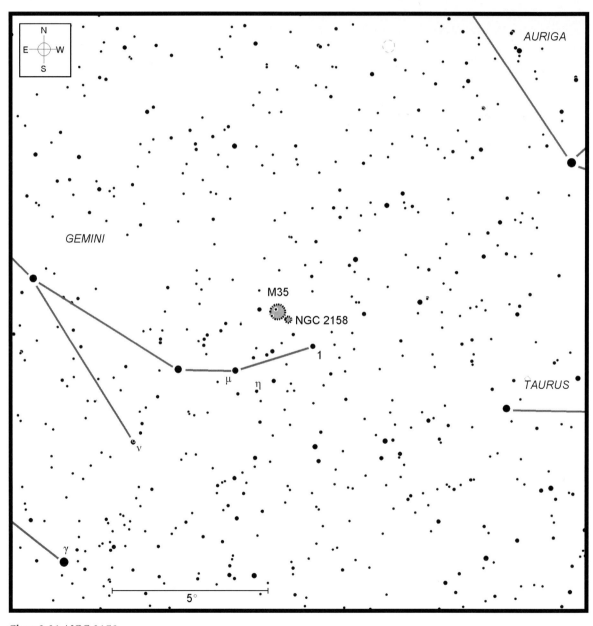

Chart 3.21 NGC 2158

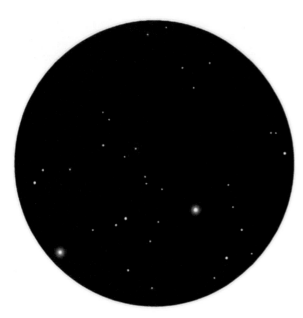

Figure 3.21 NGC 2158

M35 in Gemini is without a doubt one of this season's most spectacular open clusters through binoculars. You'll find it near the foot of the twin brother, Castor. Because of that position, I think of M35 as the "Soccer Ball Cluster," since it looks like Castor is about to kick it. More than 200 stars lie within, with over two dozen shining brighter than 10th magnitude.

Spotting M35 by eye is a very good test for the darkness of a winter night, and in fact, is included in Chapter 2 as naked-eye Challenge 15. It's a simple task

to find it through binoculars, even with less than ideal sky conditions.

Once you've spotted M35, try your luck with NGC 2158, a second open cluster found just $\frac{1}{4}°$ to the southwest. In my book *Touring the Universe through Binoculars*, I recalled my attempts at seeing NGC 2158 through binoculars this way: "Try as I might, I have never been able to see NGC 2158 in 7×50 binoculars, although it should be visible if it is actually that bright. Instead, I find it requires at least 11×80's to be seen."

Well, that was more than 20 years ago, and in those two decades, I've come to see the cluster several times on dark nights through my 10×50s (Figure 3.21). It takes just a little extra effort – okay, maybe a lot of extra effort. A special night that is free of any clouds and light pollution is also a must to make out its small, grayish blur just off the southwestern edge of M35. While 50-mm binoculars can resolve as many as a half dozen stars within M35, nary a flicker will be seen in NGC 2158. That's because none of its 1,000 stars shine brighter than 14th magnitude.

NGC 2158 is one of the most unusual open clusters in the winter sky. Its great distance, estimated at 13,000 light years or more than four times the distance to M35, puts it very near the edge of our Milky Way Galaxy. Studies also reveal it to be one of the oldest open clusters known. Judging by its unusually high number of mature yellow and red stars, NGC 2158 may be more than one billion years old. M35, consisting primarily of young blue-white stars, is only about 150 million years old.

44   NGC 2237

Target	Type	RA	Dec.	Constellation	Magnitude	Size	Chart
NGC 2237	Emission nebula	06 31.7	+05 04.0	Monoceros	–	80′×60′	3.22

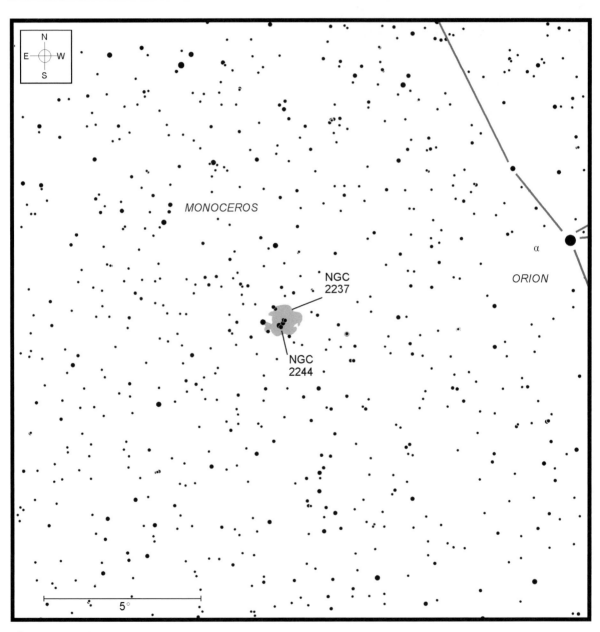

Chart 3.22 NGC 2237

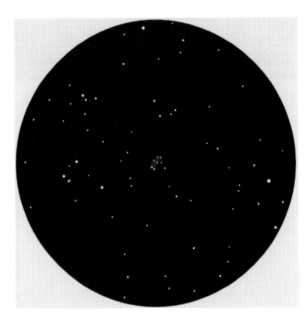

Figure 3.22 NGC 2237

Of the three dozen open star clusters within Monoceros, NGC 2244 is the largest and brightest. You might even spy it by eye alone on very dark, clear winter nights as a faint blur about halfway between Procyon [Alpha (α) Canis Minoris] and Orion's left (western) shoulder, Bellatrix [Gamma (γ) Orionis]. Aim your binoculars at that point and you'll easily count half a dozen 6th- and 7th-magnitude cluster stars set in a distinctive rectangular pattern. The brightest point in the rectangle is the 6th-magnitude yellowish star 12 Monocerotis. One hundred stars belong to NGC 2244, all lying about 4,900 light years away from our Solar System and covering about a Moon's diameter in our sky. Given that distance, NGC 2244 must span about 70 light years.

If you're viewing under country skies, you might also notice a faint haze encircling NGC 2244. That's the dim glow of the famous Rosette Nebula (NGC 2237), a huge wreath-shaped cloud that engulfs the star cluster. The Rosette measures more than a degree in diameter, but can be very difficult to spot. Actually, the Rosette is quite bright, but unfortunately, like all emission nebulae, it shines primarily in the red end of the spectrum, where the human eye is not terribly sensitive.

That means that, while it is easy to record with fairly short photographic exposures, it is a tough visual test. I've been able to glimpse it as a ghostly, broken ring surrounding the cluster through 10×50 binoculars on clear winter nights (Figure 3.22), but under average skies, it stays hidden among the stardust. Narrowband nebula filters, like the Lumicon UHC and Orion UltraBlock, help bring out the cloud, but you will need two for viewing through binoculars. Thus equipped, the Rosette is a remarkable sight through binoculars, but take your time. What appears to be a dull haze at first will evolve into a mottled wreath of light as you concentrate on it for several minutes. Look for irregular regions throughout the cloud, notably toward the southeastern edge, where a hook-shaped arm appears to curl off from the main body.

Several skilled observers, including the late Walter Scott Houston, have seen not only the open cluster with just their eyes, but also the Rosette itself, by peering through nebula filters. To try this test for yourself, hold the filter square to your eye and then slowly tilt it back and forth just a bit. This "rocking technique," as Houston reported in one of his Deep-Sky Wonders columns in *Sky & Telescope* magazine, causes the nebula to flicker in and out of view.

45 NGC 2403

Target	Type	RA	Dec.	Constellation	Magnitude	Size	Chart
NGC 2403	Galaxy	07 36.9	+65 36.2	Camelopardalis	8.9b	22.1′×12.4′	3.23

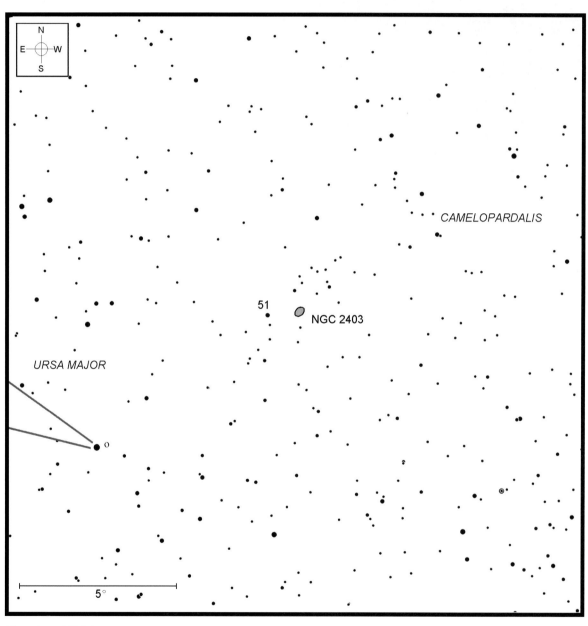

Chart 3.23 NGC 2403

NGC 2403 is one of the brightest galaxies in the northern celestial hemisphere that was missed by Messier and Méchain. Although their omission may have been due to the galaxy's sparse surroundings, NGC 2403 is actually not that hard to find. Begin at Muscida [Omicron (o) Ursae Majoris], the nose of the Great Bear. From there, slip about 5° northwest to 6th-magnitude 51 Camelopardalis. NGC 2403 lies in wait just 1° further to the west.

Recently I revisited NGC 2403 through my 10×50 binoculars. I could just make out its tiny oval glow against the background sky from my suburban backyard here on Long Island. Under darker skies, however, it has come through as a dim splotch in 7×35 binoculars, while the 10×50s uncovered the galaxy's round, diffuse core centered in an elongated halo, shown in Figure 3.23. As always, there is no substitute for a dark sky.

Great detail in this striking broad-armed spiral galaxy can be glimpsed when viewed through moderate-size amateur telescopes. A 10-inch instrument hints at spiral-arm structure toward the galaxy's western edge, a trait that is more readily confirmed in 12-inch and larger telescopes. These same scopes also reveal a very faint nebulous "star" within one of the spiral arms. Higher power confirms that this is not a star at all but rather a huge hydrogen-II region separately cataloged as NGC 2404. Photographs reveal this is only the brightest of many H II regions and clusters of stars sprinkled across the galaxy.

Figure 3.23 NGC 2403

Hovering above the northeastern horizon at this time of year is the obscure constellation Camelopardalis the Giraffe. Though the human eye alone reveals little more than a void populated by a scattering of 4th-magnitude and fainter stars, binoculars begin to unleash some of the beast's latent wonders. One of the Giraffe's few hidden treasures visible through binoculars is NGC 2403, a spectacular spiral galaxy tilted nearly face-on to our perspective.

46   Apollo landing sites

Target	Type	Best lunar phases	Chart
Apollo landing sites	Solar System	varies with landing site	3.24

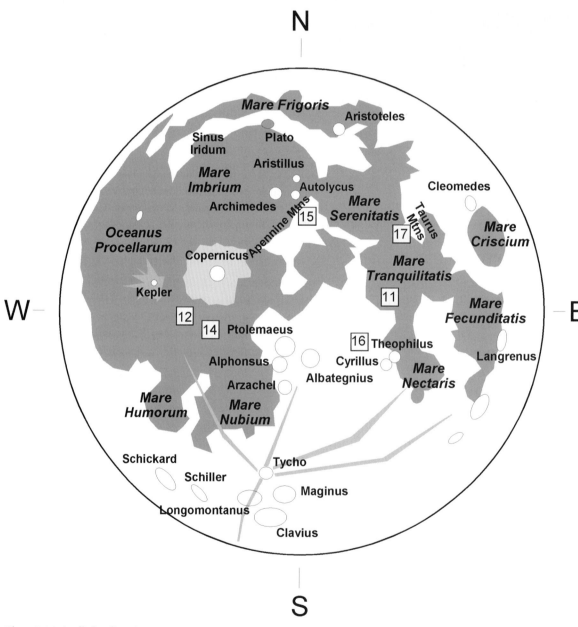

Chart 3.24 Apollo landing sites

Figure 3.24 Apollo 11 descent stage from Lunar Reconnaissance Orbiter

Between July 1969 and December 1972, six teams of United States astronauts ventured across the gap between Earth and Moon to land and walk on that distant world. Have you ever visited their landing sites? If not, let's do so now.

We begin with Apollo 11, "Tranquility Base." The dark gray outline of the Sea of Tranquility, Mare Tranquilitatis, looks almost perfectly round from our earthly vantage point. The best time to view Mare Tranquilitatis is during the waxing crescent phases, five to six days after New Moon. Apollo 11 landed near the southwestern shore (lower left edge).

Apollo 12 landed on Oceanus Procellarum (the Ocean of Storms) in November 1969. Dominating the waxing gibbous phases, the Ocean of Storms covers more than one million square miles of lunar terrain. The mission's exact landing site lies south of the prominent crater Copernicus, which sees sunrise two nights after First Quarter. Watch as sunlight first bathes the crater's sharply defined walls, catching the strong central mountain peak before sliding down to the crater floor. Mark your calendar to come back in a few nights when the brilliant ray system of Copernicus explodes into view against the darker background of the mare. Its starburst pattern is unmistakable through even the most modest binoculars.

The hilly region known as Fra Mauro was the February 1971 landing site of Apollo 14. Fra Mauro is found near the southeastern shore of the Ocean of Storms, to the east, or right, of Apollo 12.

July 1971 saw Apollo 15 touch down near the Apennine Mountains, which mark the southeastern edge of the Sea of Rains, Mare Imbrium. The arc of the Apennines is found just south of the prominent triangle of craters formed by Aristillus, Autolycus, and Archimedes. All three lie near the Moon's terminator, or sunrise line, the night after First Quarter.

Apollo 16 landed near the crater Descartes in the highlands south of the Sea of Tranquility in April 1972. The craters Theophilus and Cyrillus are to the east (right) of the landing site, while Albategnius is roughly an equal distance to its west, or left. The night before First Quarter is perfect for viewing this area. Just west of Albategnius, three more striking craters that almost touch each other's borders – Ptolemaeus, Alphonsus, and Arzachel – see sunrise the following evening.

Apollo 17's landing in December 1972 signaled an end to the Apollo era. We find its site near the Taurus Mountains, which form the eastern rim of the Sea of Serenity, Mare Serenitatis. The best time to view this area is during the waxing crescent phases.

Plans are in the making for a return to the Moon, but why wait? We can relive the magic of the historic Apollo program tonight with just our binoculars.

4
Small-scope challenges
Giant binoculars, 3- to 5-inch telescopes

Telescopes in the 3- to 5-inch aperture range can be divided into three wildly diverse groups. On one hand, we have an overabundance of small instruments that are marketed toward unsuspecting newcomers to the hobby. These are often sold through department stores, consumer-club warehouses, toy stores, hobby shops, online auctions, and other mass-market outlets like television shopping networks. Two underlying traits are shared by the vast majority of these instruments: mediocre optical quality and shaky mounts.

On the other hand, we have some nice, middle-of-the-road economy instruments that serve their owners well. These instruments excel as grab-and-go telescopes, allowing their users to spring outside at a moment's notice without the burden of carrying massive mountings and huge tube assemblies.

Finally, we have some of the finest instruments sold today. Small-aperture refractors from well-known companies like Tele Vue, Takahashi, and Astro-Physics delight and entertain amateurs who are looking for the highest-quality optics, top-notch mechanics, and superior portability. They share the same spur-of-the-moment freedom as the middle group, while delivering top-notch images.

Giant binoculars, defined here as anything 70 mm or greater in aperture and 10× or more magnification, are also included in this chapter's mix. Owing to their lower magnifications, some of the challenges here are beyond their capability, while others are made to order.

Regardless of which instrument you are using, the challenge objects gathered here will test its limits, and yours.

47   NGC 2959, NGC 2976, and NGC 3077

	Target	Type	RA	Dec.	Constellation	Magnitude	Size	Chart
47a	NGC 2959	Galaxy	09 45.1	+68 35.7	Ursa Major	13.6p	1.3′×1.2′	4.1
47b	NGC 2976	Galaxy	09 47.3	+67 55.1	Ursa Major	10.8b (note 1)	5.9′×2.6′	4.1
47c	NGC 3077	Galaxy	10 03.4	+68 44.0	Ursa Major	9.9	5.5′×4.0′	4.1

Note 1. Magnitude numerical values followed by "b" denote blue magnitudes based on UBV photometric studies; They may be as much as two magnitudes lower than how an object appears visually.

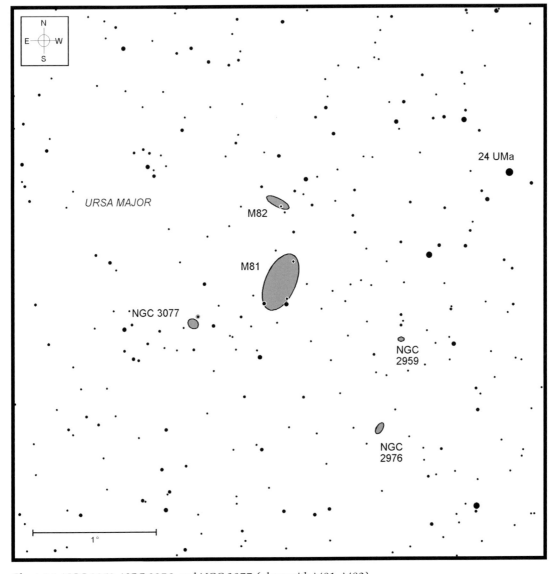

Chart 4.1 NGC 2959, NGC 2976, and NGC 3077 (along with M81, M82)

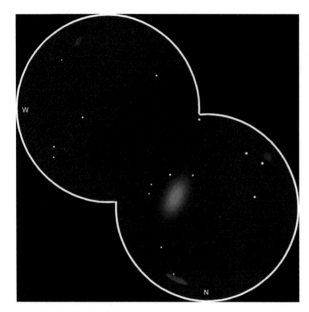

Figure 4.1 M81, M82, NGC 3077

M81 and M82 form perhaps the most famous pair of galaxies north of the celestial equator. The former was already discussed as naked-eye Challenge 1 in Chapter 2, while seeing the latter was offered as binocular Challenge 22 in Chapter 3. This time, we are back once again to visit two of their immediate neighbors. Bode missed these fainter companions when he discovered M81 and M82 in 1774, as did Pierre Méchain when he independently rediscovered this dynamic duo in 1779. They even evaded Charles Messier when he added the brighter pairing to his catalog in 1781. Two more decades would pass before William Herschel would discover their dim glows, yet both NGC 2976 and NGC 3077 can be spotted through small backyard telescopes given good skies.

NGC 3077, an odd looking 10th-magnitude elliptical galaxy, is the brighter of the two. Look for it about 45′ to the southeast of M81, just beyond a 10th-magnitude field star. As shown in Figure 4.1, NGC 3077 looks like a round, fuzzy patch of gray light that is best described as "featureless." Look carefully with averted vision, however, and you should spot a faint stellar nucleus just peaking out of the galaxy's center.

Long-exposure photographs show several opaque dust clouds along the edge of the galaxy that radiate away from the galaxy's core like spokes on a bicycle wheel. Elliptical galaxies are typically void of nebulosity, but NGC 3077's unusual appearance is probably due to its gravitational interplay with M81. Evidence of this is clearly visible in radio images, which reveal long filamentary threads of gas swirling between the two.

NGC 2976 is set about $1\frac{1}{2}°$ southwest of M81. Rated a full magnitude fainter than NGC 3077, NGC 2976 appears as little more than a dim, oval glimmer through 3- to 5-inch telescopes. Like NGC 3077, NGC 2976 appears riddled by gravitational warping caused by its proximity to M81. Although classified as a spiral galaxy, its nearly edge-on tilt to our line of sight coupled with these distortions blur its spiral arms in photographs.

There is still more to the M81 family of galaxies. Indeed, more than 30 individual systems belong to the brood, although many are found far from M81's immediate vicinity. NGC 2403, a bright spiral and the second most massive member in the group, lies 14° to the west in Camelopardalis, while NGC 4236, a barred spiral, is in Draco, a distant 21° to the east. Both are visible through 3- to 5-inch telescopes, even through moderate light pollution. Another groupie, the irregular galaxy NGC 2366, is also bright enough to be seen through your telescope as a small smudge some 4° north-northwest of NGC 2403.

| 48 | **M109** | | | | | | |
Target	Type	RA	Dec.	Constellation	Magnitude	Size	Chart
M109	Galaxy	11 57.6	+53 22.5	Ursa Major	10.6b	7.6′×4.6′	4.2

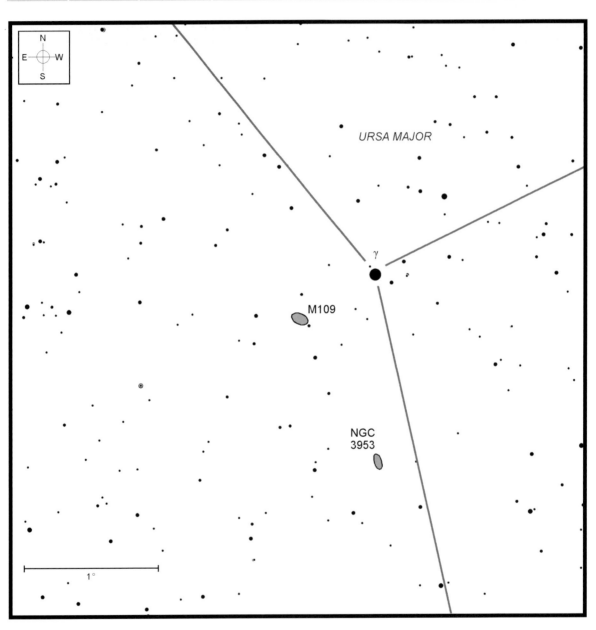

Chart 4.2 M109

The galaxy we know today as M109, cross-identified as NGC 3992 in John Dreyer's 1888 *New General Catalogue of Nebulae and Clusters of Stars*, was first spotted by Messier's contemporary, Pierre Méchain, on March 12, 1781. Although this was after Messier had submitted his original list of 103 objects for publication in *Connoissance des Temps*, Méchain reported his new find to Messier as "close to Gamma in the Great Bear." Messier did not live to see a second edition of his catalog, but objects 104 through 110 have been added posthumously by others. M109 joined the exclusive club of Messier objects in 1953, when astronomy historian Owen Gingerich noted Messier's observations of six additional "Méchain objects," now known as M104 through M109.

Not only is M109's history a little cloudy, spotting it represents one of the toughest challenges in this chapter. Indeed, many a seasoned observer has trouble seeing M109 through much larger telescopes. The low magnifications of giant binoculars further confound the situation.

M109, a nearly face-on barred spiral galaxy, lies just 38′ southeast of Phecda [Gamma (γ) Ursae Majoris], the star marking the southeastern corner of the bowl. At magnitude 2.4, Phecda's starlight easily washes out the dim glow of M109 at low power, especially if viewing with less than perfectly clean optics.

That's only part of the problem, however. M109 is one of those objects that, by their nature, have a very low surface brightness. The open structure of M109's spiral disk make it so dim that spotting it through anything smaller than a 6-inch telescope is all but impossible. As a result, smaller instruments reduce M109 to only its central nucleus, which appears as little more than a dim point (Figure 4.2).

These two facts led the creators of the Astronomical League's Binocular Messier Club to list M109 as a challenge object for an 80-mm binocular. Through my pair of 16×70 binoculars, it only shows up as a dim field "star" with perhaps the slightest hint of fuzziness. Increasing to a 20×80 binocular helps to single out the

Figure 4.2 M109

galaxy from among the few field stars in its immediate area.

The higher magnifications possible through my 4-inch *f*/10 refractor help to isolate M109's dim glow from the background. At 102×, the galaxy's nucleus appears decidedly lopsided, elongated roughly east-northeast/south-southwest. With averted vision, I can also detect a subtle, somewhat mottled hint of the galaxy's central bar protruding in the same direction, but any trace of the spiral arms that curl away from the ends of that bar remain in the realm of larger apertures and/or more skillful eyes.

While you are in the area, try to spot NGC 3953, another barred spiral set 1.4° due south of Phecda. Some suggested that Messier may have missed Méchain's reported galaxy and actually saw NGC 3953 instead. Although that conjecture is generally dismissed today, NGC 3953 is often spotted first by observers seeking M109 because of its slightly greater surface brightness.

49   M40, NGC 4290, and NGC 4284

	Target	Type	RA	Dec.	Constellation	Magnitude	Size	Chart
49a	NGC 4284	Galaxy	12 20.2	+58 05.6	Ursa Major	14.3p	2.5′×1.1′	4.3
49b	NGC 4290	Galaxy	12 20.8	+58 05.6	Ursa Major	12.7p	2.3′×1.5′	4.3
49c	M40	Double star	12 22.2	+58 05.0	Ursa Major	9.0, 9.3	50″	4.3

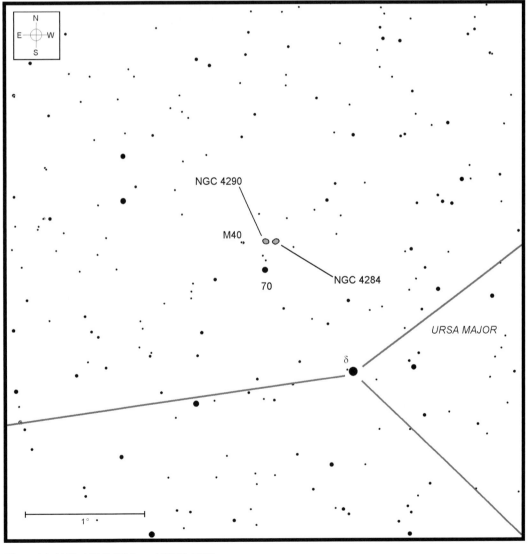

Chart 4.3 *M40, NGC 4290, and NGC 4284*

Most agree that the Messier catalog of deep-sky objects stands as the finest single compilation of star clusters, nebulae, and galaxies visible from the northern hemisphere. When it is time to single out the finest of the list's 109 entries, however, we often have trouble agreeing. Is it the Orion Nebula, M42; the Great Globular Cluster, M13; or maybe the Ring Nebula, M57? One thing is for certain – you'll never find Messier's 40th entry on anyone's "finest" list.

We all make mistakes, and M40 was one of Messier's. The story goes that Johann Hevelius, a noted observer from Dantzig, Germany, reported seeing a "nebula" near the star Megrez [Delta (δ) Ursae Majoris] in 1660. Four years later, try as he might, Messier could not repeat Hevelius's observation. All he found were a pair of close-set 9th-magnitude stars. Messier noted on October 24:

I searched for the nebula above the tail of the Great Bear, which is indicated in the [Hevelius] book *Figure of the Stars*, second edition . . . I have found, by means of this position, two stars very near to each other and of equal brightness, about the 9th magnitude, placed at the beginning of the tail of Ursa Major: one has difficulty to distinguish them with an ordinary [nonachromatic] refractor of 6 feet [focal length]. There is reason to presume that Hevelius mistook these two stars for a nebula.

For reasons lost to history, Messier decided to include the pair in his catalog, even though he knew well that they were just two stars.

Hevelius's legacy was resurrected again in 1863 when Friedrich Winnecke rediscovered the double star from Pulkovo Observatory in St. Petersburg, Russia. He marked down its position and, not knowing of the previous observations, subsequently included it as the fourth listing in his double star inventory, *Doppelsternmessungen* (Double Star Measurements). As a result, M40 is often cross-listed as Winnecke 4. More recent observations based on data from the European Space Agency's astrometric satellite Hipparcos suggest that the two components may actually be just an optical alignment – an optical double star – not a true binary pair.

Call it M40 or Winnecke 4 as you prefer, we are looking at a pair of stars shining at magnitudes 9.0 and 9.3 and separated by about 50″, and slowly widening. Both are resolvable through giant binoculars and,

Figure 4.3 M40, NGC 4290

Messier's description to the contrary, offer little visual challenge for 3- to 5-inch apertures. Look for M40 about $\frac{1}{2}°$ northeast of 5.5-magnitude 70 Ursae Majoris, itself a degree northeast of Megrez.

Although M40 leaves something to be desired, another object looms nearby that proves a worthy challenger for this size-class instrument. NGC 4290, a small barred spiral galaxy, is just 11′ to the west of M40. As portrayed in Figure 4.3, both easily fit into the same 85× field of my 4-inch *f*/10 refractor, with NGC 4290 looking like a faint, perhaps slightly oval blur. The galaxy, rated at magnitude 12.7(p), is visible with difficulty through my refractor from my suburban backyard, but it is clear from darker sites.

If you have good luck with NGC 4290, then try to find another, even fainter galaxy just to its west. Even under the darkest skies, NGC 4284 is a very difficult object through 5-inch (and even larger) telescopes. Shining at only magnitude 14.3(p), its presence is only whispered through my 6-inch Schmidt–Cassegrain from a site where the naked-eye limiting magnitude is slightly better than 6.0. As a visual clue that you are looking in the right place, NGC 4284 lies very near two 13th-magnitude field stars, one to its east, and the other to its south. Good luck!

50 Markarian's Chain

Target	Type	RA	Dec.	Constellation	Magnitude	Size	Chart
Markarian's Chain	Galaxy group	12 28	+13	Coma–Virgo	–	1.5°	4.4

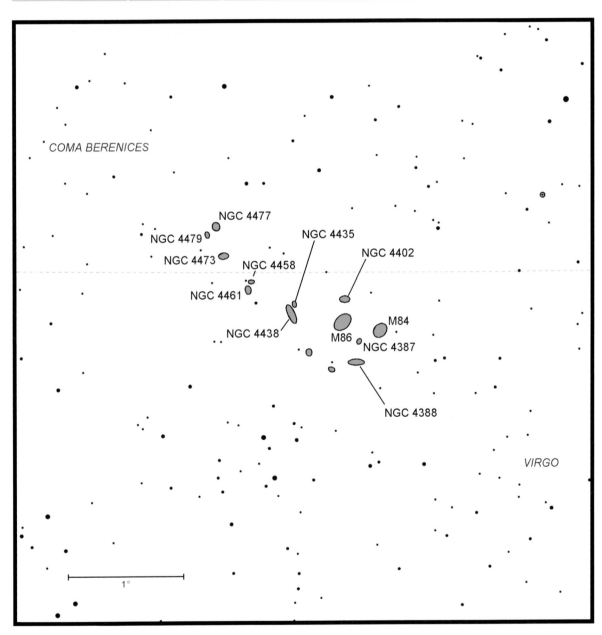

Chart 4.4 Markarian's Chain

Table 4.1 *Markarian's (expanded) chain*

Object	RA	Dec	Mag	Size
M84	12 25.1	+12° 53.3'	9.2	6'×6'
NGC 4387	12 25.7	+12° 48.6'	13.0	2'×1'
NGC 4388	12 25.8	+12° 39.7'	11.8	8'×1'
M86	12 26.2	+12° 56.8'	8.9	9'×6'
NGC 4402	12 26.1	+12° 26.1'	12.6	5'×1'
NGC 4435	12 27.7	+13° 04.7'	10.8	3'×2'
NGC 4438	12 27.8	+13° 00.5'	10	9'×3'
NGC 4458	12 28.9	+13° 14.5'	11.8	2'×2'
NGC 4461	12 29.1	+13° 11.0'	11.1	4'×1'
NGC 4473	12 29.8	+13° 25.8'	10.2	5'×3'
NGC 4477	12 30.0	+13° 38.2'	10.4	4'×3'
NGC 4479	12 30.3	+13° 34.7'	13.4	2'×1'

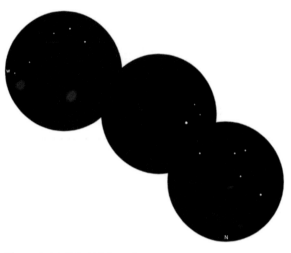

Figure 4.4 M84, M86, et al.

Aim your telescope anywhere in the large, seemingly empty gap between the stars Denebola [Beta (β) Leonis] and Vindemiatrix [Epsilon (ε) Virginis] and, given sharp eyes and a dark sky, you are bound to see one or more faint splotches of light somewhere in the eyepiece's field of view. You've entered the Coma–Virgo Realm of Galaxies, a collection of upwards of 2,000 separate galactic systems. Talk about a population explosion!

While there are many possibilities from which to choose, let's concentrate on a centralized line of galaxies known as Markarian's Chain. The name comes from the Russian astrophysicist Benjamin Markarian (1913–1985), who published an article entitled "Physical Chain of Galaxies in Virgo and its Dynamic Instability" in the December 1961 issue of the *Astronomical Journal*. In the article, he discussed how this $1\frac{1}{2}°$ arc of galaxies could not have just formed serendipitously. Instead, he "arrived at the conclusion that the chain of galaxies in the Virgo cluster is not a chance grouping but a real physical system."

Markarian's original chain had eight galactic links but, by veering off course a little to include some immediate neighbors, the list grows quickly to more than a dozen. How many of those listed in Table 4.1 can you spot?

Two galaxies, M84 and M86, dominate the western end of the chain and make a good jumping-off point. Together, they were detailed in the last chapter as binocular Challenges 23a and 23b, respectively. There,

just seeing them was the test. Now, with larger apertures and more magnification, subtle differences begin to appear. M84, an E1 elliptical galaxy, shows a slightly oval disk that's elongated northwest/southeast and punctuated by a brighter central core. Its neighbor, M86, also looks to be an elliptical galaxy, although its larger, slightly dimmer disk seems more elongated. It was that appearance that led astronomers to classify it initially as an E3 elliptical. Modern studies, however, now identify it as an S0 or lenticular galaxy, the "missing link" between ellipticals and spirals in the galaxy morphology classification system.

Do you ever get the feeling that you're being watched? You might when you're viewing M84 and M86, if your skies are dark enough. A pair of faint galaxies to their south joins the two Messier objects to create sort of an intergalactic face (Figure 4.4). The tiny elliptical galaxy, NGC 4387, marks the nose, while the edge-on spiral NGC 4388 forms the thin mouth. Neither galaxy shines brighter than 11th magnitude, however, so spotting the face can be tough. Finally, the 12th-magnitude edge-on spiral NGC 4402 looks like a raised eyebrow to the north of M86, as if our face is wondering "hey, what are *you* looking at?"

Although M84 and M86 are the eyes of our intergalactic face, the next two links in Markarian's Chain, NGCs 4435 and 4438, actually bear the nickname The Eyes. This soubriquet goes back more than half a century, to an article entitled "Adventuring in the Virgo Cloud" that appeared in the February 1955 issue of *Sky & Telescope* magazine. The article was

accompanied by a chart created by its author, Leland
Copeland, that affixed whimsical labels to some of the
galactic groupings. To Copeland, NGC 4435 and 4438
looked a pair of eyes staring back at him.

NGC 4435 and 4438 are also collectively known as
Arp 120 for their listing in American astronomer Halton
Arp's *Atlas of Peculiar Galaxies*. The catalog's 338 entries
comprise a fascinating collection of interacting and
merging galaxies. In the case of entry 120, Arp believed
that both galaxies were close enough to distort one
another because of their complex gravitational
interplay. Streamers of hot material extending away
from NGC 4438 add evidence to a much more recent
study using the Chandra X-ray Observatory that
concluded both galaxies had collided with each other
about 100 million years ago.

NGC 4435, the northern "eye," impresses me as
slightly brighter than NGC 4438, even though it is listed
as half a magnitude dimmer. That disparity is likely
because of its smaller size, which concentrates its light
more effectively. Both appear oval, with their long axes
oriented due north/south.

Hopping another link along the chain to the
east-northeast brings NGC 4458 and NGC 4461 into
view. The former looks nearly circular, while the latter
displays a distinctly elongated disk that will probably
require averted vision to be appreciated fully.

The chain crosses into Coma Berenices before
arriving at NGC 4473. Even if you have had some
difficulty spotting some of the other galaxies here
through your telescope or large binoculars, NGC 4473
should come clean fairly easily. Its elliptical disk,
oriented approximately east/west, surrounds a stellar
core that should be evident at about 80×.

Finally, we arrive at NGC 4477, the last link in
Markarian's Chain. Like NGC 4473, this should be a
fairly easy catch through 3- to 5-inch telescopes. Look
for a small, foggy patch engulfing a brighter central
core. Lying right next door, NGC 4479 offers an even
more difficult test than its neighbor. Though not part of
Markarian's original list, this 13th-magnitude galaxy
offers up a worthy challenge for small-scope observers.
Crank up the magnification to 100× or more, if
conditions permit, to maximize image contrast.

51	3C 273

Target	Type	RA	Dec.	Constellation	Magnitude	Size	Chart
3C 273	Quasar	12 29.1	+02 03.2	Virgo	13.0	stellar	4.5

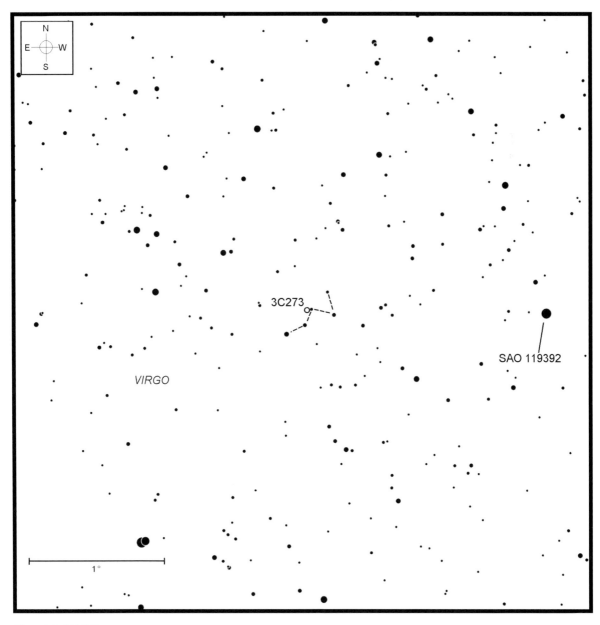

Chart 4.5 3C 273

Figure 4.5 3C 273

Whenever my neighbor Joe sees me at my telescope, he'll come over and ask "so, how far can you see with that thing?" Every time! You've also probably met someone like Joe. Well, unless you have a double-digit telescope, your answer should probably be "2 billion light years." That's the distance to 3C 273, the sky's brightest quasi-stellar object, or quasar for short, and the first of its type to be detected. The unusual designation indicates that it's the 273rd entry in the *Third Cambridge Catalogue of Radio Sources*, which was compiled with the historic radio telescope at Jodrell Bank, England, and published in 1959.

Like many other quasars, 3C 273 looks like just another dim, slightly bluish star. Looks can be deceiving. Given its tremendous distance, its total light output is an incredible 2 trillion times that of our Sun. What's even more amazing is that this enormous power source takes up a volume of space no larger than our Solar System! Photographs also show a jet of material blasting outward at nearly the speed of light, further indicating that something pretty powerful is happening here.

The trick to seeing 3C 273 through small telescopes is to wait for the best nights, when the air is dry and the sky is crystal clear. Even then, picking it out from all of the other equally dim stars can be a challenge. To try your luck, aim your telescope toward the naked-eye star Eta (η) Virginis and then, with your lowest-power, widest-field eyepiece in place, nudge about $1\frac{1}{2}°$ northeastward to the reddish star SS Virginis, which fluctuates between magnitudes 6.0 and 9.6. From here, head about $1\frac{1}{4}°$ due north to 7th-magnitude SAO 119392. The quasar lies $\frac{3}{4}°$ directly east of this last star, right next to a 13th-magnitude star that some could confuse for 3C 273 itself. The quasar's blue color, however, will help set it apart from any stellar neighbors, as should the "mini Cassiopeia" asterism shown on the chart.

Like many quasars, 3C 273 is known to vary in brightness. Although it usually shines at magnitude 13.0, it has brightened by as much as a full magnitude in the past. But it can also slip a little unexpectedly, slowly dimming to magnitude 13.2 or so before returning to normal brightness.

So, Joe, my answer is still 3C 273, and I'm sticking with it!

52 NGC 5195

Target	Type	RA	Dec.	Constellation	Magnitude	Size	Chart
NGC 5195	Galaxy	13 30.0	+47 16.4	Canes Venatici	10.5b	5.8′×4.6′	4.6

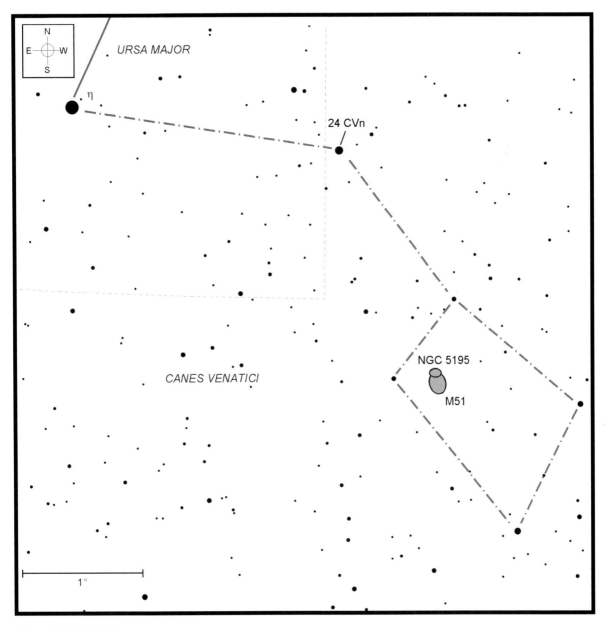

Chart 4.6 NGC 5195

Of all the galaxies in the spring sky, my favorite has to
be M51, the magnificent Whirlpool Galaxy in Canes
Venatici. M51 is one of those objects that keeps on
giving no matter what size or type of telescope (or
binocular) you are using. Even through modest 7×35
binoculars, its 8th-magnitude disk can be spotted
against the surrounding stars. Look for its dim glow
$3\frac{1}{2}^{\circ}$ southwest of Alkaid [Eta (η) Ursae Majoris], just
inside the eastern corner of a rectangle of four faint
stars.

With a bit more aperture and magnification, we can
also spot M51's close friend, NGC 5195. Although
Messier discovered M51 in October 1773, he apparently
failed to notice its sidekick. Some $7\frac{1}{2}$ years later, on
March 21, 1781, Messier's contemporary Pierre
Méchain noticed for the first time that Messier's 51st
catalog entry showed not one, but rather two
concentrations of light. Returning later that year,
Messier described his object as "double, each has a
bright center, which are separated 4'35". The two
'atmospheres' touch each other; the one is even fainter
than the other." It therefore appears that Messier did
indeed see NGC 5195, but did not realize that it was a
separate object deserving a separate numerical entry in
his catalog.

Both M51 and NGC 5195 suffer badly in light
pollution. While both merge into a single glow that is
relatively easy to see through 50-mm binoculars on
dark nights, light pollution can obliterate them through
larger instruments. Even under the darkest skies, 11× or
12× is needed to sever NGC 5195 from its domineering
companion. I can identify each individually with little
trouble through my 16×70 and 20×80 binoculars, but
have never been able to in 10×70s.

Switching to a 4.5-inch *f*/4 reflector, only 27× adds
the impression of the luminous bridge of material that

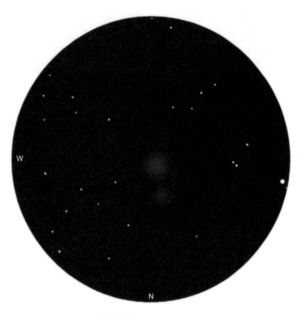

Figure 4.6 M51 and NGC 5195

curls from one galaxy to the next, while 91× reveals a
faint, nearly stellar nucleus in each galaxy (Figure 4.6).
This galactic umbilical cord connecting both galaxies is
probably the remnant of a close encounter that
occurred several million years ago. Not only did the
gravitational interplay between smaller NGC 5195
distort its already tumultuous internal structure, but the
repercussions also seem to have enhanced the spiral
structure in the far more massive M51. Indeed, some
gifted observers report seeing subtle blotches in the
spiral arm halo through telescopes as small as 4 inches
in aperture. For the vast majority of amateurs, seeing
spiral structure remains a challenge for 6- to 9.25-inch
class instruments, discussed in Chapter 5.

53 Izar

Target	Type	RA	Dec.	Constellation	Magnitude	Size	Chart
Izar [Epsilon (ε) Boötis]	Binary star	14 45.0	+27 04.5	Boötes	2.5/5.0	2.8″	4.7

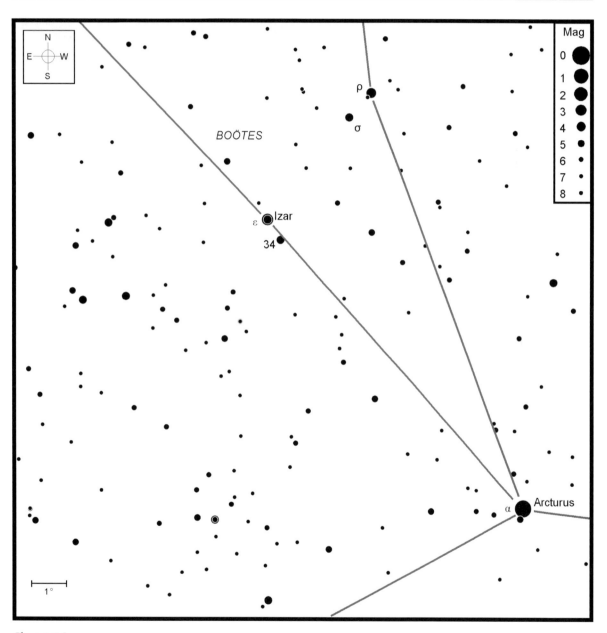

Chart 4.7 Izar

Finding this next challenge object is no challenge at all
unless you are trying to starhop to it from the inner city.
That can be tough, but, for everyone else, Izar [Epsilon
(ε) Boötes] is visible easily by eye to the northeast of
brilliant Arcturus as one of six stars that make up the
constellation's distinctive kite shape. Swing your
telescope its way and it still looks like a single star, as it
does to the naked eye. So, what's the attraction?

Izar is one of the more challenging binary stars in
the northern spring sky. Binoculars readily show it to be
accompanied by 34 Boötes, a red, 5th-magnitude sun
39″ to its southwest. They form an attractive pairing,
even though their affiliation is strictly a chance,
line-of-sight alignment.

The real challenge here is to see Izar's actual
companion, 5th-magnitude Izar B. Izar B is 2.8″ away
from the system's primary star, Izar A. Observers
usually describe Izar A as looking either yellowish or
pale orange, while Izar B ranges from bluish to sea
green or emerald.

Two stars separated by 2.8″ sound as though they
are wide enough to drive a truck through until you
consider the circumstances. The problem is due to the
optical properties of light. As discussed in Chapter 1,
because light travels in waves, stars never appear as
perfect points of light through telescopes. Instead, a star
will focus to a small central disk – the Airy disk –
encircled by one or more dim rings called diffraction
rings. In the aperture range set for this chapter's
challenges, Izar A's first diffraction ring is
approximately 3 arc-seconds away from its Airy disk. As
a result, the companion lies nearly superimposed on
the primary's first diffraction ring, lowering image
contrast and creating the challenge.

With this in mind, I headed out with my 4-inch
refractor to my suburban backyard a few years ago, to
see what I could see. Popping in a 7-mm eyepiece
(143×), I could readily split the pair, with Izar B
appearing as a bright spot on A's diffraction ring
(Figure 4.7). It helped that this particular instrument
has good image contrast and that the seeing was steady.
Had I been viewing through a similar-size telescope
with an obstructed optical design (that is, a reflector or
catadioptric), then the resulting lower contrast could
have prevented me from seeing the companion.

Like many refractors, this particular instrument
came with an objective lens cap that has a smaller

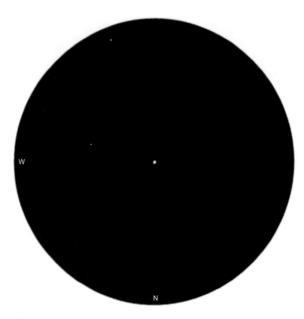

Figure 4.7 Izar

central opening. In this case, the smaller opening,
which is normally covered by a plastic dust cap,
measures 1.75 inches (45 mm) across. Could I still see
Izar B if I effectively stopped down my refractor to a
create 45-mm *f*/22 instrument? Not expecting much in
the way of success, I removed the dust cap, placed the
lens cover over the objective, and aimed at the star
again. To my surprise, I could see both suns fairly
easily, the companion now just inside the first
diffraction ring. Incidentally, to make this as "blind" a
test as possible, I purposely conducted these
observations before looking up the companion's
position angle. Sometimes, knowing where something
is will cause you to see it even if it is not visible –
witness the plethora of Martian canal observations at
the end of the nineteenth century! Afterward, when I
went back inside to check, sure enough it was right
where I saw the companion, northwest of
Izar A.

To be successful at spotting Izar B, you will need
high-quality optics, both in your telescope and in the
eyepiece, and steady seeing. Light pollution is of little or
no concern once Izar is in view, since both stars are so
bright.

54   NGC 6369

Target	Type	RA	Dec.	Constellation	Magnitude	Size	Chart
NGC 6369	Planetary nebula	17 29.3	−23 45.6	Ophiuchus	12.9p	38″	4.8

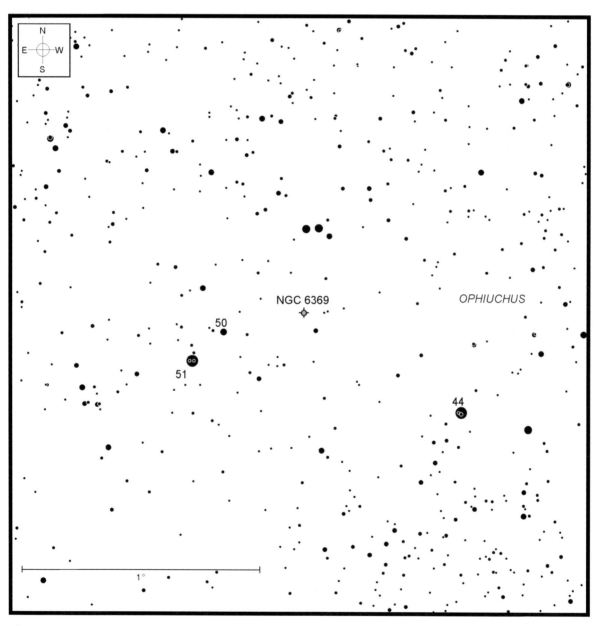

Chart 4.8 NGC 6369

When he accidentally found Uranus among the stars of Gemini in March 1781, William Herschel opened up our Solar System. Until that instant, it was thought that our Sun's planetary family contained only six members, ending at Saturn. But, with Uranus suddenly joining the group, the prospect for even more planets – perhaps many more – took the astronomical world by storm. Herschel, subsequently enjoying the fruits of his unexpected discovery as the new Astronomer Royal at the Royal Greenwich Observatory in England, led the charge to find them.

Within a month of commencing his systematic effort of scanning the skies, Herschel found another "planet," this time in Aquarius. No doubt he judiciously marked its position on his charts so that he could return again a few nights later to see just how far his new world had moved against the starry backdrop. But, when he returned, he immediately realized that it hadn't moved at all; instead, it had stayed anchored in place. Other discoveries of similar objects that looked like tiny versions of Uranus soon followed, but all also remained stationary among the stars. Because their small disks looked like distant planetary spheres, Herschel coined the phrase "planetary nebula." Even though it is one of the greatest astronomical misnomers of all times, since it says nothing of the true nature of these ghostly objects, the term stuck.

Herschel went on to discover 33 planetary nebulae[1] scattered across the sky. One of the more interesting and, at the same time, challenging of Herschel's planetaries to view through 3- to 5-inch instruments is NGC 6369 in southern Ophiuchus. Nicknamed the "Little Ghost Nebula," NGC 6369 is an example of a ring-type planetary nebula, a faint version of M57. That is, if you can find it.

Zeroing in on NGC 6369's exact location isn't too difficult, since it lies just half a degree northwest of 51 Ophiuchi. To get there, first locate 3rd-magnitude Theta (θ) Ophiuchi, about 12° east of Antares. Hop a little more than a degree northeast from Theta to 44 Ophiuchi, and then another degree east-northeast to 5th-magnitude 51 Ophiuchi.

---

[1] William Herschel originally classified 79 objects as planetary nebulae, but only 20 proved to be correct. The others were subsequently reclassified as other types of objects. However, 13 other objects that he had categorized as other types of objects were actually planetary nebulae. Therefore, whether or not he knew it, Herschel discovered 33 planetary nebulae.

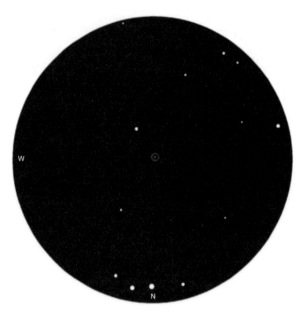

Figure 4.8 NGC 6369

You should take that final step from 51 Oph to NGC 6369 using a low-power eyepiece, but then switch to at least 120× to see the nebula's disk. NGC 6369 measures about half the apparent diameter of Jupiter. The Little Ghost looks like as a small, wraithlike disk of grayish light (Figure 4.8) through my own 4-inch refractor at 142×, as if floating in the ether once thought to flood the cosmos. When the air is especially transparent, the nebula's pale turquoise hue becomes clearer, but resolving its distinctive ring shape remains difficult even under the best conditions. The nebula's delicate color results from the ionizing radiation that is generated by the progenitor star and streams through the tenuous cloud. In the process, electrons are ripped away from their parent hydrogen and oxygen atoms, causing the cloud to glow colorfully.

Like its more famous kinfolk, M57, the remnant of the star that spawned the nebula remains invisible in all but the largest backyard telescopes. NGC 6369's central star, a white dwarf measuring perhaps no larger than our planet in diameter, glows weakly at 16th magnitude and is a challenge to spot through far larger instruments than those here. Those who successfully spot the star will notice that it is not exactly centered, but rather is offset slightly.

55 Barnard's Star

Target	Type	RA	Dec.	Constellation	Magnitude	Size	Chart
Barnard's Star	Star with high proper motion	17 57.6	+04 41.6	Ophiuchus	9.5	stellar	4.9

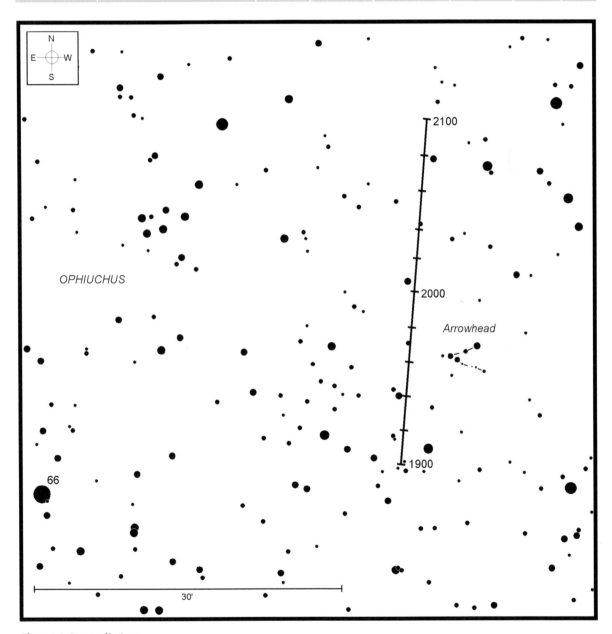

Chart 4.9 Barnard's Star

The curtain opened on Challenge 55 in September 1916, when a pair of articles written by Edward Emerson Barnard appeared in the journals *Nature* and the *Astronomical Journal*. Both recounted Barnard's discovery of a faint star in the constellation Ophiuchus that appeared completely unremarkable except for the fact that its proper motion was faster than any other star ever found. *Proper motion* is the apparent angular distance that an object shifts against the background celestial sphere over a specified period of time.

All stars exhibit proper motion to a degree because of their motion caused by the Milky Way's slow rotation as well as their own gravitational interaction with nearby stars. As a rule, a star's annual proper motion is limited to fractions of an arc-second. That's why the constellations we see and enjoy today appear as they did when they were first identified by the first civilizations thousands of years ago.

There are, however, exceptions to this rule. Barnard's initial computations found that "his" star had a proper motion of 10.3 arc-seconds per year, far exceeding any other star discovered up to that point. No doubt it was this curious behavior that led stellar astronomers to investigate this star's other characteristics to see if any contributed to Barnard's Star's rapid transit. Their studies showed it to be a rather ordinary red dwarf sun, spectral class M4V to be exact, lying just 6 light years from us and closing. Right now, it shines weakly at magnitude 9.5, but, as the millennia pass, it will be growing brighter steadily, if only subtly, as it draws closer to our Solar System. According to *Burnham's Celestial Handbook*, in about 8,000 years Barnard's Star will have closed to less than 4 light years away from us. In the process, its annual proper motion will have accelerated to 25 arc-seconds and it will have grown in brightness to about magnitude 8.6.

That's magnitude +8.6, incidentally. Even though Barnard's Star will be our Solar System's closest

neighbor in that distant era, its small size and low luminosity will keep it well below naked-eye limit. With a mass only 16% of the Sun and an estimated diameter of 140,000 miles (224,000 km), Barnard's Star radiates only 0.04% as much energy as our star. As a result, it will never outshine some of our other neighbors like Sirius or Alpha Centauri.

Red dwarf stars are by far the most common type of star in our universe, accounting for approximately 75% of the population. Although their low masses keep them below naked-eye visibility, their typical life spans extend well beyond that of more massive stars like the Sun. While our star is expected to exist for a total of 10 to 11 billion years, red dwarfs like Barnard's Star may go on for a trillion years or more.

Barnard's Star received a great deal of attention in the 1960s when Dutch astronomer Peter van de Kamp claimed to have discovered a giant planet (or planets) in orbit. Van de Kamp's claim was accepted for years, but was ultimately proven wrong by later studies. However, the prospect of a nearby solar system caused wide speculation back then. What would the planets look like? Were they inhabited? Could we visit them? The last question was studied closely by the British Interplanetary Society from 1973 to 1978 as part of their Project Daedalus program. Project Daedalus was to be an unmanned scientific research mission propelled by a two-stage nuclear fusion engine. Designed to accelerate to 10% of the speed of light, the spacecraft would take an estimated 50 years to reach its destination.

Use Chart 4.9 to find Barnard's Star. With your widest field eyepiece in place, position 5th-magnitude 66 Ophiuchi along the eastern edge of the view and look toward the opposite side for an arrowhead-shaped asterism of faint stars that points eastward. Barnard's Star lies less than 10 arc-minutes to the arrow's northeast, near an 11th-magnitude sun.

56 NGC 6781

Target	Type	RA	Dec.	Constellation	Magnitude	Size	Chart
NGC 6781	Planetary nebula	19 18.4	+06 32.3	Aquila	11.8p	108″	4.10

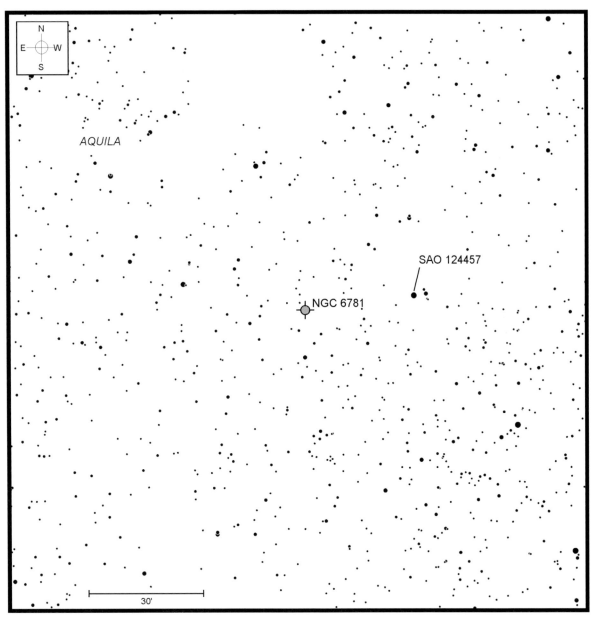

Chart 4.10 NGC 6781

No fewer than 65 planetary nebulae are scattered throughout the star clouds of Aquila the Eagle. Most are so small and so dim that they are nearly indistinguishable from the stars that surround them, but there are exceptions. NGC 6781 is one of them. Not only is it the second brightest planetary in the constellation, it is also the second largest in appearance; its disk measures 109 arc-seconds across. This makes it easy to tell apart from a star, although at the same time also lowers its effective surface brightness.

To find NGC 6781, aim exactly halfway between Altair [Alpha (α) Aquilae] and Mu (μ) Aquilae. Both should squeeze into the same finderscope field, with Altair to the northeast and Mu to the southwest. Now, adjust your aim so that Mu crosses the field and ends up in the northeast corner where Altair was a moment ago. Look toward the southwestern edge, where Mu was originally, and you should find the 6.7-magnitude star SAO 124457. NGC 6781 can be found just half a degree to its east.

NGC 6781 is visible through my 4-inch refractor as a faint, featureless disk at just 39×. From my light-polluted suburban backyard, a narrowband filter is needed to make it out but, under darker skies, it is visible in the same telescope without a filter. No matter the sky conditions, however, narrowband and oxygen-III filters are needed to see some of the nebula's structure. NGC 6781 is not a homogenous disk as it may appear at first. Look carefully and you should see how the northeastern edge looks noticeably brighter than the southwestern edge, as shown in Figure 4.9. Photographs bear this out, as well. The effect is

Figure 4.9 NGC 6781

probably the result of our viewing the nebula at an angle slightly off its axis of expansion. At the same time, the center of the disk may look still darker than the nebula's outermost circumference. NGC 6781 is fashioned in the style of M57, the Ring Nebula in Lyra, although its annulus structure is not nearly as pronounced. The white dwarf star at the center of it all – actually, it appears slightly off-center – shines feebly only at 16th magnitude, well beyond the reach of most backyard telescopes.

57 NGC 6803

★★

Target	Type	RA	Dec.	Constellation	Magnitude	Size	Chart
NGC 6803	Planetary nebula	19 31.3	+10 03.3	Aquila	11.3p	6″	4.11

58 NGC 6804

★★

Target	Type	RA	Dec.	Constellation	Magnitude	Size	Chart
NGC 6804	Planetary nebula	19 31.6	+09 13.5	Aquila	12.2p	35″	4.11

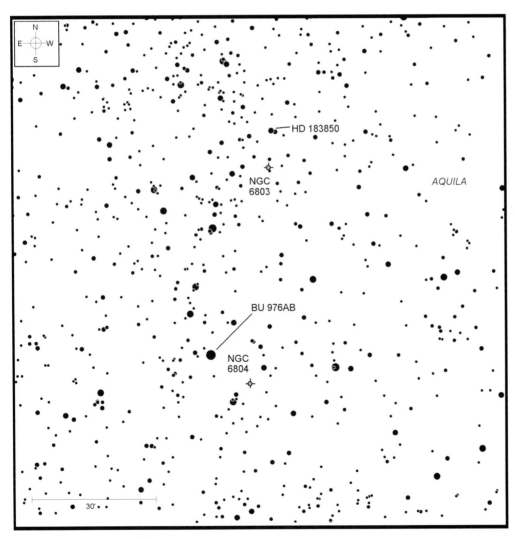

Chart 4.11 NGC 6803 and NGC 6804

Here's two for the price of one for you. Our next stop is
a pair of challenges found within 1° of each other in the
constellation Aquila the Eagle. Both of these planetary
nebulae present interesting tests for smaller apertures,
each in its own way. In fact, despite their closeness to
one another in our sky, NGC 6803 and NGC 6804 have
no physical relationship and couldn't be farther apart
in terms of appearance.

Let's begin with NGC 6803, found a little less than
4° west of Tarazed [Gamma (γ) Aquilae]. The American
astronomer Edward Pickering (1846–1919) was first to
lay eyes on this tiny target on September 17, 1882, using
the 15-inch refractor at Harvard College Observatory.
Pickering is most famous for his work determining
characteristics of stars by studying their spectra. Truth
be told, much of Pickering's acclaim was due largely to
the computational work performed by more than a
dozen female astronomers who assisted him. Known in
certain circles as "Pickering's Harem," his team of
assistants included Annie Jump Cannon, Henrietta
Swan Leavitt, Antonia Maury, and even his former
maid, Williamina Fleming. Each went on to make many
important contributions to the science in her own
right.

You'll find NGC 6803 at the end of a meandering
line of eight 8th-magnitude stars, just 10′ south of the
9th-magnitude double star HD 183850. The problem is
telling NGC 6803 apart from the rich surrounding star
field, since its tiny disk measures a scant 6″ across. Even
viewing at 200× through my 4-inch refractor on a night
of exceptionally steady seeing, it is still tough to tell
which is the planetary and which are just faint
surrounding stars without a little help.

Time to call in a narrowband nebula filter, such as a
Lumicon UHC or Orion UltraBlock. An oxygen-III
(O-III) filter also works well, although it tends to dim
the field far more than a UHC.

For minuscule planetaries like NGC 6803, however,
don't simply screw the filter into your eyepiece barrel.
Instead, center your telescope on the field suspected of
containing NGC 6803, hold your filter in between the
eyepiece and your eye, and take a careful look. By
alternately moving the filter in and out of the optical
train, you will see the planetary "blink." Stars, which are
broadband emission objects, will dim more noticeably
than the planetary, which focuses its energy emissions
only in a narrow portion of the visible spectrum. Do
this back and forth rapidly, checking each stellar point
as you go, and the planetary will have no choice but to

Figure 4.10 NGC 6804

reveal itself. After you capture NGC 6803, screw the
nebula filter into your eyepiece's barrel to see if you can
make out its disk.

Leave the filter in place as we move on to part 2 of
this challenge. From NGC 6803, slide 20′ southeast to a
7th-magnitude star, and then another 30′ due south to
the binary star BU 976AB, a close-set pair of
6th-magnitude suns that are a nice resolution test for
3-inch telescopes. NGC 6804 is just 11′ to their
southwest.

While NGC 6803 is challenging for its tininess,
NGC 6804 measures 35″ in diameter. That's easily large
enough to be distinguishable through my 4-inch
refractor at 100× (Figure 4.10). My notes recall that I
saw it "first with averted vision, then directly; a faint,
homogenous disk of grayish light floating within a
distinctive kite-shaped asterism." Larger apertures add
several faint stars immediately around the planetary,
producing a ghostly, faux three-dimensional effect that
is quite captivating. These same instruments may also
show the nebula's 14th-magnitude central star.

It was probably this appearance, with the nebula
framed by those unrelated stars that led William
Herschel to misclassify NGC 6804 as an open cluster
when he discovered it in August 1791. Only after it was
scrutinized more closely through the 100-inch reflector
at Mount Wilson Observatory by Francis Pease in 1917
was its true nature discovered.

59 Campbell's Hydrogen Star

★
★

Target	Type	RA	Dec.	Constellation	Magnitude	Size	Chart
Campbell's Hydrogen Star	Planetary nebula	19 34.8	+30 31.0	Cygnus	9.6p	35″	4.12

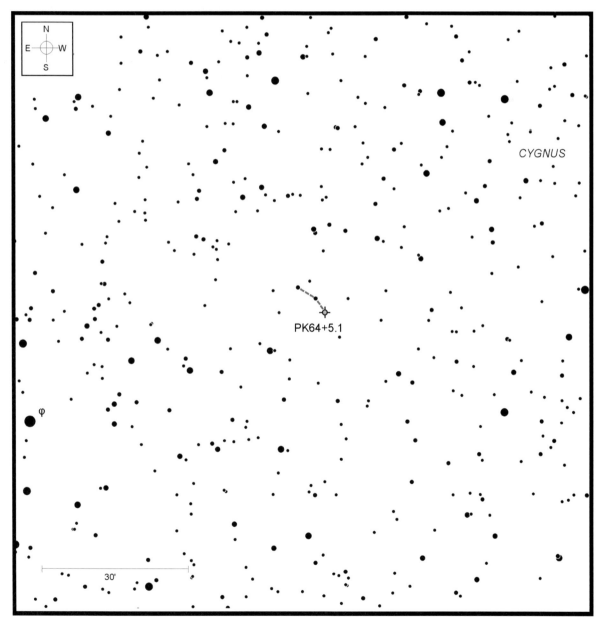

Chart 4.12 Campbell's Hydrogen Star

The surprisingly bright planetary nebula Campbell's Hydrogen Star is located just $2\frac{1}{2}°$ north of Albireo, but few people have ever seen it. The Herschels missed it, and it is missing from the *New General Catalogue*. American astronomer William Wallace Campbell spotted this unusual starlike object through a visual spectroscope at Lick Observatory in 1893. He could tell immediately from its spectrum that, despite its starlike appearance, he was not seeing an ordinary star at all. Instead, he had spotted an uncharted planetary nebula.

Although we know this little-observed target best by its nickname, Campbell's Hydrogen Star, it is more often labeled on charts as either PK64+5.1 or Henize 2-438. The former is its designation in Perek and Kohoutek's *Catalogue of Galactic Planetary Nebulae* (Academia Publishing, 1967), while the latter is from Karl Henize's paper "Observations of Southern Planetary Nebulae," which appeared in the *Astrophysical Journal Supplement* [Vol. 14 (1967), p. 125].

Campbell's Star is $2\frac{1}{2}°$ north of my favorite double star, Albireo [Beta (β) Cygni], but, to find it, I usually prefer to cast off from Phi (φ) Cygni, a little way northeast along the Swan's neck. Heading 1° due west and $\frac{1}{3}°$ due north of Phi puts you in the right place. There, look for a diagonal line of three equally spaced 10th-magnitude stars. The planetary nebula is actually the line's southwestern star.

Part of the problem in identifying Campbell's Star is that the nebula's 10th-magnitude central star

Figure 4.11 Campbell's Star

completely overpowers the tiny, 7.5 arc-second disk at low and medium powers. At 200×, my 4-inch refractor shows the star surrounded by the nebula's subtle glow, as shown in Figure 4.11, but only after studying it intently with averted vision. Little benefit is realized by "blinking" the nebula with either a narrowband or an oxygen-III nebula filter, although there is some modest improvement using a hydrogen-beta filter. If you happen to have one, be sure to give it a try.

60 NGC 6822

Target	Type	RA	Dec.	Constellation	Magnitude	Size	Chart
NGC 6822	Galaxy	19 44.9	−14 48.2	Sagittarius	9.3b	15.6′×13.5′	4.13

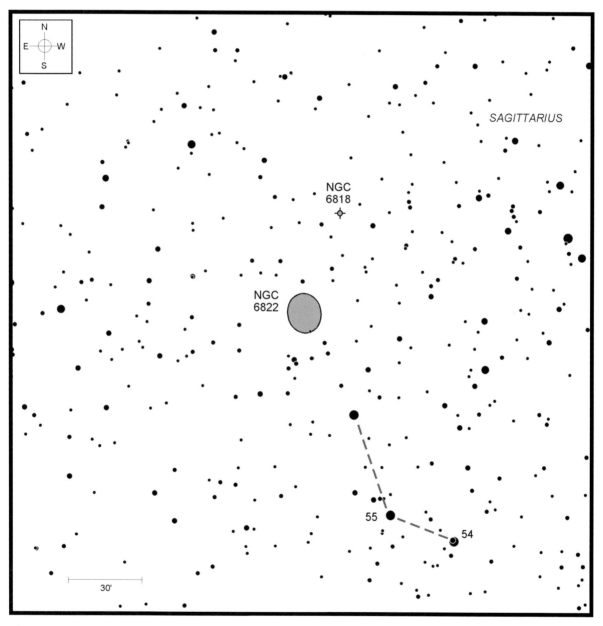

Chart 4.13 NGC 6822

Our next challenge may well be the most difficult in this chapter. In fact, there are readers who could argue, quite successfully, that it is one of the most difficult in this entire book. What is this scourge of so many amateur astronomers? NGC 6822, a dwarf barred irregular galaxy in Sagittarius. You may recognize it better by its nickname, Barnard's Galaxy.

The Barnard of Barnard's Galaxy is none other than Edward Emerson Barnard, whom we have already met numerous times in Chapters 2 and 3 for his study of dark nebulae. During one of his observing exploits with the 5-inch Byrne refractor at Vanderbilt University's observatory in Nashville, Tennessee, Barnard stumbled upon a dim, irregular glow among the stars of eastern Sagittarius. Thinking it to be a nebula that had strayed from the pack in western Sagittarius, he marked its position and published his find. Four years later, John Dreyer included Barnard's object as the 6,822nd entry in his *New General Catalogue*.

The nickname Barnard's Galaxy wasn't bestowed on NGC 6822 until some time after Edwin Hubble completed a photographic study of this unusual object in 1924. Hubble's groundbreaking work would prove without doubt that NGC 6822, like the Andromeda "Nebula," lay far beyond the edge of our Milky Way. He showed that both were separate island universes, which ultimately led to our modern view of the universe. NGC 6822 is now known to be a member of the Milky Way's Local Group of galaxies.

The easiest way to find NGC 6822 is to first locate a neighbor. That intermediate stop is a real gem, too. NGC 6818, known as the Little Gem, is an easy planetary nebula to see through just about any backyard telescope. From Nunki [Sigma (σ) Sagittarii] in the Teapot's handle, scan northeastward to an asterism known as the "teaspoon" formed by Pi (π), Omicron (o), Xi (ξ), Rho-1 (ρ1), and Rho-2 (ρ2) Sagittarii. Follow the spoon's handle about a finder field farther northeast to an arc of three 5th-magnitude stars, including 54 and 55 Sagittarii. NGC 6818 is a little over a degree to their north. Its disk appears uniform through 3- to 5-inch telescopes, while larger apertures show it as resembling the "CBS eye," the logo of the CBS television network.

With an eyepiece yielding no more than 40× in place, look for a very faint smudge of light about 40

Figure 4.12 NGC 6822

arc-minutes south-southeast of NGC 6818. Do you see it? If so, congratulations; you've spotted Barnard's Galaxy (Figure 4.12). Don't be surprised if it takes a few attempts. Barnard's Galaxy is easily hidden by any haze or light pollution, especially when viewed from mid-northern latitudes because of its southerly declination. Its light is also dimmed by intervening clouds of cosmic dust that litter the plane of our galaxy located nearby. If you can't find it at first, take heart. Remember, William and John Herschel never saw it either!

Once you spot the galaxy, see if you can make out any of the rich structural details that are so plainly evident in photographs. Although much of the galaxy's light is concentrated in a bar running north–south, its overall shape resembles a gently curved triangle that is widest toward the north and concave toward the west. Look for slightly brighter patches to the north and south of center. One, known as IC 1308, is tucked just inside the galaxy's northeastern edge. Try using the "blink technique" with either a narrowband or oxygen-III filter; review the discussion under NGC 6803 (small-scope Challenge 57) earlier in this chapter.

61   61 Cygni   ★

Target	Type	RA	Dec.	Constellation	Magnitude	Separation	Chart
61 Cygni	Binary star	21 06.9	+38 44.8	Cygnus	5.2, 6.0	30″	4.14

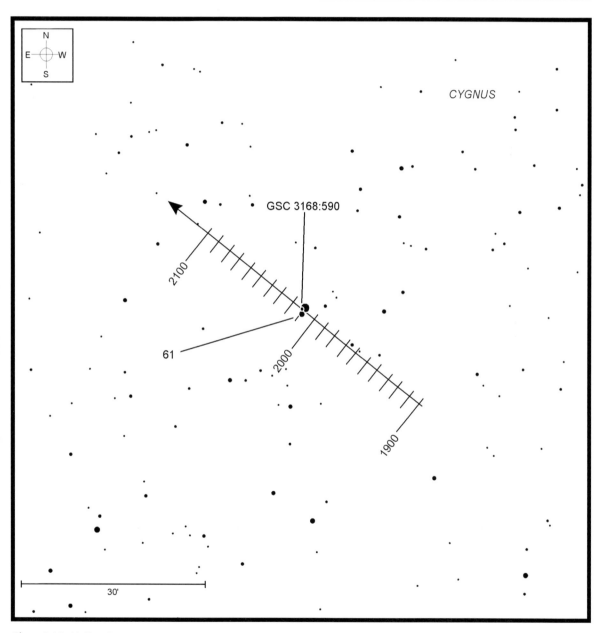

Chart 4.14 61 Cygni

Like Barnard's Star, 61 Cygni has an unusually high rate of proper motion, causing it to move visibly against the more distant stellar backdrop over the course of relatively few years. Although Barnard's Star wins the race in terms of speed, 61 Cygni is no slouch either. At present, 61 Cyg has a proper motion of more than 5 arc-seconds per year.

The Italian astronomer Giuseppe Piazzi (1746–1826), who is also credited with discovering the first asteroid (excuse me, "dwarf planet"), Ceres, was also the first to notice 61 Cyg's rapid motion after completing a 10-year study in 1804. Piazzi called it the "Flying Star," a name that sticks to this day. Curiously, Piazzi made no mention of the fact that 61 Cyg is a binary star, however, even though both stellar members must have been visible through his telescope. It wasn't until 1830 that German astronomer Friedrich von Struve (1793–1864) announced that 61 Cyg was a binary system. Eight years later, German astronomer Friedrich Bessel (1784–1846) measured 61 Cyg's parallax shift, becoming the first to use that trigonometric method to calculate a star's distance. His estimate of 10.4 light years is impressively close to the modern value of 11.2 light years.

We now know that 61 Cygni is a pair of orange main sequence stars, both of which are smaller, cooler, and older than our Sun. The primary sun, 61 Cygni A, shines at magnitude 5.2, while 61 Cygni B shines at magnitude 6.0. Separated by about 30 arc-seconds, the stars appear far enough apart to be resolvable through $8\times$ binoculars.

The real challenge presented by 61 Cygni is not in splitting the binary, but rather in monitoring its proper motion over the course of several years. Chart 4.14 shows the pair's path from 1900 to 2100. Notice how 61 Cygni A and B will pass to either side of an 11th-magnitude field star between now and 2015. Predictions show that the star GSC 3168:590 will lie between the 61 components in 2011, creating the look of a triple star system for a brief period. The 61 pair will continue on, of course, while GSC 3168:590 will be left behind. Use the chart here to follow the stars' progress by marking their precise locations every year or so. Doing so will let you see for yourself, just as Piazzi saw more than 200 years ago, that 61 Cygni truly is the flying star.

62 IC 5146 and Barnard 168

	Target	Type	RA	Dec.	Constellation	Magnitude	Size	Chart
62a	Barnard 168	Dark nebula	21 53.3	+47 16	Cygnus	–	100′×10′	4.15
62b	IC 5146	Emission nebula	21 53.5	+47 15.7	Cygnus	–	11.0′×10.0′	4.15

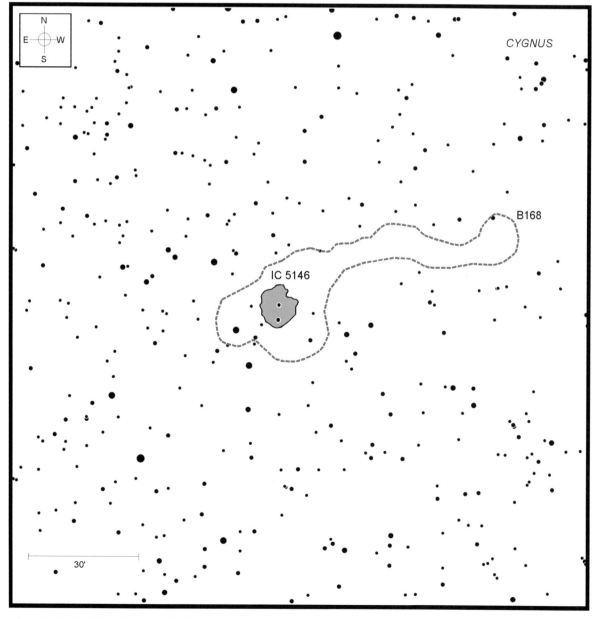

Chart 4.15 IC 5146 and Barnard 168

As a group, emission nebulae, or hydrogen-II (H II) regions, are the most difficult deep-sky objects to see visually. The problem is that they emanate light in very narrow segments of the visible spectrum, with their brightest emissions in the red wavelengths. That's a problem, since the human eye is all but color blind to red light under dim light conditions.

Arguably, the only objects more difficult to spot than emission nebulae are the opaque profiles of dark nebulae. These cosmic dust clouds are themselves invisible; we only see their silhouettes against the starry backdrop. No starry backdrop, no dark nebula; it's that simple. That brings us to this double challenge in Cygnus. IC 5146, known to many by its nickname, the Cocoon Nebula, is a taxing patch of glowing gas, while Barnard 168 is a thin, sinuous lane of darkness that seems to start at the bright nebula and extend far to its northwest.

To spot this celestial odd couple for yourself, begin at the bright open cluster M39 to the northeast of Deneb [Alpha (α) Cygni]. Famous as a bright, loose congregation of stars and covering an area of sky as large as the Full Moon, M39 is best appreciated at very low powers. Be sure to take a moment to enjoy the view.

From M39, steer your telescope $2\frac{1}{2}°$ east-northeast to 4th-magnitude Pi-2 (π2) Cygni, and then slowly scan southward, watching for the starry background to drop off abruptly. That will be Barnard 168. Because of its length – more than a degree tip-to-tip – Barnard 168 is best appreciated with binoculars. My 16×70 binoculars reveal a winding stream of black ink flowing through a valley in the stars, as recorded in Figure 4.13.

By following the dark cloud to its eastern end, you will come to a pair of 9.5-magnitude stars. Both are engulfed in the subtle clouds of the Cocoon. Armed with a 22-mm Tele Vue Panoptic eyepiece (46×), my 4-inch refractor can only muster the slightest hint of the nebula itself, looking like an oval glow surrounding those stars.

So-called nebula filters prove only moderately successful with the Cocoon. The biggest boost, modest as it is, through my 4-inch refractor is with a narrowband filter. A hydrogen-beta line filter also has a positive effect on the Cocoon, but only in larger apertures. A hydrogen-beta filter on my 4-inch scope renders the nebula invisible. Surprisingly, an oxygen-III filter, considered by experienced observers to be the most useful filter of all on emission nebulae, proves

Figure 4.13 IC 5146

worthless with IC 5146, regardless of telescope aperture.

The question of who discovered IC 5146 is the subject of some debate. Most references state that Thomas E. Espin was first to spot it on August 13, 1899. Espin was a British clergyman and astronomer who specialized in the study of binary stars using his observatory in Tow Law, a small town in the Wear Valley district of County Durham, England. Some dispute Espin's role as discoverer, however. While his is probably the first visual observation of the Cocoon, it was actually discovered photographically by Edward Emerson Barnard on October 11, 1893, using the 6-inch Willard lens at Lick Observatory.

Some sources claim that IC 5146 is a star cluster, not a nebula. In fact, a cluster of more than 100 young stars is embedded within the Cocoon Nebula. Espin's original notes, however, refer to his discovery as a "faint glow about 8 arc-minutes [across], well seen each night." The entry in Dreyer's *Index Catalogue* echoes Espin's words, referring to IC 5146 as "pretty bright, very large, irregularly faint, magnitude 9.5 star in the middle."

Credit for the first separate mention of the Cocoon's cluster actually belongs to the Swedish astronomer Per Collinder. Collinder's 1931 listing of open star clusters includes it as Collinder 470. According to Brent Archinal and Steven Hynes in their book *Star Clusters* (Willmann-Bell, 2003), Collinder himself apparently caused the confusion when he incorrectly cross-labeled the cluster as IC 5146. This error has since been carried over to many other references, including the first edition of the popular *Uranometria 2000.0* star atlas (Tirion and Cragin). Modern studies show that 110 stars belong to Collinder 470, including the two 9.5-magnitude stars embedded within the Cocoon's clouds. Unfortunately, most of the other cluster stars are far too faint to be visible through amateur telescopes.

★ ★
★ ★ ★
★

63   NGC 147 and NGC 185

	Target	Type	RA	Dec.	Constellation	Magnitude	Size	Chart
63a	NGC 147	Galaxy	00 33.2	+48 30.5	Cassiopeia	10.5b	13.2′×7.7′	4.16
63b	NGC 185	Galaxy	00 39.0	+48 20.2	Cassiopeia	10.1b	11.9′×10.1′	4.16

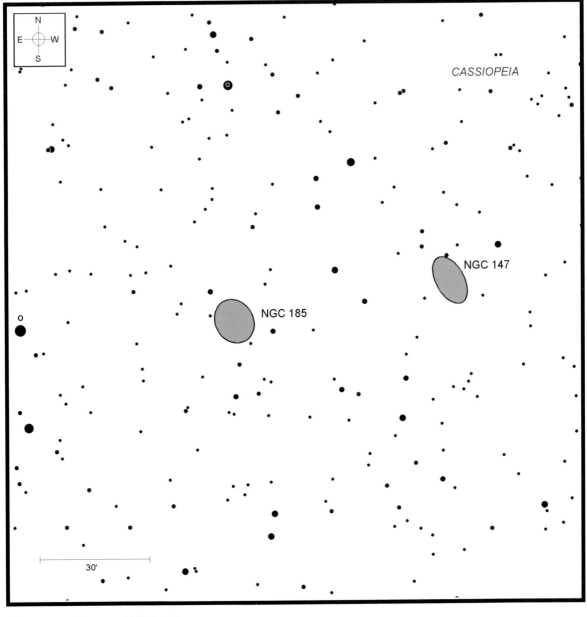

Chart 4.16 NGC 147 and NGC 185

The Local Group is a galactic conglomerate of more than 40 member systems. Three spiral galaxies – the Milky Way, the Andromeda Galaxy M31, and the Triangulum Spiral M33 – dominate the collection, but a horde of smaller irregular and elliptical systems are also scattered throughout.

At least four of those less massive galaxies are gravitationally bound to M31. Two – M32 and M110 – were already discussed in Chapter 2 as binocular Challenge 35. Most amateurs are familiar with them, since they lie in the same field of view of the parent Andromeda Galaxy. But two others, designated as NGC 147 and NGC 185, are not nearly as well known. Both are several degrees to M31's north, across the border in Cassiopeia, and feature smaller, fainter disks that are much more challenging to see.

To find them, aim your telescope or binoculars exactly halfway between M31 and Shedir [Alpha (α) Cassiopeiae], the brightest star in the Cassiopeia W, where you should find a line of three 5th-magnitude stars aligned almost exactly north–south. Using a low-power eyepiece, focus on the trio's northernmost star, Omicron (o) Cassiopeiae. NGC 185 is just 1° west of Omicron, close enough so that both just fit into the same field of a 26-mm Plössl eyepiece through my 4-inch refractor.

To see the galaxy clearly, however, move the star out of view. Then, NGC 185's slightly elongated disk can be spotted squeezed inside a triangle of 8th- and 9th-magnitude stars. Although the full extent of this

Figure 4.14 NGC 147 and NGC 185

dwarf E3 galaxy spans $12' \times 10'$ in photographs, the outer boundary is far too faint to be seen visually, at least in this aperture range (Figure 4.14).

Bump another degree farther west and you should discover NGC 147. Although it has almost the same magnitude value as NGC 185, the surface brightness of this dim elliptical galaxy is noticeably lower, making it more difficult to pick out. Again, expect to see a small, dim, featureless blur.

64 NGC 188

Target	Type	RA	Dec.	Constellation	Magnitude	Size	Chart
NGC 188	Open cluster	00 47.5	+85 14.5	Cepheus	8.1	13′	4.17

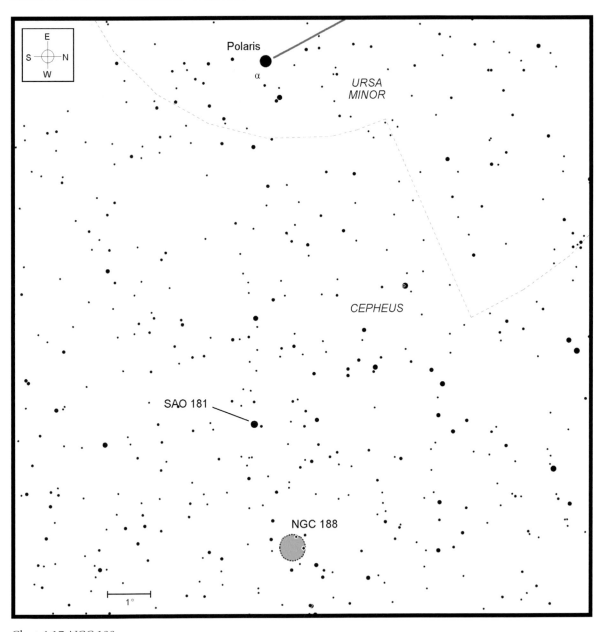

Chart 4.17 NGC 188

Although most open star clusters reside along, or at least near, the plane of the Milky Way, a few renegades have strayed into more sparse surroundings. Certainly, the Pleiades (naked-eye Challenge 13) comes to mind, as does the Coma Berenices Star Cluster (naked-eye Challenge 2). We now add a third cluster to that list with NGC 188. Although not nearly as showy as the other two, NGC 188's lonely neighborhood in northern Cepheus makes it stand out on star atlases, if not in the sky itself.

NGC 188 lies only 4° from the north celestial pole, keeping it above most of our northern horizons throughout the year. That's a plus, but also a minus. Its location in such a sparse area discourages many an amateur from even trying to spot it, especially if using an equatorially mounted instrument. Trying to aim at something so close to the celestial pole, where a polar-aligned telescope has to be twisted around the mount's right ascension axis at wild angles, is an exercise that only a contortionist can appreciate.

The best estimates place as many as 550 stars ranging in brightness from 12th to below 18th magnitude within NGC 188. Together, their collective starlight merges into a soft, 8th-magnitude glow that spans 13 arc-minutes, or about half a Moon's diameter in our sky.

Don't let that "bright" magnitude rating fool you, however. Lacking a central concentration causes the surface brightness of NGC 188 to plummet. Factor in even the slightest light pollution, haze, or cloud and you have an object that is both difficult to zero in on as well as find once in the right spot.

That's probably why the acclaimed observer William Herschel missed it as he surveyed the sky in the eighteenth century. NGC 188 was first spotted by his son John, himself an outstanding visual astronomer, in November 1831. The junior Herschel described his find as "very large, pretty rich, 150 to 200 stars of 10th to 18th magnitude; more than fills the field."

To find NGC 188, begin by centering on Polaris. Offset Polaris so that your finder field is shifted toward Gamma (γ) Cephei, the northern tip of the constellation Cepheus. In between, 3° from Polaris, lies 4th-magnitude SAO 181. NGC 188 is just 1° to its south. Under exceptional skies, some astronomers have reported seeing the cluster's gentle glow through 50-mm, and even smaller, binoculars. The best I can

Figure 4.15 NGC 188

claim, however, is spotting a very subtle hint through 16×70 binoculars under the dark Vermont skies from the Stellafane convention many years ago (Figure 4.15). On another night from a less ideal spot, NGC 188 appeared as a very faint, circular glow peppered by feeble points of light through my 4-inch refractor at 45×. A couple of those stars are closely paired, which adds to the target's visual interest.

NGC 188 is unique in more ways than its unusual location. Looking at its individual stars reveals that its hottest main sequence star is spectral class F2, with many others rated as spectral classes G and K. That's in sharp contrast to what we find in most star clusters, which are made up of spectral type O, B, and A stars. All of these are far hotter and more massive. They live life in the fast lane, consuming their hydrogen fuel at a furious rate, only to evolve off the main sequence in several million years. Lower-mass stars like our Sun and those in NGC 188 survive for billions of years, as they conserve their hydrogen far more judiciously. Therefore, from these studies, we know that NGC 188 is unusually old for an open cluster. Current estimates put it at 5 billion years old. That makes it the second oldest open star cluster north of the celestial equator. Only NGC 6791 in Lyra is older.

| 65 | **NGC 300** | | | | | | | |

Target	Type	RA	Dec.	Constellation	Magnitude	Size	Chart
NGC 300	Galaxy	00 54.9	−37 41.0	Sculptor	8.7b	22.1′×16.6′	4.18

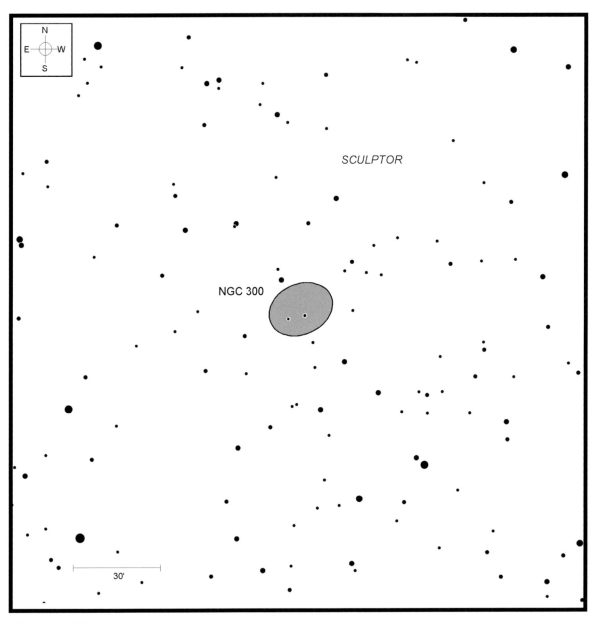

Chart 4.18 NGC 300

Have you heard of NGC 300, the Southern Pinwheel Galaxy? If it were positioned high in our autumn sky in a prominent constellation, you certainly would have. In fact, NGC 300, an Sc spiral tilted nearly face-on to our view, would be one of the season's showpieces, especially through large backyard telescopes. Because it lies in the far southern sky, nestled among the faint stars of Sculptor, it remains the purview of diehard deep-sky fanatics only.

NGC 300 presents an interesting challenge for more than one reason. First, yes, its position in unfamiliar terrain makes it difficult for starhoppers. The brightest star in the area, Ankaa [Alpha (α) Phoenicis], shines only at magnitude 2.4 and is not on anyone's "wow, look at that" list. Therefore, the first challenge to finding NGC 300 is finding Ankaa, our starhop's starting point. I always begin about 55° to its north, at the Great Square of Pegasus. Extend the eastern side of the square, from Alpheratz [Alpha (α) Pegasi] to Algenib [Gamma (γ) Pegasi], straight south for 34° to 2nd-magnitude Deneb Kaitos [Beta (β) Ceti], and then for another 25° southward to Ankaa. Binoculars will help to trace the route. Once there, shift your finderscope's aim 5°, or about one field, northeast to the wide pair of 6th-magnitude stars Lambda-1 and -2 (λ1 and λ2) Sculptoris. NGC 300 is just 2° to their east-northeast.

Even when you are aimed exactly at it, NGC 300 remains a demanding target because of its very low surface brightness. Like two other difficult objects, M33 in Triangulum (naked-eye Challenge 12) and M74 in Pisces (binocular Challenge 39), the light from NGC 300's S-shaped spiral arms softly diffuses away from a nearly stellar core. As a result, even though it is typically rated at 8th magnitude, NGC 300's surface brightness plummets to nearly 14th. On one hand, the large apparent size of NGC 300's spiral disk, 22.1′×16.6′, means that a low-power eyepiece is best for segregating the galaxy from its surroundings. On the other hand, low-power eyepieces and their proportionally large exit pupils typically lower the contrast of an object and the surrounding sky. If the target's surface brightness is low to begin with – well, you can see where this is going.

To see for myself what combination would be best for spotting NGC 300, I set up my 4.5-inch f/4 reflector. Armed with several eyepieces ranging in focal length from 5 to 26 mm, I aimed toward the galaxy and took a look. Going back and forth between eyepieces, I found the most aesthetically pleasing view came at 38× and is

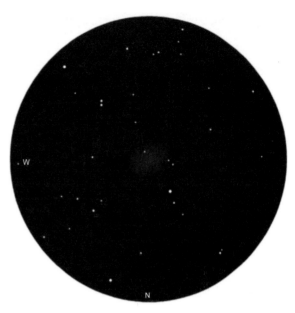

Figure 4.16 NGC 300

captured in Figure 4.16. The field was large enough to take in the dim glow of the galaxy as well as reveal three of the Milky Way stars floating in front. The galaxy's dim oval disk could be seen with averted vision, but looking directly at it caused it to vanish, save for the faint glow of its central core. Higher magnifications caused the galaxy to disappear from view entirely.

The foreground stars misled early observers into thinking they were looking at a distant star cluster. In fact, the galaxy's discoverer, James Dunlop, described his find in his 1827 compilation *A Catalogue of Nebulae and Clusters of Stars in the Southern Hemisphere* as a large, faint nebula that was "easily resolvable into exceedingly minute stars, with four or five stars of more considerable magnitude." Seven years later, John Herschel, who is sometimes incorrectly credited with NGC 300's discovery, also saw "several centres of condensation." Some of those are the field stars noted above, but NGC 300 is also home to several hydrogen II regions sprinkled across its disk. The most prominent are found just south of the galactic core as well as near the northwestern edge of the spiral disk.

Most references place NGC 300 in the Sculptor Group of Galaxies, thought to be some 4 million light years away. Some recent studies, however, cast doubt on this. Instead, these suggest that it is actually a member of the Milky Way's Local Group. If so, then it is considerably closer to home.

★
★ ★
★

66   NGC 404

Target	Type	RA	Dec.	Constellation	Magnitude	Size	Chart
NGC 404	Galaxy	01 09.5	+35 43.1	Andromeda	11.2b	3.4′×3.4′	4.19

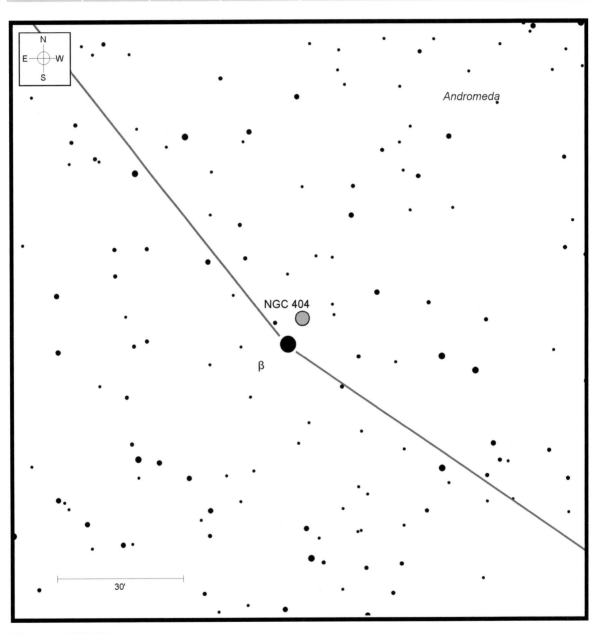

Chart 4.19 NGC 404

Did you know that I discovered a comet in the fall of 1973? I was out with my 8-inch Newtonian reflector, just hopping around the autumn sky, when I noticed stars weren't focusing sharply. Thinking the telescope's collimation was off, I aimed at a nearby bright star to check whether the silhouette of the secondary mirror was centered correctly in the star's out-of-focus image. After I tweaked things a bit, all appeared well, so I focused on that star to check things before moving on. Lo and behold, I saw a dim blur of light right next to the star! Checking things further, it wasn't an internal reflection or an optical aberration. Whatever I was seeing was real! And it wasn't on my star atlas, which at the time was Antonin Becvar's *Skalnate Pleso Atlas of the Heavens* (Sky Publishing, 1969).

My excitement was dashed when I recalled reading something in Walter Scott Houston's Deep-Sky Wonders column about a little-observed galaxy in Andromeda. It turns out that I "discovered" NGC 404, a far-off S0 lenticular galaxy that happens to be just 8′ away from my test star of choice that night, 2nd-magnitude Mirach [Beta (β) Andromeda]. So much for my immortality.

NGC 404, nicknamed Mirach's Ghost for obvious reasons, has since gone on to be a favorite little treasure of my own. Since it appears so close to Mirach, locating this tiny lenticular galaxy is easy enough. Aim your scope toward Mirach and – BANG – you're there. But now comes the problem of *seeing* NGC 404. Rated at magnitude 11.2, the galaxy is more than 4,300 times fainter than the star. As a result, even the slightest bit of sky haze or optical contamination by dust will scatter the star's light across the field and sweep away the galaxy's delicate image.

So, we need a strategy. Conventional wisdom suggests that in order to find a tough object that is so near an overwhelmingly bright distraction, we need to segregate one from the other. Select an eyepiece that produces a high enough magnification to do just that,

Figure 4.17 NGC 404

and move Mirach just out of view while hunting down the galaxy. For my own 4-inch *f*/9.8 refractor, I found that the best combination to spot the galaxy was a 12-mm Plössl eyepiece (Figure 4.17). Even though the field measured more than 30′ across, it was just narrow enough to let me move the star off to one side and spot the galaxy. If seeing conditions permit, try the same set-up, but add a high-quality 2× Barlow lens into the mix. That extra oomph should make seeing the galaxy a little easier, but only if the field is in sharp focus. Use Mirach to check that, but then wait a few moments after it has been moved out of view, so that your eye becomes dark-adapted again.

As for my discovery, well, it seems that William Herschel beat me to it by 189 years; he bumped into NGC 404 in 1784. I wonder if he was checking his telescope's collimation on Mirach at the time?

67 NGC 604

Target	Type	RA	Dec.	Constellation	Magnitude	Size	Chart
NGC 604	Emission nebula/stellar association in galaxy	01 34.5	+30 47.0	Triangulum	11.5b	1.5'	4.20

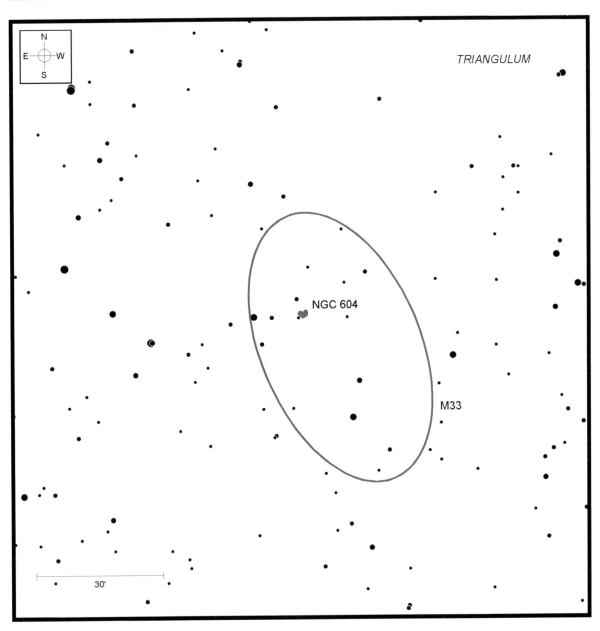

Chart 4.20 NGC 604

The spiral galaxy M33 in Triangulum is one of those
unusual targets that can actually be more difficult to see
through larger telescopes than through smaller ones
because of its huge apparent size. Although it was more
than four decades ago now, I can still recall my own
frustration at trying to find its elusive glow through my
first telescope, a long-gone 4-inch Newtonian reflector.
Yet today, I see it routinely through binoculars as small
as 7×35 in size. Indeed, many use its naked-eye
visibility (naked-eye Challenge 12) as a gauge for
judging the darkness of an autumn night.

It may come as a surprise to find out that a few of
M33's own deep-sky objects can be found in 3- to
5-inch telescopes, as well as through giant binoculars.
NGC 604 is a huge hydrogen II region within one of
M33's spiral arms. Its small glow lies about 11′
northeast of the galaxy's central nucleus.

William Herschel was first to lay eyes on this distant
cloud of glowing hydrogen gas when he noticed it on
September 11, 1784. He subsequently listed it as
"H-III-150," his short-hand way of cataloging it as the
150th entry in his compilation of "very faint nebulae."
Herschel and others went on to catalog many other H II
regions and star clouds within M33. Many of those are
discussed in the next chapter as medium-scope
Challenge 147. From our earthly vantage point, NGC
604 only spans 1.5′ and shines at magnitude 11.5b
(about 9th magnitude visually). While that may sound
unimpressive at first, consider what we are viewing.
Edge-to-edge, NGC 604 actually extends over 1,500
light years. By comparison, M42 is estimated to extend
over some 30 light years. If we were observing NGC 604
from 1,600 light years away, the distance to M42, the
nebula would cover some 50° of sky! Imagine that
sight. We would see billowing clouds of ionized
hydrogen gas entangled like tendrils of an octopus and
enveloping more than 200 recently formed stars. Each

Figure 4.18 NGC 604

of those brilliant stars exceeds our own Sun's mass by
some 15 and 60 times. And there are many more stars
to come from NGC 604. One estimate claims that there
is enough interstellar matter to create tens of thousands
of individual stars.

With good sky conditions, NGC 604 appears as a
tiny oval smudge of light near the edge of the spiral arm
halo. At 100×, my 4-inch refractor shows a distinctly
brighter nucleus (Figure 4.18) but none of the 200
individual stars in the cloud's cluster. Those stars barely
break 15th magnitude, well beyond the grasp of our
telescopes here. The largest backyard telescopes may
just be able to glimpse a few, however. We will return to
M33 in the next chapter as we go hunting for other
deep-sky objects in that galaxy far, far away.

68   NGC 1360

Target	Type	RA	Dec.	Constellation	Magnitude	Size	Chart
NGC 1360	Planetary nebula	03 33.3	−25 52.2	Fornax	9.6p	6.4′	4.21

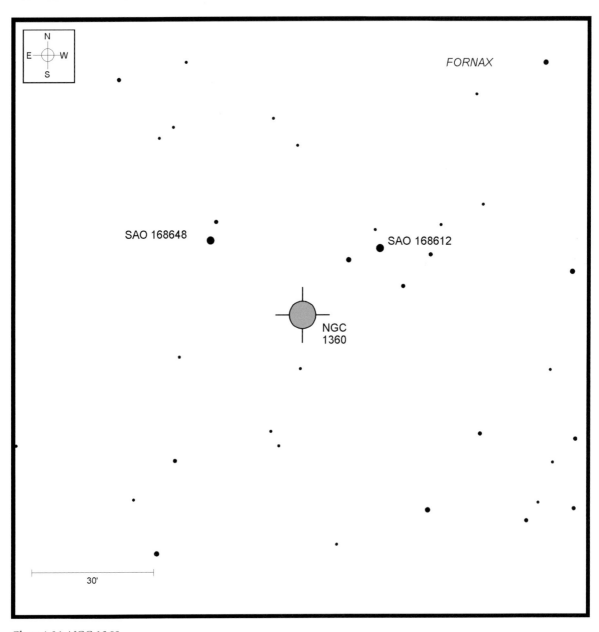

Chart 4.21 NGC 1360

Many stargazers consider Fornax, the Furnace, to be a constellation of the deep south, and therefore, invisible from mid-northern latitudes. While it is true that Fornax scrapes the southern horizon on early winter evenings, it does so at much the same altitude as Scorpius does during the summer. If you can see Scorpius from your observing site, you can also see Fornax six months later.

The real reason so few of us take notice of Fornax is not because of its southerly location, but rather its lack of luster. The constellation's brightest star, Alpha (α) Fornacis, shines at a relatively dim magnitude 3.9. The two other primary stars that contribute to the constellation's formal pattern, Beta (β) and Nu (ν) Fornacis, are both below 4th magnitude. To most of us, those few dim stars are not much to look at, but, to the inventive eye of Nicolas Louis de Lacaille, they formed a furnace. Lacaille's furnace was not the type you would use to heat your house, however. To him, this was Fornax Chemica, a small heating unit used by the chemists of his day to heat chemicals during experiments.

Admittedly, the furnace may not look so hot to naked-eye stargazers. It does, however, hold many amazing deep-sky sights, including one of the most unusual planetary nebulae in the sky. Shortly after it was discovered in 1857 by the American comet hunter Lewis Swift, that nebula, NGC 1360, became an object of mystery and intrigue to those trying to classify it. Some suggested it was an unusual emission nebula, while others thought it was a planetary nebula. Even after decisive studies were conducted in the 1940s by Rudolph Minkowski at Mount Wilson Observatory in California, many still found it a curiosity.

Part of that curiosity probably stemmed from NGC 1360's odd appearance. The internal structure displayed by most planetary nebulae is the result of strong, swirling streams of charged particles from their embedded white dwarf progenitor stars. These stellar winds hollow out the central portion of the nebula and create denser outer levels, or shells.

NGC 1360 does not show a characteristic central void. Instead, it appears all mixed up. The October 2004 issue of the *Astronomical Journal* (Vol. 128, pp. 1711–15) reported on the research results conducted by Daniel Goldman and his colleagues of the Department of Astronomy, University of Illinois at Urbana-Champaign. Their studies find that:

There exist planetary nebulae that do not possess morphological features that suggest the presence of wind-wind interactions. NGC 1360 is such a planetary nebula. Its surface brightness does not dip deeply at the center or rise steeply at the limb to indicate a hollow-shell structure.

They concluded that the lack of a sharp internal edge to NGC 1360 is due to the absence of fast stellar winds. A later study published in the March 20, 2008 *Astrophysical Journal* (Vol. 676, pp. 402–7) by M. T. Garcia-Diaz and others from the Instituto de Astronomia, Universidad Nacional Autonoma de Mexico concluded that the "fast stellar wind from the central star [in NGC 1360] has died away at least a few thousand years ago and a back-filling process has modified its structure producing a smooth, nearly featureless and elongated high excitation nebula." Why this is so remains a mystery.

Clearly, Swift's discovery continues to intrigue stellar astronomers, just as it intrigues us, if for different reasons. While NGC 1360 is bright enough to be seen through large binoculars and small telescopes with relative ease, pinpointing its location in the emptiness of the early winter sky can be difficult. Therefore, our challenge is not to understand why NGC 1360 looks like it does. Our challenge is to find this unusual egg-shaped cloud in the first place.

Of course, one way to overcome this challenge is simply to use a Go-To telescope. Punch in "NGC 1360" and you're there without any fuss or muss. But what challenge is there in that? Therefore, I challenge you to find NGC 1360 without any aid whatsoever save for your finderscope and a star atlas; that is, a star atlas other than the second edition of *Sky Atlas 2000.0*. NGC 1360 was omitted from Chart #18, where it should be plotted, but don't worry, we'll find it together.

I prefer to start at Lepus, the Hare, just south of mighty Orion. Extend a line from Delta (δ) to Epsilon (ε) Leporis and follow it toward the west for about 17°. Through your finderscope or binoculars, look for a trapezoidal pattern formed from Tau-6 (τ6), Tau-7 (τ7), Tau-8 (τ8), and Tau-9 (τ9) Eridani. Once there, extend a line from Tau-9 through Tau-8, continuing for about 4° westward to a close-set pair of 6th-magnitude stars, SAO 168612 and SAO 168648. NGC 1360 lies just south of the halfway point between those two stars. In fact, all three may just squeeze into a low-power field.

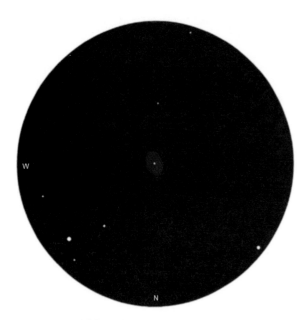

Figure 4.19 NGC 1360

Keep in mind that you are not looking for a small disk of light, but rather a large glowing cloud. To put things in perspective, the largest, brightest planetary nebula north of the celestial equator is M27, the Dumbbell Nebula in Vulpecula (binocular Challenge 30c). M27 shines at magnitude 7.4 and measures $8' \times 6'$. By contrast, NGC 1360 spans 6′, nearly identical in apparent size. At 9th magnitude, however, it is also some 4 times fainter.

Through my 4-inch refractor, NGC 1360 looks like an unusually symmetrical oval cloud of faintly greenish light resembling a cosmic egg (Figure 4.19). The central star, shining at 11th magnitude, is just visible in this aperture, but is quite prominent in larger instruments. At first glance, the cloud will look perfectly uniform. Take a closer look, however, and a very subtle, almost spiral-like structure becomes evident. A narrowband filter helps to bring this out, but unless the eye is trained to spot delicate details – a talent only gained by years of experience – then this will probably pass undetected.

69 Pleiades nebulosity

★
★ ★
★

Target	Type	RA	Dec.	Constellation	Magnitude	Size	Chart
Pleiades nebulosity	Reflection nebula	03 46	+24 10	Taurus	n/a	70′×60′	4.22

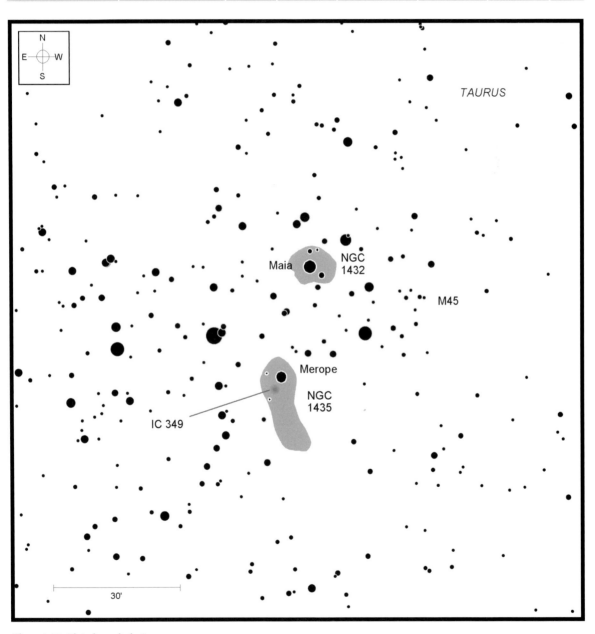

Chart 4.22 Pleiades nebulosity

The Pleiades have inspired many a stargazer over the ages. Perhaps no lovelier a literary image has ever been painted of their glistening appearance than that penned by Alfred Lord Tennyson in "Locksley Hall" (1842):

> Many a night I saw the Pleiads,
> rising thro' the mellow shade,
> glitter like a swarm of fire-flies
> tangled in a silver braid.

In a case of life imitating art, Tennyson's allusion to silvery braids unintentionally predicted Ernst Wilhelm Tempel's discovery on October 19, 1859. That night, while observing through a 4-inch refractor from Venice, Italy, Tempel spotted a dim cloud extending southward from the Pleiad Merope. He likened the nebula's appearance to "a stain of breath on a mirror." Tempel's Nebula, as it is nicknamed, was subsequently listed by John Herschel in his *General Catalogue of Nebulae and Clusters* as GC 768, and then later as NGC 1435 in John Dreyer's *New General Catalogue*.

   We now know that the entire cluster is bathed in the gentle glow of a reflection nebula passing through the region. The striking blue color seen so vividly in photographs comes from starlight reflecting off myriad grains of interstellar dust that engulf the cluster.

   For years, it was assumed that the nebulosity was the material left over from the interstellar cloud that begat the cluster some 100 million years ago. While that would appear to be a reasonable conclusion, looks can be deceiving. Studies conducted at the end of the twentieth century concluded that the clouds and the cluster are moving at different speeds and in different directions. These facts point to two completely different origins. It just happens to be that, right now, the vivid blue dust clouds seen in photographs are passing through the same region of space as the cluster. They will continue on their independent courses, parting company sometime in the far distant future.

   Visually spotting the reflection nebulae that surround the Pleiades takes determination; clean, well-collimated optics; and a top-notch night. The slightest interference, whether from the Moon, light pollution, or a passing cloud, will probably render them invisible. Yet, under ideal conditions, NGC 1435 can be seen through 70-mm aperture binoculars as a fan-shaped wisp of very dim light extending southward from the star (Figure 4.20). The shape and direction of NGC 1435 are dead giveaways. Haze and ill-mannered optics can cause nebulous glows to appear around stars.

Figure 4.20 NGC 1435

Unless you specifically see the glow around Merope fanning away to the star's south, much as a comet's tail extends away from its coma, then you are probably seeing something much more local and insidious in origin. To make sure you're seeing the real thing, check the nearby Hyades cluster, where there is no trace of nebulosity.

   Once you confirm that what you are seeing is actually the Merope Nebula and not an optical illusion, try to spot other portions of the Pleiades dust cloud. NGC 1432, discovered in 1875, surrounds Maia at the northwest corner of the Pleiades "bowl" and is considerably fainter than Merope's Nebula. Unlike NGC 1435, which extends to only one side of its host star, NGC 1432 diffuses more or less uniformly around Maia. This makes it more difficult to tell apart from starlight scattering inside a telescope.

   Other patches enveloping Alcyone, Electra, Celaeno, and Taygeta were discovered in 1880. All are fainter still. Contrary to what you might expect, using a larger aperture will *not* necessarily improve their visibility. Indeed, the seventeenth-century Prussian comet discoverer Heinrich Louis d'Arrest once noted that "here are nebulae, invisible or barely seen in great telescopes which can be easily seen in their finders." While I might contest his use of the word "easily," his sentiment remains true to this day.

One final section of the Pleiades dust cloud, however, will benefit from a large aperture. IC 349 is a small, segregated portion of the Merope Nebula that appeared slightly brighter in photographs taken by Edward Emerson Barnard in 1890. Spotting it visually will probably take at least a 12-inch telescope and a trained eye. Look for a small patch within NGC 1435 just 36″ south and 9″ east of Merope.

One of the few to view IC 349 successfully was Sue French, author of the Deep-Sky Wonders column in *Sky & Telescope* magazine (2004 to present). She notes that, in order to isolate IC 349 from the overwhelmingly bright star, she rotated her 14.5-inch Newtonian reflector in such a way that the diffraction spikes from the secondary mirror's spider mount straddled the target. She then used an eyepiece outfitted with an occulting bar to block Merope's intense glare.

70   NGC 1535

Target	Type	RA	Dec.	Constellation	Magnitude	Size	Chart
NGC 1535	Planetary nebula	04 14.3	−12 44.4	Eridanus	9.6p	60.0″	4.23

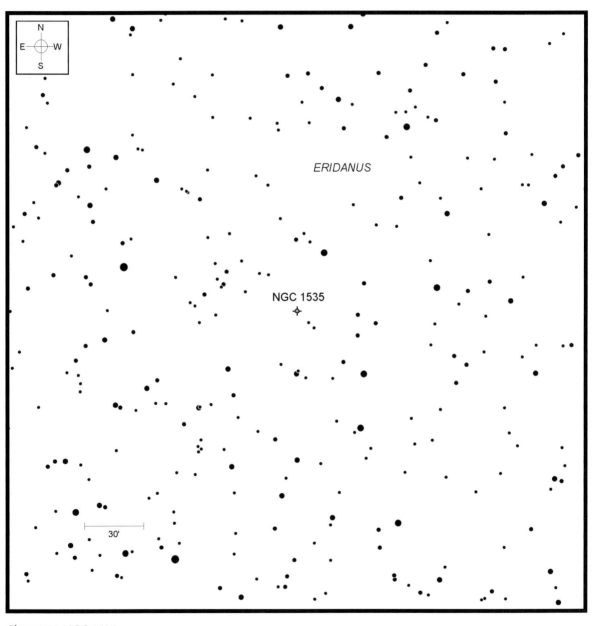

Chart 4.23 NGC 1535

Planetary nebula NGC 1535 is a victim of circumstance. Take a look at its facts. Its bluish disk spans 60.0″, which is quite large as planetaries go, and shines as brightly as the Ring Nebula in Lyra. Its central star glows at magnitude 11.6, creating a surreal scene resembling a disembodied human eye, giving rise to the nickname "Cleopatra's Eye." Those in the know rate NGC 1535 as one of the sky's finest planetary nebulae. Yet this enticing target remains unknown to most backyard stargazers.

NGC 1535's anonymity is due in large part to its empty surroundings within the vague constellation Eridanus the River. I suspect that most amateurs, at least those who prefer to starhop, feel that taking the extra time to zero in on NGC 1535 is just too much effort. The closest bright star, Rigel, lies a distant 20° to the east. None of the nearby stars shine brighter than 4th magnitude. This makes NGC 1535 a challenge to find, but not impossible.

Kick off your hunt from Rigel [Beta (β) Orionis]. With your finderscope, scan about $1\frac{1}{2}°$ to the star's west-southwest to a short arc of three stars aligned east–west. The brightest star shines at 4th magnitude and is labeled Lambda (λ) Eridani. Continue westward past the arc for 6°, or about a finderscope field, to the stars 55 and 56 Eridani, and another half-field further west to reddish 47 Eridani. Are we there yet? No, but we're getting close. Keep moving westward another finder field to Omicron-2 (o2) Eridani and stop. From Omicron-2, shift $2\frac{1}{2}°$ due south to 5th-magnitude 39 Eridani, then finally, another $2\frac{1}{2}°$ to NGC 1535.

NGC 1535 should be immediately obvious if it's in a telescope's field of view – no hunting required. Even at only 38×, my 4-inch refractor easily displays the planetary's pale steel-blue disk nestled in a sparsely populated field (Figure 4.21). Compare its appearance to that of Neptune and, apart from the disparity in

Figure 4.21 NGC 1535

color, it's easy to see why early observers mistook these vaporous disks for distant, undiscovered planets.

Many people mistakenly think they are seeing the central star at low magnification, when in fact, they are seeing the nebula's bright central core. We need more power! Thankfully, NGC 1535 takes magnification well, since 200× or more is needed to isolate the star from the bright annulus that immediately surrounds it. In his book *Hidden Treasures*, author Stephen O'Meara notes his success through a 4-inch refractor operating at 504×, an incredible magnification for any telescope, but especially that size. My local seeing conditions never let me approach those crazy high numbers, but I have had success seeing the central star through my 4-inch at 248×. To increase the odds of success, try looking at the nebula with direct vision to better suppress the nebula's glow.

71   NGC 1851

Target	Type	RA	Dec.	Constellation	Magnitude	Size	Chart
NGC 1851	Globular cluster	05 14.1	−40 02.8	Columba	7.1	12.0′	4.24

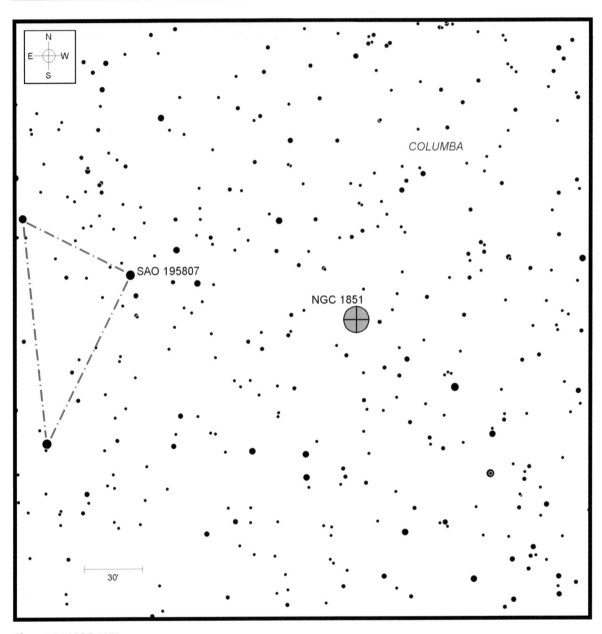

Chart 4.24 NGC 1851

The challenge faced by observers in mid-northern latitudes when it comes to spotting globular cluster NGC 1851 is its location in the sky, some 40° south of the celestial equator. As a result, it never crests more than 20° above the southern horizon from most of North America. Our southern hemisphere colleagues remark that they easily see NGC 1851 through 10×50 binoculars and that it is one of the showpieces of January and February skies. Our view, however, is hampered by horizon-hugging haze and light pollution. The result is merely a dim glimmer of the globular's true self.

The easiest way that I have found to locate Columba is to imagine a large equilateral triangle between Rigel, Sirius, and the dove's brightest star, Phact [Alpha (α) Columbae]. Each is separated from the others by between 23° and 26°. Aim your binoculars or finderscope at Phact, and then look to its southeast for Wazn [Beta (β) Columbae] and to its southwest for Epsilon (ε) Columbae. Shifting about a field due south of Epsilon should bring a right triangle of three 6th-magnitude stars into view. The star at the triangle's right angle, SAO 195807, glows with a reddish tinge and is a little less than 2° to the east-northeast of NGC 1851.

NGC 1851 is a globular cluster for all telescopes and all magnifications. At 45×, my 4-inch refractor reveals an unresolved, nebulous glow punctuated by a brighter, almost starlike central core (Figure 4.22). In many ways, it resembles the coma of a tailless comet. But

Figure 4.22 NGC 1851

then, increase the magnification and that illusion begins to change. At 143×, the edges of the cluster begin to take on a graininess, as if on the brink of resolution. A 4-inch telescope doesn't have quite enough oomph to take that final step, but upping the aperture just 2 inches will show some of NGC 1851's 13th-magnitude stars around its fringe.

72 Horsehead Nebula

Target	Type	RA	Dec.	Constellation	Magnitude	Size	Chart
Horsehead Nebula	Dark nebula	05 41.0	−02 27.7	Orion	n/a	4′	4.25

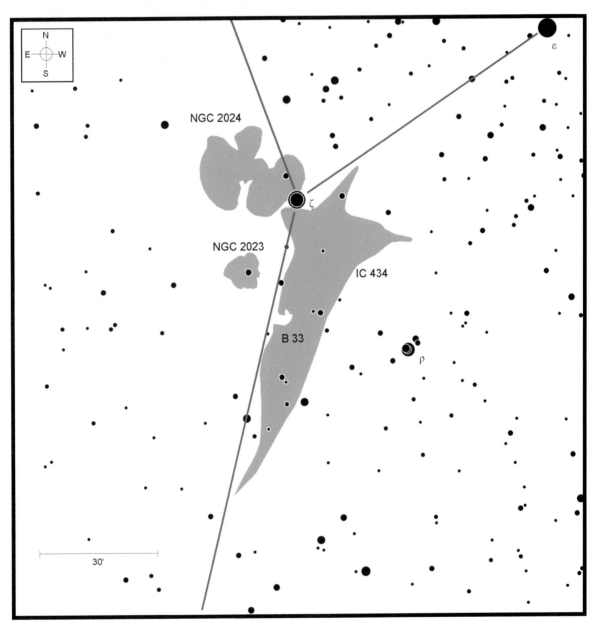

Chart 4.25 Norshead Nebula

For this next challenge object, we should really have strains of the "Mission: Impossible" theme song playing in the background. Of all the deep-sky objects in the winter sky, none carries the mystique of the dark nebula Barnard 33, better known as the Horsehead Nebula. The Horsehead is located 1° due south of Orion's easternmost belt star, Alnitak [Delta (δ) Orionis], making it very easy to pinpoint. But, as easy as it is to locate, this "night mare" is a nightmare to see.

The problem is not the object, but rather our eyes. The human eye is a marvelous tool with an incredible range. We can adapt to almost any lighting condition, from very bright to very dark and still find our way around. But when it comes to dim, red deep-sky objects, it's almost worthless. The Horsehead is visible only because it is situated in front of the dim, red emission nebula, IC 434. And IC 434 is, to all intents and purposes, invisible unless viewed under very dark skies or by using nebula filters, or both.

In my book *Touring the Universe through Binoculars* (John Wiley and Sons, 1990), I stated that "the Horsehead Nebula . . . is too small and faint to be visible in binoculars." I reasoned that it's tough enough to find it through large backyard telescopes, let alone binoculars. But at the 1991 Winter Star Party in the Florida Keys, after spending some time observing alongside Tom Lorenzin, a talented astronomy author and observer from North Carolina, I found out that I was wrong. Tom showed me that the Horsehead is indeed visible in giant binoculars!

Here's how we did it. First, we were in an ideal observing site. The Keys' crystal-clear skies and Orion's height above the horizon certainly made a big difference. Next, he taped a pair of hydrogen-beta line filters to the eyepieces of his 10×70 Fujinon binoculars. Finally, we made sure that Alnitak was just outside the northern edge of the field, while nearby Sigma (σ) Orionis was toward the western edge. Then, with a detailed chart of the area at our side, we looked for a close-set pair of 8th-and 9th-magnitude stars near the

Figure 4.23 Horsehead Nebula

center of the field. These coincide with the leading edge of IC 434 and are just west of the Horsehead. With Tom's help, it took me only a few minutes to pick out both nebulae. Of course, the Horsehead was very small, looking like a thumb viewed from a few dozen feet away, but it was unmistakably there. More recently, and under less ideal conditions, the Horsehead has appeared through my 4-inch refractor, as captured in Figure 4.23.

This is not to say that seeing the Horsehead is a simple task in larger instruments. Nothing could be further from the truth. Indeed, I can rarely see it from my own backyard observatory using my 18-inch reflector with a hydrogen-beta filter in place. So, regardless of the telescope you're using, the Horsehead Nebula is a difficult challenge, one that will test your instrument's quality, the clarity and darkness of your night sky, as well as your skills as an observer.

73 Jonckheere 900

Target	Type	RA	Dec.	Constellation	Magnitude	Size	Chart
Jonckheere 900	Planetary nebula	06 26.0	+17 47.4	Gemini	12.4p	9.0″	4.26

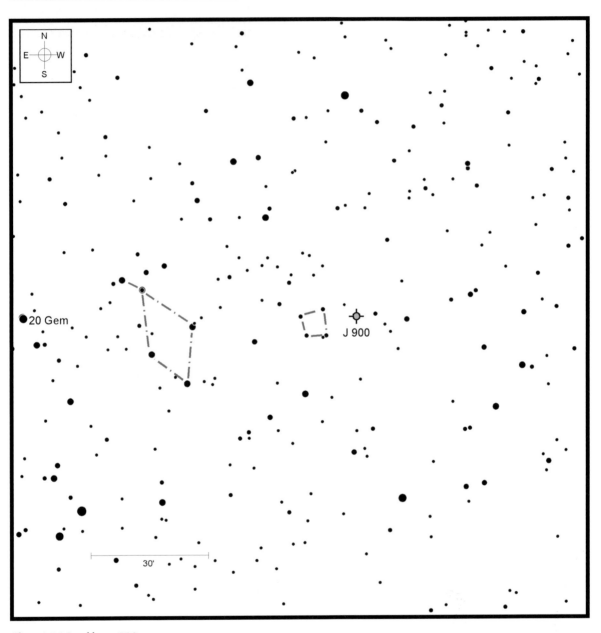

Chart 4.26 Jonckheere 900

Few amateur astronomers are familiar with the name Robert Jonckheere. Jonckheere was a French double-star observer who conducted research at a number of observatories over his six-decade career, including the Strasbourg Observatory in France, the Royal Greenwich Observatory in England, as well as the McDonald Observatory in Texas. His life's work culminated with the 1962 publication of his *General Catalogue of 3,350 Double Stars*, an expansion of his earlier *Catalogue and Measures of Double Stars discovered visually from 1905 to 1916 within 105° of the North Pole and under 5″ Separation.*

Of those 3,350 double stars, the one listed here is particularly unique. In 1912, using a 13-inch refractor at the University of Lille's observatory in France, Jonckheere discovered a vague, blurry pairing embedded in a planetary nebula. Jonckheere announced his discovery the following year in *Astronomische Nachrichten*[2] and included it as entry number 900 in his catalog.

Strangely, subsequent observations by Edward Emerson Barnard through the 40-inch refractor at Yerkes Observatory in 1913 and 1915 turned up the planetary, but showed no sign of the embedded binary star. Barnard described the planetary as "a brightish, ill defined, bluish white disc, possibly a little brighter in the preceding part . . . there is no central condensation and no trace of the central double star."[3] That's odd, since Jonckheere, a seasoned observer, clearly described seeing two stars within the nebula. How could he make such an error?

Jonckheere's planetary nebula, abbreviated as J 900 in most references, lies within the constellation Gemini and is bright enough to be visible through 4-inch telescopes aimed its way. Zeroing in on J 900 is a simple task thanks to its prominent location near the feet of the Gemini twins. Beginning at Alhena [Gamma (γ) Geminorum], hop northwestward to 23 Gem and then onward to 20 Gem, both shining at 7th magnitude. Continue due west to an asterism of 8th-magnitude stars in the shape of an upside-down kite, flying half a degree west of 20 Gem, and then onward another 45′

Figure 4.24 Jonckheere 900

westward to a rhombus of 9th-magnitude suns. J 900 is found 10′ west of those four stars. An unrelated 12.5-magnitude star lies just 11″ southwest of J 900, which at first blush looks like a wide double star. This illusion was not what Jonckheere reported in his discovery announcement, however.

The biggest challenge posed by J 900 is not from how dim it is, but rather from how small it is. Through my 4-inch refractor at 40×, J 900 disguises itself very effectively as just another faint star, only revealing itself when flashed with a narrowband or oxygen-III filter (Figure 4.24). A magnification of 200× hints that J 900 has a tiny disk, although distinguishing an exact shape is impossible. The central star, at 16th magnitude, is also well below our detection threshold.

So, what exactly did Jonckheere see? Barnard's report provides an important clue. In his description, Barnard noted that the nebula appears irregularly illuminated. When viewed through large telescopes, J 900 displays an oddly rectangular shape accented by two opposing bright lobes, one to the west and one to the east. In all likelihood, that is what Jonckheere saw. He simply misinterpreted the two brighter regions as a double star.

[2] *Astronomische Nachrichten*, Vol. 194 (1913), p. 47.
[3] *Astronomical Journal*, Vol. 30, No. 719 (1917), p. 208.

74   NGC 2419

Target	Type	RA	Dec.	Constellation	Magnitude	Size	Chart
NGC 2419	Globular cluster	07 38.1	+38 52.9	Lynx	10.3	4.6'	4.27

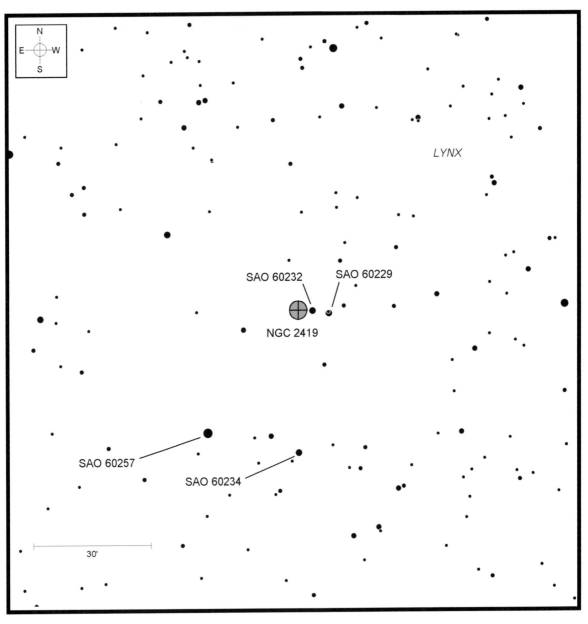

Chart 4.27 NGC 2419

Probably known better by its nickname the "Intergalactic Tramp"[4] bestowed by Harlow Shapley in 1944, NGC 2419 is unusual among winter's deep-sky objects for many reasons. First, it lies within the nondescript constellation of Lynx, a nearly starless span between the Gemini twins and the leading paw of Ursa Major.

Next, the nature of the object, a globular cluster, seems quite out of place in a region dominated by distant galaxies. Aren't globular clusters supposed to swarm around the center of the Milky Way? That's certainly where we find most, scattered throughout Sagittarius and neighboring Ophiuchus, Scorpius, and other nearby summer constellations. The fact that NGC 2419 is so far from the galactic core led many astronomers in the early twentieth century to conclude, albeit erroneously, that it is an independent system. The "Intergalactic Tramp" epithet echoes that incorrect assumption.

Like the other 150 or so globulars in the Milky Way's family, NGC 2419 does indeed orbit the center of our galaxy. Its orbital path, however, is quite unique among globulars. Unlike most other globulars, which hug the core, NGC 2419 follows a wide, eccentric track that takes an estimated three billion years to complete. At the present time, NGC 2419 is projected to be between 275,000 and 300,000 light years away from the Solar System and about the same distance away from the galactic center. That's farther away than two of our galaxy's dwarf galaxies, the Large and Small Magellanic Clouds.

Although the constellation Lynx is difficult to spot, locating NGC 2419 is not as tough as it might appear at first, thanks to its proximity to Castor [Alpha (α) Geminorum]. Center Castor in your finderscope and then slowly scan about half a field northward to the 5th-magnitude stars Omicron (o) and 70 Geminorum.

---

[4] It should be pointed out that when Shapley used the word "tramp" in 1944, its connotation was quite different than how many use the same word today. Shapley probably used the term to signify NGC 2419's slow, somewhat aimless meandering, like a hobo, rather than a comment on its lack of morality.

Figure 4.25 NGC 2419

Hop another half-field northward and you'll come to another, fainter pair of stars (6th-magnitude SAO 60257 and 7th-magnitude SAO 60234) oriented east–west. Time to switch to your telescope. You should see another tighter pair of 7th-magnitude stars (SAO 60229, a fine double star in its own right, and SAO 60232) just to the north. All four stars should fit into the same field of your widest-field eyepiece. Now, swap eyepieces to medium magnification, say 100× or so, and take another look. Can you spot a very faint, diffuse blur just to the east of those two 7th-magnitude stars? If so, you've nabbed NGC 2419 (Figure 4.25).

NGC 2419 glows dimly at magnitude 10.3 and spans about 4 arc-minutes. Since none of its stars shines brighter than 17th magnitude, NGC 2419 appears pretty much the same when viewed through all but the very largest backyard telescopes – faint and fuzzy with just the slightest suggestion of a brighter central core. To catch any hint of NGC 2419, however, requires a dark sky, so wait for one of those special winter nights when the Moon is down and the air is dry.

75 Moon: Bailly

Target	Type	Best lunar phases (days after New Moon)	Chart
Moon: Bailly	Solar System	Days 13 and 23–24	4.28

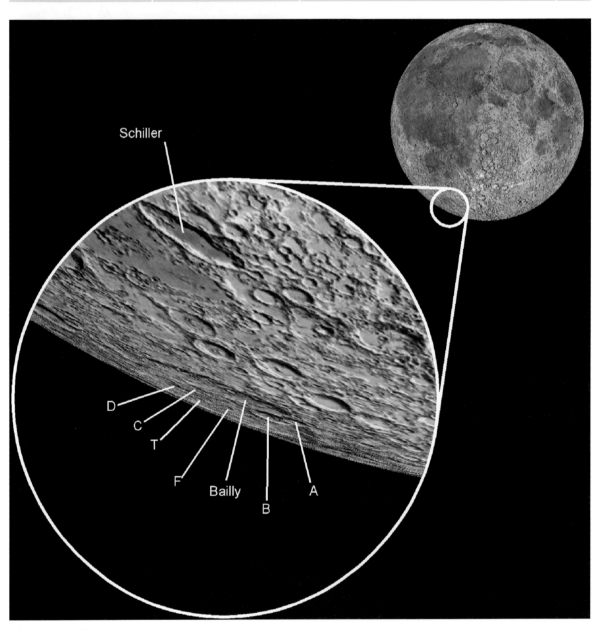

Chart 4.28 Bailly

"What's the largest crater on the Moon?" That's a common question asked at star parties. You would think that, whatever the answer is, it would be cut and dry, and that the crater would be easy to see through binoculars. Unfortunately, neither is true. Not only is that seemingly simple question difficult to answer, the most common answer is anything but easy to see.

Look up the question in most references and the answer given is Bailly, a huge impact scar measuring some 190 miles (305 km) in diameter. That's 9% of the Moon's diameter! But – and you knew there was a "but" coming – it all depends on how you define what a crater is. Many authorities today refer to Bailly as a "walled plain," a "mini-mare," or a "lunar basin." By definition, a lunar basin possesses certain key identifying features, including an interior ring of low hills that were created subsequent to the impactor striking the lunar surface. By contrast, a traditional impact crater shows a central mountain peak, which Bailly does not possess. So, is Bailly a large crater or a small impact basin? There is no general consensus.

Regardless of the definition you subscribe to, Bailly's huge girth makes it sound as though it should be easily visible. Not so. That's because Bailly lies near the southwestern limb in the Moon's libration zone. Although the same side of the Moon always faces Earth, the Moon's elliptical orbit causes it to wobble east–west and north–south ever so slightly. This effect, called *libration*, allows us to see a little over half of the lunar disk – 59% to be exact – although not all at once, of course. Since the Moon orbits Earth in an ellipse, its actual orbital speed varies. When near perigee, or closest to Earth, the Moon is moving along at 3,897 kilometers per hour, while, at apogee, its farthest point

from Earth, it is traveling at a more leisurely 3,455 kilometers per hour. As a result, when near perigee, the Moon twists a bit to expose more of its eastern limb, while, at apogee, its twist favors the western limb.

Libration is also enhanced by the 7° tilt of the Moon's rotational axis with respect to its orbital plane around Earth. As a result, the Moon's north polar region is skewed toward us for half of each lunar orbit, while the south polar region is favored for the other half.

Bailly is impossible to see unless its quadrant is tilted toward us. Adding to the problem is that the best time to look for Bailly is on the night before Full Moon, when the terminator sweeps across its huge floor. To spot Bailly under these blindingly bright conditions, isolate its area by using a reasonably high magnification. Begin at, say, 100×, but don't be surprised if you need to move up to 200×.

Alternatively, Bailly is also visible in the last few days before New Moon, when the waning crescent's terminator is signaled the coming night. Moonlight will not be nearly so bright, but the Moon's lower altitude, and the accompanying poorer seeing conditions, could be problematic.

Once you spot Bailly, take your time touring its ancient floor. Surprisingly lava-free, Bailly shows the scars of exposure to deep space for perhaps 3 billion years. Can you see some of the smaller craters that overlap or infiltrate its enormous dimensions? The two largest, known as Bailly A (22 miles or 35 km diameter) and Bailly B (38 miles or 61 km diameter), partly cover the western wall. Several smaller craters, including Bailly C, D, F, and T, lie between their namesake and the lunar limb. If you thought Bailly was difficult to spot, these last four are all on the next level still.

76   Clavius craterlets

Target	Type	Best lunar phases (days after New Moon)	Chart
Moon: Clavius craterlets	Solar System	Days 8 and 21	4.29

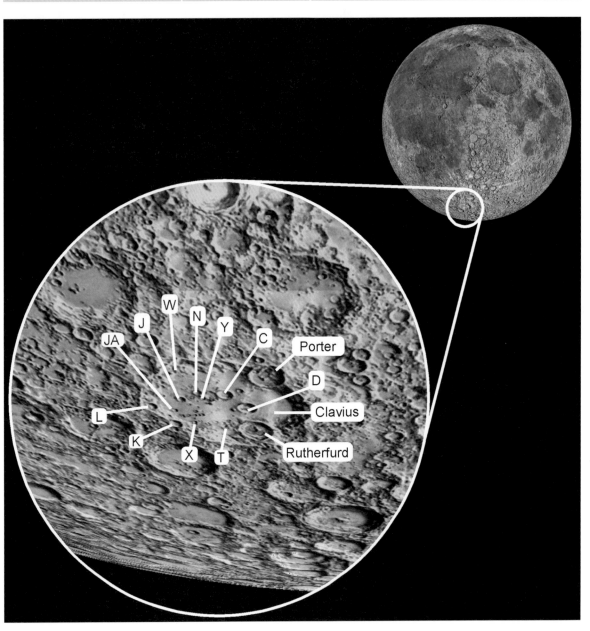

Chart 4.29 Clavius craterlets

One of my favorite lunar landmarks is monstrous Clavius. Named for the sixteenth-century German astronomer/mathematician Chistoph Klau (who was also known as Christophorus Clavius), the crater Clavius, like Bailly, is more properly classified as a walled plain because of its girth and comparatively smooth floor that was flooded with lava perhaps 3.9 billion years ago. Clavius spans 136 miles (219 km) from edge to edge, which brings it into the range of steadily held 10× binoculars, especially when it is near the terminator. That happens when the Moon is 8 days past New, as well as Day 22, the night after Last Quarter. On these nights, sunlight just catches the top of the rim, causing a bright ring to protrude into the cold, dark lunar night.

Clavius is a relatively old formation because it is riddled with more than a dozen smaller craters, or craterlets, across its floor. Spotting these craterlets is a fun challenge for small telescopes. How many can you see?

The two largest lie along opposite edges of Clavius and really stand out from the bunch. The larger of the two, Rutherfurd, can be found along the southeastern edge of the rim. Take a careful look at Rutherfurd for its terraced walls that plunge more than a mile to its rugged floor. Notice how its central peak, a telltale sign that the crater is the result of an impact, is not centered at all, but offset toward the north.

Multiple central peaks await you in Porter (formerly known as Clavius B). Named for Russell Porter, the father of amateur telescope making, this crater spans 32 miles. Its rim is not quite as sharp and distinct as Rutherfurd, which tells us that Porter is the older of the two. In fact, while Rutherfurd was probably formed during the Moon's final, light-cratering era some 1 billion to 2 billion years ago, Porter has been around for nearly as long as Clavius itself.

Two smaller craters are found riding the rim's craggy northwestern edge. Clavius L, covers 14 miles

Table 4.2 *Clavius craterlets*

Crater name	Diameter (miles)
Porter	32
Rutherfurd	31
Clavius D	16
Clavius L	14
Clavius C	12
Clavius K	12
Clavius N	8
Clavius J	7
Clavius JA	5
Clavius T	5
Clavius W	4
Clavius X	4
Clavius Y	4

(23 km), while Clavius K is 12 miles (19 km) in diameter.

Several smaller craters set in an unusually symmetrical arc scar the floor of Clavius. Clavius C (12 miles or 19 km diameter) lies almost perfectly centered, while larger Clavius D (16 miles or 26 km diameter) is offset westward toward Rutherfurd and Porter. Both should be visible in telescopes as small as 50-mm aperture.

Completing the crater arc to the east of Clavius C are a trio of smaller, equally spaced craters. Clavius N, Clavius J, and Clavius JA, measure 8, 7, and 5 miles across (13, 11, and 8 km), respectively. All three are visible through my 4-inch refractor at 150×. Can you spot them through something smaller? If so, try your luck with Clavius Y, a 4-mile (6-km) diameter pothole just to the west of Clavius N.

There are many other craterlets in and around Clavius. Chart 4.29 and Table 4.2 show several, along with their diameters. How many can you spot?

77 Moon: Hesiodus A

Target	Type	Best lunar phases (days after New Moon)	Chart
Moon: Hesiodus A	Solar System	Days 8 and 22	4.30

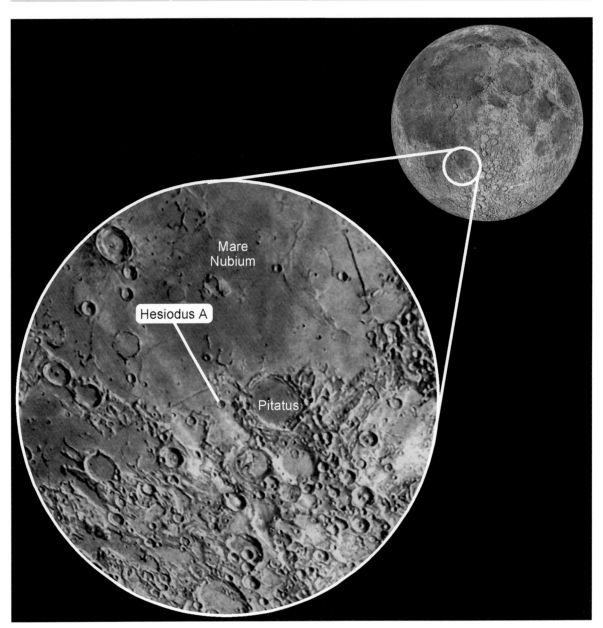

Chart 4.30 Hesiodus A

Some of the oddest classes of surface features found on the Moon are so-called concentric craters. While craters formed from impacts normally show a well-defined rim surrounding a sunken floor and accented by a central mountain peak, concentric craters also have an oddly raised inner ring that typically measures about half of the crater's outer diameter. To picture a concentric crater, imagine a donut lying in a larger bowl.

More than 50 concentric craters have been discovered on the Moon. Most are small, measuring between 1 and 8 miles (1.6–13 km) in diameter. What caused these strange occurrences remains a topic of continued study and controversy among selenologists to this day. Planetary geologist and author Charles Wood noted in one of his Exploring the Moon columns in *Sky & Telescope* magazine that it is "probably not a coincidence that these craters formed where they did; all of them occur near the edge of maria."[5] Studies

suggest that, just like the maria, the internal ring is related to volcanism. Whether the inner rings formed immediately after the impacting projectile slammed into the surface or some time later remains controversial. Wood, however, goes on to note that "it's strange that other nearby craters of similar size don't have inner rings, too."

Perhaps the most striking example of a concentric crater is Hesiodus A. Hesiodus A measures 9 miles (14.5 km) in diameter and lies just west of the much larger crater Pitatus on the south shore of Mare Nubium. The best opportunities to observe Hesiodus A's unusual interior come on the nights after First Quarter and Last Quarter. If the timing is right, watch sunrise (or sunset) over Hesiodus A. As sunlight falls along the smooth walls of the crater's bowl, the rays will strike the inner ring and cast a shadow along the opposite wall that can momentarily make the ring look as though it is floating.

---

[5] "Concentric Fractures and Craters," *Sky & Telescope*, January 2007, pp. 71–2.

★
★★
★

78 Moon: Lambert R

Target	Type	Best lunar phases (days after New Moon)	Chart
Moon: Lambert R	Solar System	Days 9 and 22	4.31

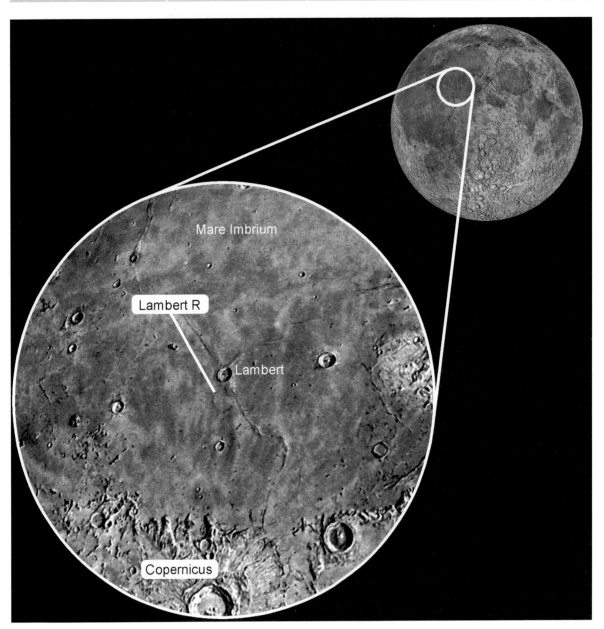

Mare Imbrium

Lambert R

Lambert

Copernicus

Chart 4.31 Lambert R

The Moon is haunted! Not by ghosts and goblins and ghouls, but by ghost craters. These craters, which probably formed during the initial period of heavy cratering some 4 billion years ago, were later washed over by the rising tide of lava that subsequently formed the maria. Today, all we see of these first-generation craters are ghostly silhouettes of their uppermost rims protruding out from the surrounding flat lava plain. What lies below remains a mystery, entombed forever in lava.

A haunting example of a ghost crater can be found nearly centered in Mare Imbrium, the Sea of Rains. As the terminator moves slowly across Imbrium, watch some 36 hours after First Quarter for sunlight to shine on the crater Lambert. Named for the eighteenth-century German mathematician Johann Heinrich Lambert, this crater measures 18 miles across, bringing it into the range of nearly all amateur telescopes. Close scrutiny reveals a spidery ejecta blanket surrounding the crater, indicating that its formation was sometime after the mare's.

Look just south of Lambert for the hint of a once-mighty crater that is now just a shadow of its former self. Can you spot a ring larger than Lambert that appears just above the surrounding surface? That's Lambert R, the "R" standing for "ruin." Lambert R, some 35 miles in diameter, would have easily dwarfed Lambert itself were it not for the intervening lava. Imagine the scene of lava encroaching on the crater's base and then rising ever higher around its steep slope, until it reached a low point in the rim. Like water breaking through a weak point in a dam, lava poured into the crater, flooding its floor. While there are probably many other craters in the area buried beneath the Imbrium lava plain, only Lambert R was tall enough for the very top of its rim to remain visible today.

79 Moon: Lunar X and Lunar V

Target	Type	Best lunar phases (days after New Moon)	Chart
Moon: Lunar X and Lunar V	Terminator shadow formations	Day 7	4.32

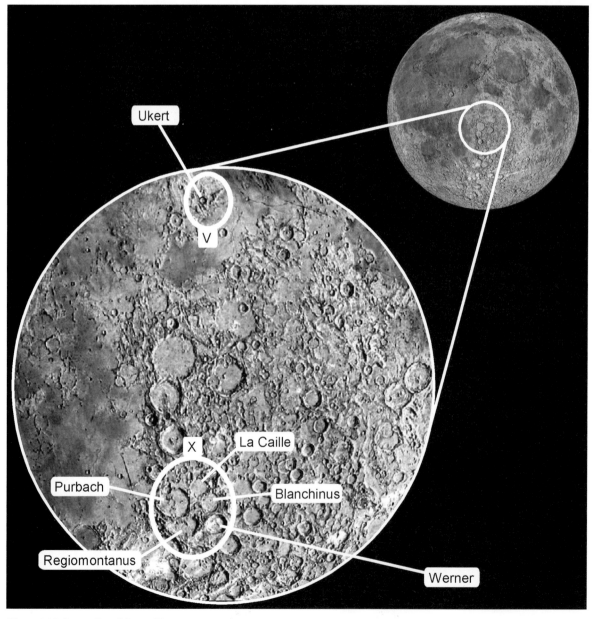

Chart 4.32 Lunar X and Lunar V

The Moon's terminator is a fascinating sight through all telescopes. Here, along the lunar sunset/sunrise line, lighting can strike familiar lunar features in very unusual ways, transforming them in ways that go unsuspected when the Sun rides high overhead.

No region of the Moon is more unusual to watch at sunrise than the area near the crater Werner. Sunlight strikes Werner and its environs on the night of First Quarter, perhaps the most popular phase among moongazers. Luna's brightness is not so overpowering that it dazzles our eyes, yet there is enough terrain showing that lets us enjoy its "magnificent desolation," as Apollo 11 astronaut Edwin Aldrin put it in 1969 when he became the second man to walk on that distant surface.

In August 2004, Canadian amateur astronomer David Chapman noticed an odd apparition along the First Quarter's terminator. Northwest of the crater Werner, which was just seeing sunlight at the time, Chapman saw an "X" seemingly floating in darkness, detached from the sunlit surface. His letter about the apparition that ran in the November/December 2004 issue of the Canadian astronomy magazine *SkyNews* started a flood of observations, photographs, sketches, and reports by others who excitedly saw the X.

It turns out that Chapman's observation of the X was not the first. The formation is well seen on the First Quarter image among Lick Observatory's famous series of high-resolution lunar photos taken decades ago. This image has appeared in countless publications, including my own book *Touring the Universe through Binoculars*. But Chapman was apparently the first person to notice it and put a name to it.

The Lunar X (Figure 4.26), also known to some as the Werner X, is easy to see as long as you are looking in the right place and at the right time. Timing is critical! The X is formed by the confluence of four lunar craters: Purbach, La Caille, Regiomontanus, and Blanchinus. Purbach forms the eastern side of the X, while Blanchinus creates the western side. La Caille forms its northern boundary, and lastly Regiomontanus marks the southern section. The X's namesake, 70-kilometer-wide Werner, does not contribute directly. Instead, as Chapman points out, "Werner is the closest well-lit crater and makes an obvious beacon for observers."

High magnification is not required to see this unique lighting effect. I've had no trouble spotting it with 20× through a 4.5-inch Newtonian reflector.

Figure 4.26 Lunar X (center) and Lunar V (top)

Indeed, it can even be seen through binoculars. But timing is important.

If the timing is right, it's fascinating to watch the Sun rise over the X, slowly unveiling its ragged form over the course of an hour or so. The first rays catch the southeastern wall of Purbach. As the Sun climbs higher in the lunar sky, the X grows as Purbach's northeastern wall is lit, eventually merging with the southeastern rim to form one side of the X. The southwestern side of Blanchinus next sees the light, followed finally by La Caille to form the X. If you are a few hours late, the shadowing effect is lost and the X illusion disappears.

While you're enjoying the X, be sure not to miss the Lunar V, which lies nearby. That's right, we have another letter of the alphabet visible at the same time just to the X's north. The V is sandwiched between Mare Vaporum to its north and Sinus Medii to its south. This puts it more or less smack dab in the center of the disk.

The so-called Lunar V is formed by low-angled sunlight spangled across several small craters. The largest, 23-km Ukert, forms a portion of the V's western edge, while a pair of intersecting ridges create the rest of the western edge as well as the eastern edge. Although the Lunar V is every bit as obvious as the Lunar X, it hasn't quite attracted the same level of attention among

Table 4.3 *Lunar X and Lunar V visibility timetable*

	2018	2019	2020	2021	2022	2023
Jan	24; 04:35	13; 12:32	1; 20:24	20; 18:12	10; 2:12	29; 00:37
Feb	22; 18:05	12; 02:13	1; 10:15	19; 08:09	8; 16:24	27; 15:02
Mar	24; 06:59	13; 15:30	1; 23:53 31; 13:05	20; 21:48	10; 06:28	29; 04:59
Apr	22; 19:18	12; 04:15	30; 01:44	19; 10:54	8; 20:00	27; 18:10
May	22; 07:08	11; 16:30	29; 13:52	18; 23:21	8; 08:49	27; 06:28
Jun	20; 18:42	10; 04:20	28; 01:36	17; 11:14	6; 20:54	25; 18:02
Jul	20; 06:15	9; 15:57	27; 13:11	16; 22:47	6; 08:25	25; 05:07
Aug	18; 18:05	8; 03:38	26; 00:54	15; 10:15	4; 19:40	23; 16:07
Sep	17; 06:25	6; 15:38	24; 12:58	13; 21:56	3; 07:00	22; 03:26
Oct	16; 19:21	6; 04:07	24; 01:32	13; 10:04	2; 18:43	21; 15:27
Nov	15; 08:49	4; 17:08	22; 14:40	11; 22:50	30; 20:17	20; 04:23
Dec	14; 22:39	4; 06:37	22; 04:17	11; 12:15	30; 10:13	19; 18:16

Note: The dates and times listed are based on calculations made with the Lunar Terminator Visualization Tool by Jim Mosher and Henrik Bondo. This useful freeware program may be downloaded from http://ltvt.wikispaces.com/LTVT.

devout lunatics. Since both are visible at the same time, why not try both?

Incidentally, Ukert displays an unusual V-shaped triangular floor after the Sun has climbed higher in its sky. Return here within a few days of Full Moon to see this unusual appearance. Finding tiny Ukert at this phase will prove very difficult, but patient searching at 100× or more should let you reel it in.

The triangular appearance of Ukert's floor has raised the eyebrows of paranormalists and ufologists for decades. Could this unusual appearance be artificial? Is Ukert an extraterrestrial construction site? I guess you'll have to judge for yourself.

Table 4.3 lists upcoming opportunities to spot both the Lunar X and the Lunar V between now and 2015. Each cell in the table lists a date followed by a time (in Universal Time) when the full X will be in view. Both will remain visible for up to two hours after the listed time.

80 Moon: Mare Marginis and Mare Smythii

★
★★
★

Target	Type	Best lunar phases (days after New Moon)	Chart
Moon: Mare Marginis and Mare Smythii	Lunar maria	Days 2–4	4.33

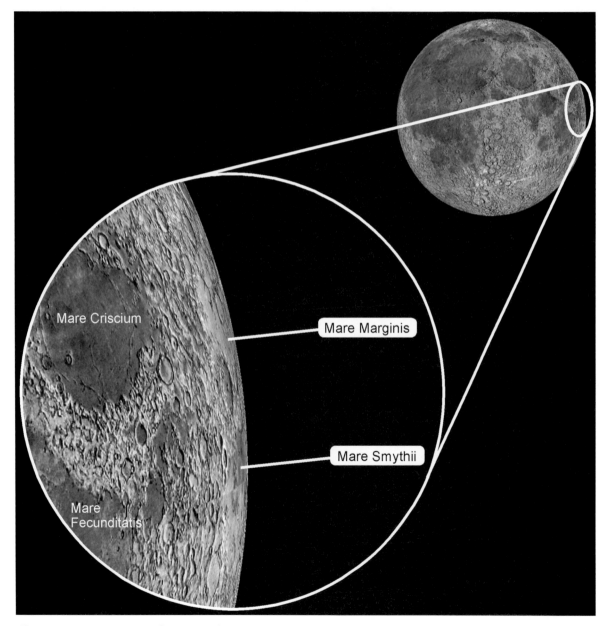

Chart 4.33 Mare Marginis and Mare Smythii

Many lunar features living on the edge of the Moon's far side can be seen from Earth thanks to libration. Challenge 80 offers up two lunar maria that live on the edge. They both take a little extra effort to spot, but, with the right timing, each can be spotted through even the smallest astronomical telescopes.

The first, aptly named Mare Marginis, lies between Mare Crisium and the lunar limb. This oddly shaped lava plain measures some 420 miles across. If the libration is especially favorable, swirls of strange, brighter material can be seen within Mare Marginis. These odd features, first noticed in photos taken during the Apollo lunar missions, seem to coincide to magnetic anomalies discovered during those same missions. Spotting the Marginis swirls takes a careful

eye looking under just the right lighting and libration conditions. The best opportunity comes when the Moon is just 2 or 3 days past New.

If you can see Mare Marginis, look along the limb to its south for another grayish smudge. That's Mare Smythii, named for British astronomer William Henry Smyth, author of the landmark nineteenth-century observing handbook, the *Bedford Catalogue*. His namesake mare measures some 450 miles in diameter and shows an unusual, two-tone floor. While the tone of most maria appears homogenous, Mare Smythii has a lighter central area bordered on either side by darker lava. The darker portions are believed to be younger − perhaps 1 to 2 billion years younger − than the lighter central portion, which may date back 3.5 billion years.

81 Moon: Mare Orientale

★
★ ★
★

Target	Type	Best lunar phases (days after New Moon)	Chart
Moon: Mare Orientale	Lunar mare	Days 13 and 26	4.34

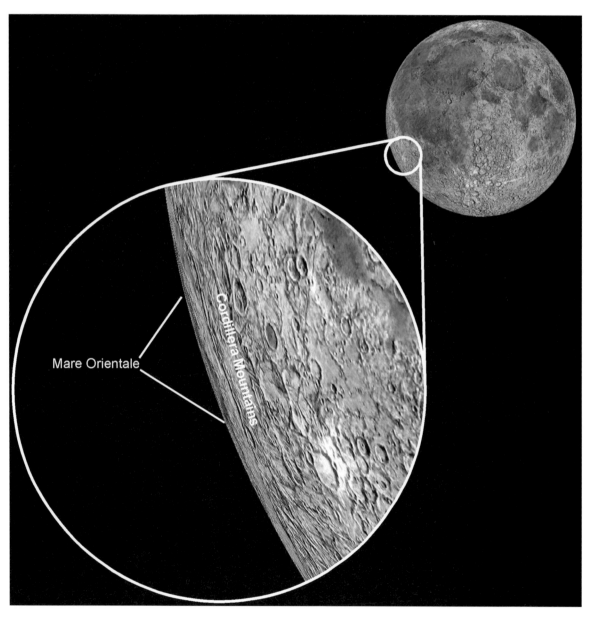

Chart 4.34 Mare Orientale

If you enjoy the test of seeing surface features that ride the lunar disk's edge, then try your luck with Mare Orientale. Mare Orientale lies along the western lunar limb, opposite Mare Marginis and Mare Smythii in the previous challenge.

Mare Orientale is the most recent impact basin on the Moon's surface, estimated to have formed about 3.8 billion years ago. Surrounded by three perfectly concentric rings of mountains, the entire feature forms a perfect bull's eye when viewed from directly overhead. The unusual three-ring system spans some 600 miles edge to edge, with the mare's circular lava plains covering the central 50%. Marking the shore of the mare are the Inner Rook Mountains, named for Lawrence Rook, a seventeenth-century British astronomer. The Outer Rook Mountains form the second set of mountains encircling the impact site and span 385 miles (620 km) in diameter. Finally, the outermost ring is formed by the Cordillera Mountains. The Cordilleras define the outermost edge of the impact basin.

Unfortunately, the Mare Orientale complex is centered 5° beyond the limb, so we never get to enjoy the full effect. The best time to look for its dark floor is within a few nights of Last Quarter, when the Sun is high in its sky and the stark lighting heightens the contrast between the mare's surface and surrounding highlands.

This is also the ideal phase to look for two smaller lava plains adjacent to Orientale: Lacus Veris and Lacus Autumni. The former lies between the Inner and Outer Rook Mountains, while the latter is found between the Outer Rooks and the Cordilleras.

Last Quarter is not a good time to look for the mountain rings, however. Instead, the best phase for spotting the Orientale mountains are the days just before Full Moon, when the Sun is just breaking their horizon and shadowing is greatest.

Lunar libration must line up with the correct lunar phases for any hope of seeing this challenge. Since that varies from month to month, it's best to use a specialized computer program to predict visibility. One of the best was written by Canadian amateur astronomer Alister Ling and is available for free from www3.telus.net/public/aling/lunarcal/lunarcal.htm. Another highly recommended freeware program, Virtual Moon Atlas by French amateurs Christian Legrand and Patrick Chevalley, is available for download from http://ap-i.net/avl/en/start.

★

82 Moon: Messier/ Messier A

Target	Type	Best lunar phases (days after New Moon)	Chart
Moon: Messier/Messier A	Lunar craters	Days 4–8 and 15–16	4.35

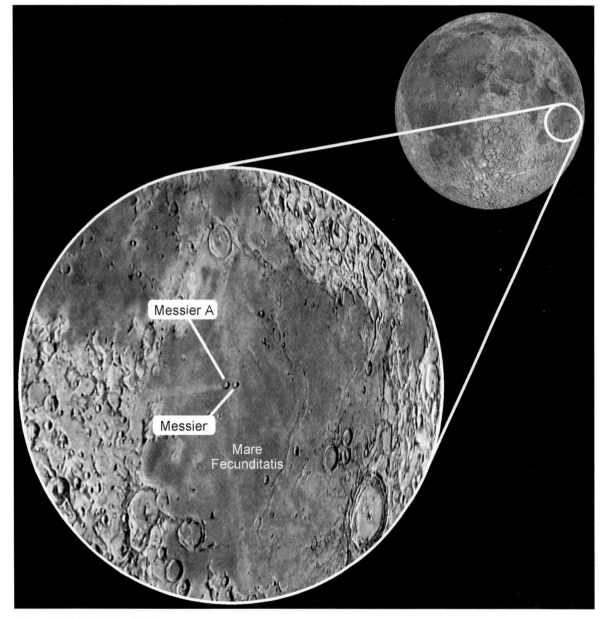

Messier A

Messier

Mare
Fecunditatis

Chart 4.35 Messier/Messier A

The double craters Messier and Messier A[6] have a story to tell. Named for the famed French comet hunter Messier, these craters are separated from one another by less than 4 miles (6 km). Both are nearly the same diameter and noticeably oval. Messier spans 5 miles by 7 miles (8 km by 11 km), while Messier A is a little larger, 7 miles by 8 miles (11 km by 13 km) across.

In addition to their similar appearance, both craters also share a common origin. A billion or so years ago, a meteoroid slammed into Mare Fecunditatis. The meteoroid itself was not necessarily special. What was unique, however, was the narrow angle at which it struck the surface. Tests conducted by Don Gault of NASA's Ames Research Center and John Wedekind of Caltech in 1978 suggest that the impact angle was probably less than 5°. The initial impact created the crater Messier, the western member of the duet. But then, just as a flat stone skips when it strikes a pond at a narrow angle, the meteoroid fractured. A large fragment skipped and was propelled another 4 miles (6 km)

downrange, until it struck the surface to create Messier A. An ejecta field from that second impact threw twin rays of bright material further downrange. This unique pair of rays, looking almost like twin comet tails, immediately calls attention to the two craters. The twin parallel rays extend westward some 90 miles (145 km), nearly to the edge of the mare.

Take a close look at Messier A. Notice how its western rim appears distended and partially upraised, almost as if there is a second rim. While that unusual appearance is not well understood, it may be the result of the meteoroid creating Messier A fracturing and ricocheting again either before or upon impact.

Although not nearly as obvious as those sweeping westward from Messier A, crater Messier also has a system of ejecta rays. Curiously, the ejecta blanket surrounding Messier fans out to the north and south of the crater itself, perpendicular to the impact. Those same tests conducted in 1978 show that not only is this possible, but it is actually very likely during a compound impact. If the lighting is right, the overall appearance of Messier, Messier A, and the craters' respective ejecta fields looks like a parrot in flight, with the fainter rays from Messier posing as the bird's outstretched wings and the twin searchlight-like beacons from Messier A its long tail feathers.

---

[6] Messier A was formerly called Pickering, after the astronomer William H. Pickering. The name was changed to Messier A in 1964 after the International Astronomical Union ruled at the time that Pickering's bizarre ideas about life on the Moon made him an inappropriate choice for a crater named in his honor. There is a crater named Pickering near the crater Hipparchus, but that is technically for Edward Pickering, William's older brother.

83 Moon: Rupes Recta (Straight Wall), Birt, and Rima Birt

★

Target	Type	Best lunar phases (days after New Moon)	Chart
Moon: Rupes Recta, Birt, and Rima Birt	Solar System	Days 8 and 22	4.36

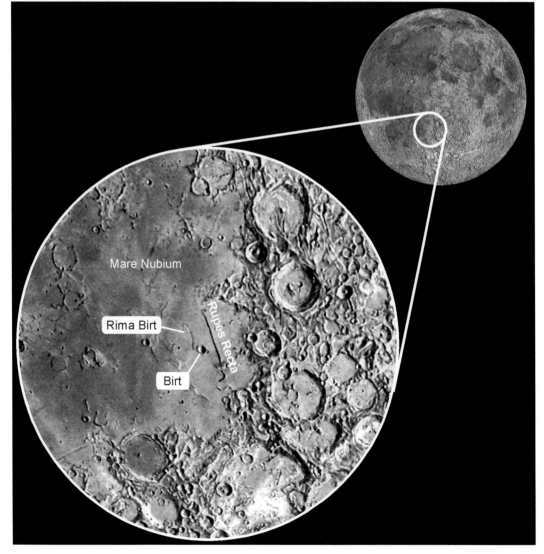

Chart 4.36 Rupes Recta, Birt, and Rima Birt

What is your favorite lunar feature? Maybe it's the mighty craters Copernicus or Tycho. Or could it be the historic Sea of Tranquility? Perhaps you enjoy visiting the rugged southern highlands around Clavius, or the Apennine and Alp Mountains.

If I had to come up with my favorite target, it would have to be a far more modest sight. I always enjoy looking for and at the Straight Wall. Best seen 8 days after New Moon, and again on Day 22, the Straight Wall is listed more properly under its official name Rupes Recta, or "Straight Fault." While fault lines on Earth are most often associated with plate tectonics, Rupes Recta was formed when a portion of Mare Nubium succumbed to subterranean pressures in the lunar crust and buckled. The area to the west of Rupes Recta sheared off and dropped some 1,000 feet (300 m) along the fault line that extends for 70 miles (113 km) from tip to tip.

Depending on when it is viewed, Rupes Recta will look like a sharp, dark line, or a bright scar in Mare Nubium. On the evening after First Quarter, we see the former appearance as the rising sunlight falling onto the surrounding mare has yet to caress the fault's westward-facing surface. As the Sun climbs higher in Nubium's sky, the shadow grows shorter, causing Rupes Recta to blend into the background.

Wait two weeks and return on the night after Last Quarter, as sunset comes to the Sea of Clouds. As the Sun sinks lower in the sky, Mare Nubium darkens in the twilight, but the face of Rupes Recta remains fully sunlit to create a dramatic bright line near the terminator.

Whether viewed at sunrise or sunset, Rupes Recta appears to live up to its popular nickname, the Straight Wall. Looks to the contrary, however, shadow studies indicate that Rupes Recta is not shear cliff at all. Instead, its face rises at a relatively gentle incline between $10°$ and $15°$. It would be a moderately steep hike in a spacesuit, but you could scale it without any of the customary mountaineering gear.

Just south of Rupes Recta lies a small clump of jumbled terrain and a half buried crater. The seventeenth-century astronomer Christiaan Huygens, credited with discovering Rupes Recta, likened the jumble and half-crater to the handle of a sword, with Rupes Recta forming the blade. Although a better allusion is that of a fencer's foil, we know the combined appearance today as Huygens's Sword. The sword is most striking at lunar sunset.

West of Rupes Recta is the perfectly round crater Birt, named for William Birt, a nineteenth-century British astronomer. Birt, spanning some 11 miles (17.7 km) in diameter, is joined by a 4-mile (6.4-km) diameter crater to its immediate east known as Birt A (although I jokingly call it Ernie).

By cranking up the magnification to at least $200\times$, can you also make out a thread-thin rille that starts just west of Birt and curves to the northwest? That's Rima Birt, a difficult-to-see tectonic feature that runs roughly parallel to the Straight Wall. Rima Birt spans some 30 miles in length, but perhaps not even a mile wide. If you look carefully, you may also notice that the rille's northern and southern termini appear slightly distended due to two small craters, Birt E and Birt F, respectively.

84 Moon: Schröter's Valley

★
★

Target	Type	Best lunar phases (days after New Moon)	Chart
Moon: Schröter's Valley	Solar System	Days 11 and 24	4.37

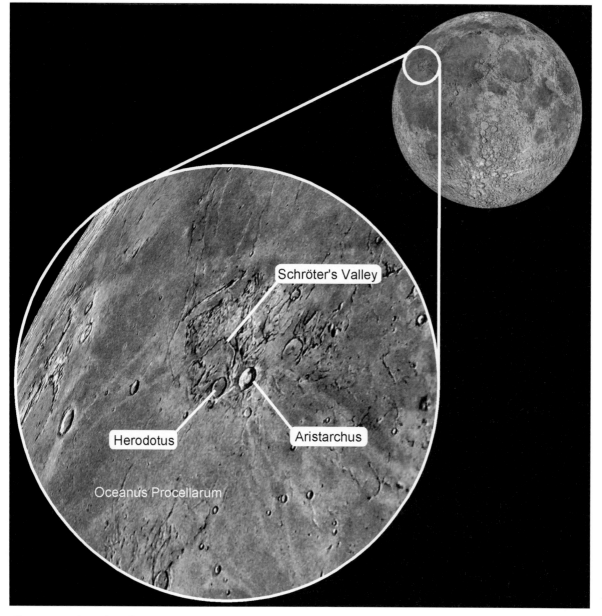

Chart 4.37 Schröter's Valley

As the terminator sweeps across the smooth surface of Oceanus Procellarum during the waxing gibbous phases, it passes over an area of jumbled terrain that looks like a cluster of islands in the otherwise tranquil "ocean." The most obvious reference point in the vicinity is the extraordinarily bright crater Aristarchus. Its sunlit floor, spanning 25 miles (40 km), and interior walls shine with an unparalleled radiance once sunlight strikes them.

Lunar dawn over Aristarchus is a wonderful time to view the area to the crater's immediate west and north. Just to the west lies another crater, Herodotus. Herodotus is just 4 miles (6.4 km) smaller in diameter than Aristarchus, although it appears far more modest than its dazzling neighbor. But it has something that Aristarchus doesn't; it has a tail! The tail curls away to Herodotus's north, twisting around toward the east as it flows through the mountainous terrain on its way toward the ocean's coastline.

We know Herodotus's tail more properly as Schröter's Valley, named for the eighteenth-century German selenographer Johannes Schröter. Although the valley looks to be connected to Herodotus, it actually originates at a 4-mile-wide (6.4-km-wide) crater some 15 miles (24 km) to the north. It then meanders 100 miles (160 km) through foothills, narrowing to less than a $\frac{1}{4}$ mile (400 m) wide at its thinnest, and eventually empties into Oceanus Procellarum.

Magnification will probably need to be increased to at least 200× to spot the tiny "source crater." If you do make it out, notice how a pile of volcanic rocks surrounding the crater looks like a cobra's head. Research suggests that the Cobra Head marks the volcanic vent from which lava welled up to cut the valley and flood a portion of Oceanus Procellarum. Technically, Schröter's Valley is not a valley, but rather a sinuous rille. Sinuous rilles abound on the Moon, with Schröter's the largest its kind.

85 Moon: Serpentine Ridge

★
★

Target	Type	Best lunar phases (days after New Moon)	Chart
Moon: Serpentine Ridge	Solar System	Days 6 and 19	4.38

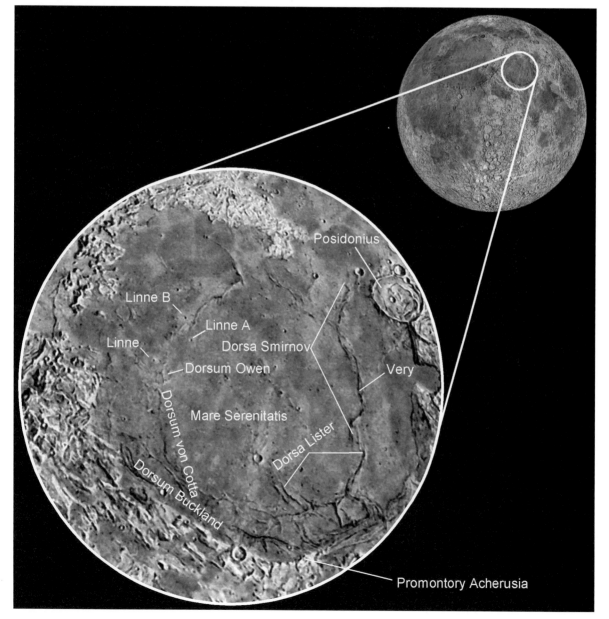

Chart 4.38 Serpentine Ridge

If you enjoy looking for different lunar features, then this challenge is for you. Rather than focus in on a crater or mountain peak, let's throw a wrinkle into it. Literally.

If you visit the Moon on the evening before First Quarter, take a careful look at Mare Serenitatis. Does it look a little wavy to you? It should, since the Sea of Serenity is anything but. Several "waves" in the form of wrinkle ridges thread their way across this sea of solidified lava. Wrinkle ridges are low, sinuous features that formed on the mare surface when the lava cooled and contracted as it filled the impact basins. Some refer to these as "lunar veins" for their resemblance to the veins in our arms and legs.

One of the most obvious and visually interesting wrinkle ridges on the Moon is the Serpentine Ridge, which parallels the eastern shore of Mare Serenitatis. The ridge's northern end starts near a little crater under the northwest wall of the magnificent walled plain Posidonius. It then wends it way southward for 310 miles, passing by the small crater Very, to end just north of the rocky Promontory Acherusia, a cape that juts out between Maria Serenitatis and Tranquilitatis. By scanning the full breadth of the Ridge using 150× or more, very subtle structural details come into view. The area near the crater Very, for instance, shows a steepness that other portions of the ridge do not.

When it came time to designate lunar features by name officially, the International Astronomical Union decided to subdivide the Serpentine Ridge into two separate entities. The northern portion is now known officially as Dorsa Smirnov, named for Russian naturalist Sergei Smirnov. The southern portion of the Serpentine Ridge is labeled on maps of the region as Dorsa Lister, after British zoologist Martin Lister.

It takes little imagination to see that the system of Serenitatis wrinkle ridges hooks further to the northwest. That extension is made up of three separately named pieces: Dorsum Buckland, followed by Dorsum von Cotta and, farthest north of the three, Dorsum Owen. The full span of the wrinkle ridge system finally turns northeastward, passing by the craters Linne, Linne A, and Linne B, to circle back toward the northern tip of the Serpentine Ridge.

86 Venus: Ashen Light

★
★★
★★
★

Target	Type
Venus: Ashen Light	Planetary lighting effect

Of all the planets, Venus puts on the most spectacular naked-eye show. Who hasn't been awestruck by the gleaming planet against an ever-darkening evening twilight? Our sister planet is a dazzler, no doubt about it.

When we look its way through our telescopes, however, our Solar System sibling proves to be very modest. We can enjoy watching the planet go through its series of phases, duplicating Galileo's historic observations of more than four centuries ago that helped to prove we live in a heliocentric solar system. Beyond that, however, Venus remains a featureless white sphere. Or does it?

Venus' atmosphere is completely opaque, whether viewing it with the largest Earth-based telescopes or through the eyes of an orbiting spacecraft. Special ultraviolet filters do let planetary astronomers view and study the circulation of the planet's clouds but, beyond that, we can only detect surface features using high-resolution imaging radar, as the Magellan mission did during its four years in orbit from 1990 to 1994. Magellan went on to map approximately 95% of the planet's surface.

Some might think that, after such a successful mission, there is nothing new under the Sun about Venus. Yet Venus continues to intrigue all who study it closely. One of the oldest and most fascinating mysteries about Venus is the so-called Ashen Light, a faint glow that illuminates the night side of Venus (Figure 4.27). In appearance, Venus' Ashen Light looks like earthshine on the crescent Moon, "the old Moon in the new Moon's arms," but is considerably fainter. While the cause of earthshine is well understood as sunlight reflecting off our planet and onto the Moon, the source of the Venusian Ashen Light remains unidentified. Venus, after all, has no moon.

The first documented sighting of the Ashen Light was made by Giovanni Riccioli, a Jesuit priest in Bologna, Italy, on January 9, 1643. Riccioli was talented

Figure 4.27 Ashen Light

observer who is also credited[7] with discovering the first binary star, Mizar, as well as spotting the first shadow transits of the Galilean satellites across Jupiter. Other observers of the period subsequently saw the Ashen Light as well.

While it might be easy to dismiss these sightings as overactive imaginations, lumping them in with Martian canals and the hunt for the mythical planet Vulcan orbiting between Mercury and the Sun, the Venusian Ashen Light refuses to go away. Since Riccioli's time, the Ashen Light has been seen innumerable times by many notable professional and amateur astronomers alike, right up to the present day. It would seem that the Ashen Light is, in fact, very real. But what causes it?

---

[7] A recent investigation of Galileo's notebook reveals that a former student of his, Benedetto Castelli, saw Mizar several decades earlier.

A variety of theories have been advanced to explain the phenomenon. Some suggest it is auroral activity within Venus' hostile atmosphere, while others believe it is caused by planet-wide lightning discharges. Another theory proposes that it arises from airglow, perhaps from heat arising from the planet's scorched surface. Adding some credibility to this last theory is the fact that, historically, Ashen Light has been seen more often when Venus appears in our evening sky, when we are facing the sunset terminator. Presumably, the surface temperature of that portion of the planet is even hotter than after it has "cooled" during the Venusian night.

Unfortunately, spacecraft observations do not support any of these theories. The Pioneer Venus orbital mission as well as the Russian Venera 11 and 12 landers surveyed the planet for the cause of Ashen Light without success. Later, while en route to Saturn, the Cassini spacecraft swung past Venus in 1998 and 1999, searching for, but failing to find, evidence of widespread lightning. Subsequent observations, however, with the European Space Agency's Venus Express mission in 2007 have confirmed that the Venusian atmosphere generates its own lightning. Could this ultimately be the source of the Ashen Light?

Since the Ashen Light's discovery, long periods of time have passed when it went completely unnoticed. At other times, observers said it was plainly evident. Studies trying to find a pattern of its visibility have yielded little success apart from showing that the Ashen Light is more evident when Venus is visible during evening apparitions than during morning appearances.

The phase of Venus also plays a role. Most sightings have occurred during the crescent phases when Venus was less than 40% illuminated. But, because of its inherent dimness, the Ashen Light fades from view unless Venus is observed against a dark background. That adds to the problem, since by the time the sky has darkened sufficiently after sunset, Venus has slipped so low in the sky that poor seeing conditions usually degrade the view. Therefore, the best time to look for the Ashen Light aligns with Venus in a crescent phase and near its maximum declination north of the celestial equator (or maximum declination south of the celestial equator for readers in the southern hemisphere).

Even with the right conditions, however, there are no guarantees. To improve your odds, select an eyepiece that will magnify the planet's disk sufficiently large to deliver a good view, but not so large that turbulence in our atmosphere renders that view unfocusable. To reduce the glare from the sunlit portion of the disk, place a piece of opaque tape, such as black photographic tape found at art supply and photography stores, across half of the eyepiece's focal plane, which is typically at the barrel's field stop. That way, Venus's bright disk can be hidden behind the tape, while the planet's night side is still in view.

If you can see the planet's disk before the sunlit portion comes into view, you are the newest member of an elite group of planetary observers who have seen the Ashen Light.

87 Eye of Mars

Target	Type
Eye of Mars	Planetary surface feature

No other planet in our Solar System appears so enticing, yet proves so frustrating, through backyard telescopes as the Red Planet, Mars. On the one hand, the planet's thin carbon dioxide atmosphere affords us a nearly cloud-free, round-the-clock view of its sun-drenched surface. On the other hand, however, the planet's small size coupled with its distance away conspire to shrink the planet's disk to no more than 25″ across at its best. Usually, Mars appears far smaller than that. As a result, whatever surface details are visible through our telescopes prove small, vague, and tenuous, at best.

This contradictory set of conditions undoubtedly led to some of the controversial surface features that early Mars observers claimed to see. Without a doubt, the best-known case of Martian illusions has to be the widespread misconception that the planet is covered in a web of thread-thin canals. Many references attribute the "discovery" of Martian canals to the Italian astronomer Giovanni Schiaparelli. Viewing Mars in 1877, Schiaparelli saw what he interpreted as dark, thin lines stretching across the lighter areas of the planet's surface and connecting the darker regions. He described these vague markings as "canali," which in Italian, means channels or grooves. Once his observations, published in 1878, reached the ears of English-speaking astronomers, canali was mistranslated to mean "canals," which of course, are artificial waterways constructed by intelligent beings. Suddenly, the hunt for the Martians was on!

Actually, Schiaparelli was not the first person to see "canali." At least half a dozen observers recorded linear features on Mars as far back as 1840. In 1867, Richard A. Proctor published a map of Mars based largely on observations and drawings by William Dawes (of "Dawes' Limit" fame, discussed in Chapter 1). Proctor presumed that the darker parts of the planet were seas and the reddish tracts continents, and proceeded to name several features after English astronomers, such as Dawes Ocean, Herschel Continent, and Terby Sea.

Schiaparelli's 1878 report also included a map of Mars, showing far more detail than Proctor's, which contained several fanciful errors. To correct these errors, Schiaparelli decided to abandon any names previously assigned and instead create his own references based on biblical and mythological entities. Terby Sea, for instance, became Solis Lacus. For the most part, the names we still use when discussing features on Mars are those assigned by Schiaparelli. That is, minus the canals, of course.

While we may chuckle today at the thought of canals crisscrossing the planet, many of the surface features that perplexed generations of astronomers continue to intrigue observers today. Even with robotic spacecraft scurrying about the surface of the Red Planet or in orbit high above, Mars still beckons backyard planet watchers. There are many striking features across the Martian surface, from the fork-shaped Sinus Meridiani (or what Proctor had christened Dawes's Forked Bay) to the dark wedge of Syrtis Major (formerly Kaiser Sea).

Since it was first detected in the nineteenth century, the region Solis Lacus, located at Martian longitude 85° west and Martian latitude 26° south, has puzzled observers. Nicknamed the Eye of Mars, or Oculus, for its cyclopic appearance, this feature has been observed to undergo dramatic changes in size and appearance. Normally, Solis Lacus appears as a dark, elliptical feature measuring some 500 miles east-to-west by 300 miles north-to-south, surrounded by a brighter region known as Thaumasia. Together, they resemble a human eye, almost as if Mars is looking back at us.

Since Schiaparelli first drew a detailed view of Solis Lacus in 1877, observers have watched it go through a variety of changes, as Figure 4.28 demonstrates. Schiaparelli's original drawing recorded a dark, segmented viaduct across Thaumasia, connecting the "eye" to Mare Erythraeum to the south. Within 30 years, others recorded not one, but several thin straits radiating outward from Solis Lacus, bridging the gap

Table 4.4 *Martian oppositions 2018–2045*

Solar opposition	Closest approach to Earth			
Date	Date	Distance (AU)	Distance (millions km)	Apparent diameter (″)
2018 Jul 27	Jul 31	0.38496	57.59	24.31
2020 Oct 13	Oct 6	0.41492	62.07	22.56
2022 Dec 8	Dec 1	0.54447	81.45	17.19
2025 Jan 16	Jan 12	0.64228	96.08	14.57
2027 Feb 19	Feb 20	0.67792	101.42	13.81
2029 Mar 25	Mar 29	0.64722	96.82	14.46
2031 May 4	May 12	0.55336	82.78	16.91
2033 Jun 27	Jul 5	0.42302	63.28	22.13
2035 Sep 15	Sep 11	0.38041	56.91	24.61
2037 Nov 19	Nov 11	0.49358	73.84	18.96
2040 Jan 2	Dec 28 (2039)	0.61092	91.39	15.32
2042 Feb 6	Feb 5	0.67174	100.49	13.93
2044 Mar 11	Mar 14	0.66708	99.79	14.03

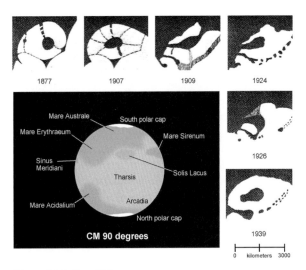

Figure 4.28 Eye of Mars

between it and the mare, as if the Eye is bloodshot. As the early twentieth century wore on, Solis Lacus continued to morph from oval to circular, blending in part into Mare Erythraeum before separating again. By the 1971 opposition of Mars, it had shrunk in size and faded in darkness, only to experience resurgence two years later. As the twenty-first century opened, the Eye was dark again, although not as large as it had appeared in the past.

The cause of these variations is probably due to dust storms that rage across the Red Planet. The powdery Martian soil can be picked up by high winds and swept across plains and down into basins. As this material is blown about, darker, subsurface regions are alternately exposed and covered up, accounting for what was once interpreted as growth of seasonal vegetation.

Mars is only ideally placed for observation when the planet is near opposition, when the Earth–Mars distance is least. Table 4.4 lists upcoming Martian oppositions.

Some Martian oppositions are better than others. An *aphelic* opposition occurs at or near Martian aphelion, when the Red Planet comes no closer than 50 to 61 million miles (81 to 98 million km) to Earth. During these comparatively poor viewing periods, Mars, which measures 4,219 miles (6,794 km) in diameter, will measure no more than 14″ across. In more favorable years, when Mars reaches opposition at or near perihelion, the planet will be less than 35 million miles (56 million km) from Earth and will appear about 25″ in diameter. These are called *perihelic* oppositions. The Martian oppositions in 2018 and 2020 will be perihelic, affording observers some prime Mars-watching, while those in 2010, 2012, and 2014 will be aphelic.

In general, dedicated planet watchers prefer refractors and long-focus reflectors because they usually produce the highest image contrast. Short-focus

Newtonians and most catadioptric telescopes yield lower image contrast owing to their large central obstructions. And since magnifications in excess of $200\times$ are usually needed to see fine details, be sure to use a high-quality eyepiece. Popular super-wide-field eyepieces are wonderful for panoramic views of star fields and broad nebulae, but they are often surpassed by simpler conventional eyepieces, such as orthoscopics and Plössls, for planetary observing. Finally, many observers report great success using color filters to enhance various features on Mars. For Solis Lacus, try an orange (Wratten #21) or red (#23A or #25) filter to increase the contrast of the dark eye against the surrounding bright region.

88   Jupiter's Great Red Spot

Target	Type	Best time
Great Red Spot	Cyclonic storm in Jupiter's atmosphere	Jupiter near opposition

"The extended forecast for the next three centuries: stormy weather." That could have been the weather prediction issued in the middle of the seventeenth century, not for here on Earth, but rather, for Jupiter. The storm is the Great Red Spot and the prognosticator could have been Robert Hooke (1635–1703). Historical records show that Hooke, a British scientist who did research in many sciences, including physics, biology, and astronomy, recorded an oval feature in the planet's atmosphere in 1664.

Many credit Hooke's historic observation as the first sighting of this huge swirling region in Jupiter's southern hemisphere. Others are not so sure. History also records that Giovanni Cassini, whom we meet again in the next challenge, recorded what many interpret as the Red Spot in 1665. His subsequent drawings, repeated in 1672 and 1677, show a dark, oval marking along the southern edge of Jupiter's South Equatorial Belt, where the Red Spot resides today. Hooke's observation is not quite as compelling as Cassini's. While there is no doubt that he saw an atmospheric storm system, some authorities believe that Hooke did not see what we know today as the Great Red Spot, but rather a now-defunct atmospheric feature. Regardless, we do know that Hooke used his sighting to measure Jupiter's sub-10-hour rate of rotation for the first time.

The first detailed observation and drawing of the Great Spot was made on September 5, 1831, by German amateur astronomer Samuel Heinrich Schwabe (1789–1875). His drawing portrays the so-called "Red Spot Hollow," an apparently empty region of the South Equatorial Belt where the Red Spot normally resides.

Schwabe saw the Spot during one of its paler periods. The Red Spot is fickle indeed. Its appearance can range from reddish to an anemic orange, beige, salmon, and, at times, even white. It was especially prominent during the period from 1878 to 1881. In the past half-century, the Red Spot appeared most pronounced from 1961 to 1966, 1968 to 1975, 1989 to 1990, and again from 1992 to 1993.

There seems to be no rhyme nor reason for these changes; at least, none that has been detected so far. What causes the Red Spot to fluctuate in color, or for that matter what causes the spot in the first place, remains unknown. Laboratory experiments suggest that the crimson hue may come from a complex brew of organic molecules sprinkled with a dash of red phosphorus and perhaps other sulfur compounds.

The Red Spot is great indeed, varying in size from 15,000 to 25,000 miles (24,000 to 40,000 km) lengthwise east–west by 7,500 miles to 8,600 miles wide (12,000 to 14,000 km) north–south. By contrast, Earth is only 7,921 miles (12,756 km) in diameter. Based on infrared observations conducted both by passing spacecraft and Earth-based telescopes, as well as the apparent direction of its counterclockwise rotation, astronomers believe that, unlike low-pressure hurricanes and cyclones on Earth, the Red Spot is a huge *high*-pressure region, an "anti-cyclone." The cloud tops that we see are higher and colder than the surrounding regions. But why the Red Spot has persisted for more than 350 years remains a mystery.

Despite these huge dimensions, spotting the Great Red Spot (Figure 4.29) can prove quite challenging for large backyard telescopes. Or it can be a simple matter to see through the smallest. It all depends on what the Red Spot is doing. When it is especially prominent, as it was when I first became interested in astronomy back in 1969, I could spot it through an inexpensive 2.4-inch refractor. But, a few years later, it evaded my 8-inch reflector save for the empty hollow.

Figure 4.29 Jupiter's Great Red Spot

Even when at its best, the Red Spot can use all the help it can get through small apertures. Good seeing and high-quality optics are musts, but color filters will also come in very handy. To enhance a feature's contrast, use complementary colors; in other words, colors that, when combined with that of the feature we seek, will turn it dark. In the case of a reddish object, such as the Great Red Spot (GRS) and reddish brown belts, try blue filters, like Wratten #82A (light blue), #80A (medium blue), or #38A (blue). And, as unlikely as it sounds, I have also had success using narrowband light-pollution filters (e.g., Lumicon UHC or Orion UltraBlock).

Of course, the finest instruments won't show the Red Spot when it is located on the far side of the planet. Jupiter rotates so quickly – the Red Spot takes just under 10 hours to make a complete circuit – that if you miss it early in the evening, it might come back around again before you pack up for the night. How can you tell beforehand when to spot the Spot? You could just try hit-and-miss, but for those who prefer to plan observing opportunities, *Sky & Telescope* magazine has a "Red Spot Calculator" on their website, www.skyandtelescope.com. Simply enter the date and your local time zone and the calculator will list the next three times that the Red Spot will transit the planet's central meridian.

89 Saturn: Cassini's Division

Target	Type	Best time
Saturn: Cassini's Division	Division in Saturn's rings	Rings near maximum tilt

One of the biggest thrills that all stargazers share is our first view of Saturn through a telescope. Remember the excitement that you felt when you saw its magnificent rings for the first time *live and in person*? For me, that epiphanic moment came through a friend's 6-inch reflector more than 40 years ago. I was mesmerized and couldn't take my eyes off it. That was it; I was hooked on Saturn for life.

Viewing Saturn is both captivating and challenging. While the rings can be seen with just a casual glance at 25×, spotting details within those rings takes a bit more magnification and quite a bit more effort.

The most famous feature of the Saturnian ring system has to be Cassini's Division, a dark gap that slices the rings in half longitudinally. Discovered in 1675 by the Italian-born French astronomer Giovanni Cassini, this apparent gap spans 2,920 miles (4,700 km) and separates the broader, brighter inner ring, referred to as the B ring, from the dimmer, outer A ring.

Spotting Cassini's Division is a fun challenge for the smallest amateur scopes. The best I can report is seeing Cassini's Division through a high-school friend's 50-mm refractor that he bought at Sears in about 1972. While telescopes of this ilk today are probably not up to the task, that little scope also showed us the annulus shape of M57, the Ring Nebula. That's not bad at all for that aperture.

Due to foreshortening, seeing Cassini's Division (Figure 4.30) span the breadth of the rings fully is far tougher than spotting its two ansae, where it curves around the far side of the planet.

Timing is also important, since twice in its 29.5-year orbit of the Sun, Saturn presents its rings edge-on to observers on Earth, effectively causing them to disappear. To see Cassini's Division at its best, wait until the rings are tilted near their maximum angle. That will next occur in 2016–2017, when their north face

will be seen from Earth. As time marches on, the tilt will shrink, as will your chances of seeing Cassini's Division, until the rings are edge-on in 2025. The odds improve once again afterward when the rings' southern face will be tilted earthward.

Incidentally, it is technically inaccurate to describe Cassini's Division as an "empty gap." Appearances through Earth-based telescopes notwithstanding, Cassini's Division is not empty at all. Instead, it contains modest amounts of comparatively dark material, as backlit photos taken by the Voyager 1 and Voyager 2 spacecraft first proved in the 1980s. These twin craft resolved the rings into a multitude of ringlets and gaps, looking almost like grooves in a record album. The cause of the rings' complex structure is the result of the interplay of orbital resonances set up with some of the planet's inner moons. Studies show that the material along the inner edge of Cassini's Division orbits the planet in exactly half the time it takes the moon Mimas to completely circle Saturn. As the gravity from Mimas pulls outward on that material, it keeps the density of material that gathers in the zone low.

Figure 4.30 Saturn's Cassini Division

90 Pluto

Target	Type	Best time
Finding Pluto	Dwarf planet	At or near opposition

Our final challenge in this chapter is a quintessential one. We're hunting for Pluto, the now-demoted dwarf planet discovered by Clyde Tombaugh after an exhaustive photographic search for what many had dubbed Planet X. He finally spotted his quarry among the stars of Gemini using a specially designed camera at Lowell Observatory in Flagstaff, Arizona.

Talk about a planet orbiting the Sun beyond Neptune sprang up shortly after Neptune itself was discovered in 1846. The same mathematics used successfully to find Neptune on paper before it was ever seen in the sky were applied to this new challenge without success. But that quest captured the imagination of one Percival Lowell (1855–1916), a member of the Lowell family, a prominent part of Boston aristocracy in the late nineteenth century. Lowell, a businessman, author, mathematician, and astronomer, used his personal wealth to pursue two astronomical passions: the search for Martian canals already discussed, and, later in life, the hunt for Planet X.

In 1894, Lowell built an observatory complex outside of the tiny town of Flagstaff in the Arizona Territory expressly for the purpose of studying Mars. Fifteen years' worth of observations and drawings led Lowell to publish three books about Mars between 1905 and 1908, and in the process, popularize the idea of intelligent life on the Red Planet.

Lowell spent much of his last years looking for the elusive Planet X beyond Neptune, a search that ended without success. After Lowell's death, the hunt for Planet X ended, only to be resurrected again 10 years later by Clyde Tombaugh, a young, Kansas-raised amateur astronomer who was hired by Lowell Observatory in 1929.

Tombaugh used the observatory's new 13-inch astrograph for his search. His method was simple, but very time consuming. Tombaugh would photograph very precise sections of the sky, each lying adjacent to the next. In those days of slow photographic emulsions, each exposure lasted 3 hours in order to accumulate enough light to record sufficiently faint stars. Tombaugh reasoned, correctly, that if there was a Planet X, it was probably quite dim. He also reasoned that the undiscovered planet would be at its brightest when at opposition, rising in the east as the Sun sets in the west. He, therefore, confined his search to along the ecliptic directly opposite the Sun in the sky.

After a series of photographic exposures were made over several nights, Tombaugh would return to the position of the first and start the process all over again across the next several evenings. In the end, he had pairs of precisely aligned photographs of the same areas of the sky taken several days apart.

Searching the plates for Planet X proved as laborious as photographing them in the first place. Tombaugh would place a matched pair of plates into a blink comparator, a microscope-like contraption that projected images of the plates through a central eyepiece, rapidly alternating between the first and the second. Once the plates were aligned exactly, "blinking" them would allow the viewer to spot any "stars" that are moving, and therefore, a member of the Solar System. This technique netted Tombaugh several new asteroids before it finally delivered the jackpot on February 18, 1930.

Pluto's position within the constellation Gemini, itself lying along the star-rich plane of the Milky Way, undoubtedly confounded Tombaugh's efforts, since he had to check so many faint points of light before Pluto showed itself. Unfortunately, our task, 80 years later, is even worse. Pluto will be spending the next decade leisurely meandering through the northern region of the constellation Sagittarius, home to the center of the Milky Way. Talk about a lot of stars to weed through! Identifying Pluto is difficult enough when it is in an empty part of the sky. But now, determining which very faint point is the dwarf planet will present an even greater challenge.

The only way to readily identify Pluto from surrounding field stars is with a detailed star chart of the region. The *Observer's Handbook*, published annually by the Royal Astronomical Society of Canada, includes charts showing the dwarf planet's track against the background stars around the date of opposition that particular year. Astronomical periodicals, such as *Sky & Telescope* and *Astronomy*, also publish charts showing Pluto's location in the sky. And of course, most popular star charting/planetarium software programs can also be customized to show, and print, finder charts tailor-made to your particular telescope and eyepiece field of view. Use a moderate magnification, perhaps 100× to 150×, to pick Pluto out from the surroundings.

Simply seeing a dim point where Pluto is expected to be, however, is not necessarily proof that you have seen the dwarf planet. Instead, make an accurate drawing of the star field, indicating which faint point you believe to be Pluto. Then, return to the same star field a night or two later and repeat the observation. Has your suspect moved? If so, you have seen Pluto!

5
Medium-scope challenges
6- to 9.25-inch telescopes

The first "good" telescope I ever owned was an 8-inch $f/7$ Newtonian reflector known as the RV-8. It was made by the Criterion Manufacturing Company of Hartford, Connecticut, and given to me by my parents as my sole Christmas gift in 1971. I have received uncountable gifts at Christmases since, but none will stand the test of time like that telescope. True, I had owned a couple of smaller telescopes beforehand, but the RV-8 was not just a telescope. It was a spaceship that catapulted me out into the *real* universe for the first time.

Here are a variety of challenging targets suitable for intermediate apertures that will take us both into the real, and sometimes, surreal universe.

★
★★
★

91   Zeta Cancri

Target	Type	RA	Dec.	Constellation	Magnitude	Separation	Chart
Zeta Cancri	Quadruple star	08 12.2	+17 38.9	Cancer	5.6/6.0/6.1/10.0	1″/0.3″	5.1

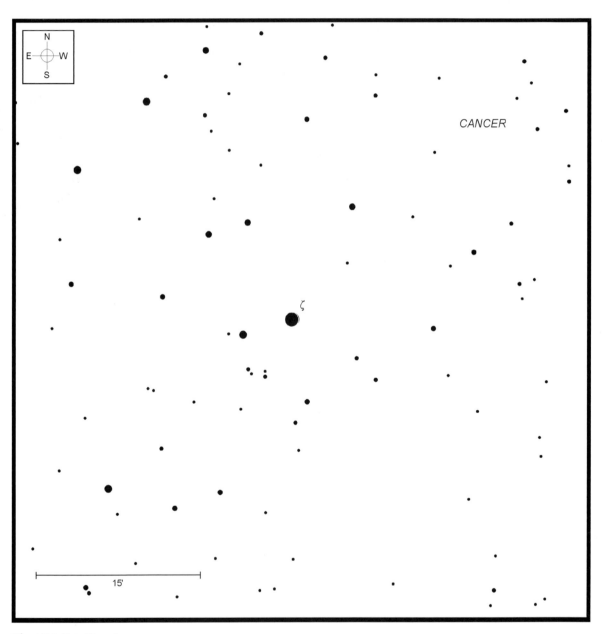

Chart 5.1 Zeta Cancri

Although it is one of the faintest constellations along the zodiac, Cancer the Crab hosts a variety of targets to test our mettle during the early spring. Spotting M44, the Beehive Cluster, by eye alone may prove very challenging for suburban observers, while the Crab's underappreciated second open cluster, M67, may also reach naked-eye visibility from more rural environs. While the constellation boasts a variety of challenging galaxies, in the test here, we will try our luck with one of the constellation's prettiest binary stars, Zeta (ζ) Cancri.

Many "best of" lists include Zeta as a spring showpiece target, so there is a good chance that you have already crossed paths. Zeta's two brightest suns, known as Zeta-1 and Zeta-2, were discovered in 1756 by German physicist/astronomer Johann Tobias Mayer. The Zetas are separated by 5 arc-seconds, which is wide enough to be resolvable through just about any amateur telescope 2 inches or greater in aperture.

In 1771, the exacting eyes of William Herschel noticed that Zeta-1 was not a solo act, but rather a tight stellar duet. Known today as Zeta Cancri A and Zeta Cancri B, these two yellow-white main-sequence stars have roughly equal luminosities and masses. They shine at magnitudes 5.6 and 6.0, respectively, and take 59.6 years to complete an orbit about their common gravitational center. During that time, their separation varies between 0.6″ at *periastron* (closest separation) and 1.2″ at *apastron* (widest separation). The last periastron occurred in 1990, with the stars' gap widening ever since. As of 2010, the two stars are separated by about an arc-second, with the gap continuing to increase until the next apastron in 2020.

Given steady seeing, a 6-inch instrument at 200× or more can resolve Zeta Cancri A and Zeta Cancri B as identical yellowish headlights nearly touching one another (Figure 5.1). As a hint, the stars are oriented approximately northeast–southwest at present (2010), although this will change as the stars continue their orbits.

By 1831, Herschel's son John noticed that Zeta-2 was wobbling ever so slightly in its orbit around Zeta-1. Although it was assumed this behavior was caused by a second star orbiting Zeta-2, this unseen companion remained unconfirmed until 2000. That year,

Figure 5.1 Zeta Cancri

photographic observations made with the Canada–France–Hawaii Telescope by J. B. Hutchings, R. F. Griffin, and F. Ménard[1] finally resolved the elusive companion. Subsequently, Zeta-2's two components have been designated as Zeta Cancri C and Zeta Cancri D.[2]

Can *any* amateur telescope possibly glimpse Zeta D? Zeta C, a yellow main-sequence star, shines at magnitude 6.1, while the newly discovered Zeta D is a weak 10th magnitude. They are separated by just 0.3 arc-seconds and have an orbital period of 17 years. That challenge may exceed even the largest backyard telescopes, although knowing the persistence of amateur astronomers I suspect it may only be a matter of time.

---

[1] J. B. Hutchings, R. F. Griffin, and F. Ménard, "Direct Observation of the Fourth Star in the Zeta Cancri System," *Publications of the Astronomical Society of the Pacific*, Vol. 112, no. 772 (2000), pp. 833–6.
[2] Although spectroscopic studies of Zeta-D reveal it to be a red dwarf, its brightness suggests we are looking at not just one, but two dwarf stars that remain too close to resolve even with today's best equipment.

92 Leo Trio 2

Target	Type	RA	Dec.	Constellation	Magnitude	Size	Chart
Leo Trio 2	Galaxy group	09 43.2	+31 55.7	Leo	–	~11′	5.2

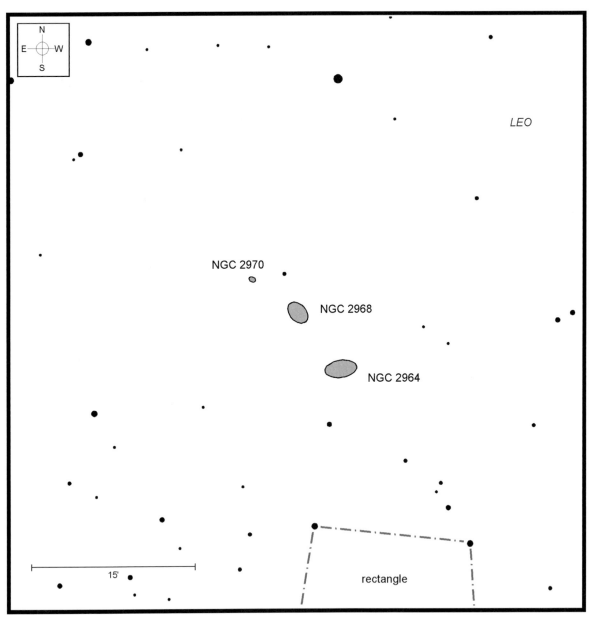

Chart 5.2 Leo Trio 2

Table 5.1 *Members of Leo Trio 2*

	Target	Type	RA	Dec.	Magnitude	Size
92a	NGC 2964	Galaxy	09 42.9	+31 50.8	12.0b	2.9′×1.5′
92b	NGC 2968	Galaxy	09 43.2	+31 55.7	12.8b	2.2′×1.5′
92c	NGC 2970	Galaxy	09 43.5	+31 58.6	14.4b	0.6′×0.4′

A nice trio of faint galaxies in northern Leo forms our next challenge. All three are tucked snuggly into the constellation's northernmost quadrant, some 7° north of the Leo "sickle."

Begin at Rasalas [Mu (μ) Leonis], the orangish star at the pointy northern tip of Leo's mane. Looking through your finderscope, scan about 5° to the northwest for the 6th-magnitude 15 Leonis. Look for a 7th-magnitude companion star just 13′ to its northwest, which helps 15 stand out from the crowd. Can you also spot a fainter, slightly orangish point (SAO 61633) about $1\frac{1}{4}°$ further to the north-northwest? If so, hop to it through your finder, and then switch to your telescope, widest-field eyepiece in place. Offset that star toward the southwestern edge of the eyepiece field until a rectangle of four 9th- and 10th-magnitude stars is centered in the field. NGC 2964, the leader of this small pack, is the same distance northeast of the rectangle as the orangish reference star is to its southwest. Table 5.1 lists the three group members.

Even though it is the brightest of the three, NGC 2964 is still a dim target in 6- to 9.25-inch scopes. Photographs reveal it to be a spiral galaxy inclined to our view some 50°. At 112× through my 8-inch reflector, NGC 2964 shows off a pale, oval glow elongated approximately east–west and surrounding a very faint, round core (Figure 5.2). I usually need averted vision to see the full span of the oval halo, but found little benefit from increasing the magnification.

NGC 2968 is a tougher catch, although it is visible through my 8-inch from under suburban skies by using averted vision. My records recall a small, very dim, featureless oval glow oriented approximately northeast–southwest. The lack of any distinguishable centralized nucleus adds to its obscurity. Although photographs record it as nearly as large as NGC 2964, it strikes my eye as perhaps only half the size. Those same photos reveal that NGC 2968, an irregular galaxy, has a

Figure 5.2 *Leo Trio 2*

pair of odd, dark, S-shaped lanes protruding from the galactic center and extending along the galaxy's major axis. Although I can see no sign of them even in my 18-inch reflector, I wonder if these lanes might be visible through larger amateur scopes.

The third and faintest member of our galactic trilogy is NGC 2970, just 5′ northeast of NGC 2968. It shines a magnitude dimmer still, and so poses a real test for medium apertures. My 8-inch can't pull it out from my light-polluted backyard, but was able to offer up a very dim glimmer from darker, rural skies. Even under the best conditions, it looks just like a very faint star. Don't feel too badly if you can't nab this last galaxy, however. William Herschel also missed it when he discovered NGC 2964 and 2968 using an 18.7-inch reflector in 1785. It took the more youthful eyes of his son, John, viewing through the same aperture in 1828 to spot it.

93 The "Antennae"

Target	Type	RA	Dec.	Constellation	Magnitude	Size	Chart
The "Antennae"	Galaxy pair	12 01.9	−18 52.8	Corvus	–	~4′	5.3

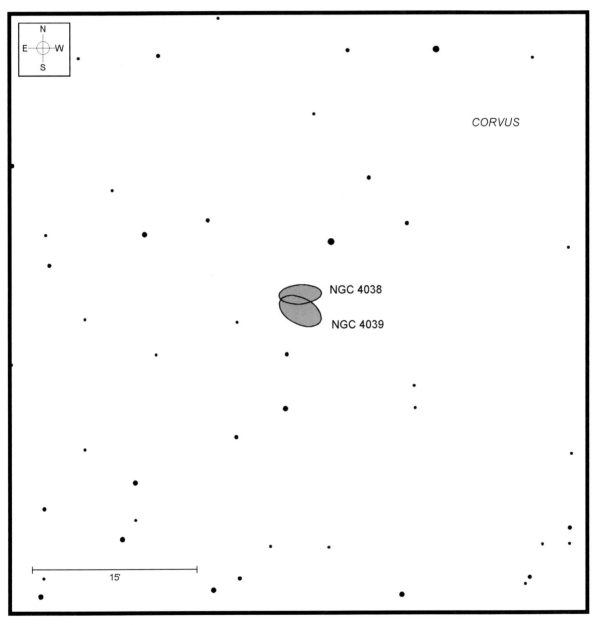

Chart 5.3 The "Antennae"

Table 5.2 *The "Antennae"*

	Target	Type	RA	Dec.	Magnitude	Size
93a	NGC 4038	Galaxy	12 01.9	−18 52.0	10.9p	3.7′ × 1.7′
93b	NGC 4039	Galaxy	12 01.9	−18 53.5	11.1p	4.0′ × 2.2′

Although the four brightest stars in Corvus, the Crow, shine no more intensely than magnitude 2.6, the constellation's distinctive trapezoidal pattern in an otherwise star-poor region of the spring sky lets it stand out well even in moderate light pollution. We will put that to good use here as we probe for one of the best-known pairs of interacting galaxies in the sky: NGC 4038 and NGC 4039, the "Antennae." Table 5.2 lists specifics.

Here, we find two galaxies in a celestial death match of tug-o'-war. Each is being yanked apart by the gravity of the other, resulting in a churning, complex mix of red hydrogen II stellar nurseries and blue clouds of newly energized stars. As some of the dust, gas, and stars from the galaxies intermix, large regions of massive-star formation are being triggered within each.

As time goes on, the momentum built up will let each galaxy escape the other's grip, only to be inexorably drawn back together in the distant future to continue their head-to-head struggle. Although there is slim chance of individual stars colliding because of their wide spacing, each galaxy will eventually be distorted beyond recognition. This struggle will continue, wrapping the galaxies around each other until the pair becomes one sometime in the far-distant future.

We can use two of the stars in the Crow's body to find our target. Trace a line from the northeastern star, Algorab [Delta (δ) Corvi], to the northwestern star, Gienah [Gamma (γ) Corvi], and continue an equal distance to the southwest. As a reference, halfway along you will pass a right triangle of 7th-magnitude stars. Continue in the same direction and you should find NGC 4038 and NGC 4039 set between two 9th-magnitude stars.

At first glance, all you may see is a single 10th-magnitude glow. What's so special about a description that could fit thousands of galaxies? Closer inspection with 100× or more reveals something different here. Look carefully and the amorphous glow will transform into a hook-shaped image with a faint extension coming off toward the south. This elongation carries its own NGC number, 4039, and glows dimly at about 11th magnitude. A dark wedge intruding from the west separates the galaxies, as portrayed in Figure 5.3.

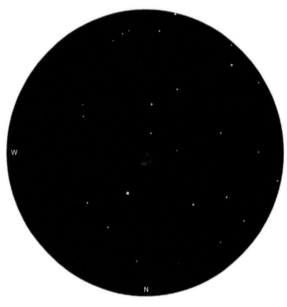

Figure 5.3 *The "Antennae"*

Given rural skies that are free of horizon-hugging haze, neither galaxy appears uniform, but looks rather clumpy. That's not an illusion. You are seeing some of the consequences of the merging process, huge starburst areas where new suns are coming into being as we watch from afar. Knots along the rim of NGC 4038, the northern galaxy in the pair, are the most obvious, although some very subtle mottling is also evident in NGC 4039.

Trying to describe the unusual appearance of this intertwined pair has pushed the imaginations of observers to the limits. The most common nickname applied to the tight twosome is the Antennae, owing to the two long, contrail-like filaments extending from both in long-exposure photographs. These "antennae" are the result of tidal forces as the two galaxies graze one another. Others prefer the monikers Ring-Tail or Rat-Tail galaxies. Visually, the pair remind many of a comma, a shrimp, or even a tadpole when viewed through medium- and large-aperture backyard telescopes.

94 NGC 4361

Target	Type	RA	Dec.	Constellation	Magnitude	Size	Chart
NGC 4361	Planetary nebula	12 24.5	−18 47.0	Corvus	10.3p	118″	5.4

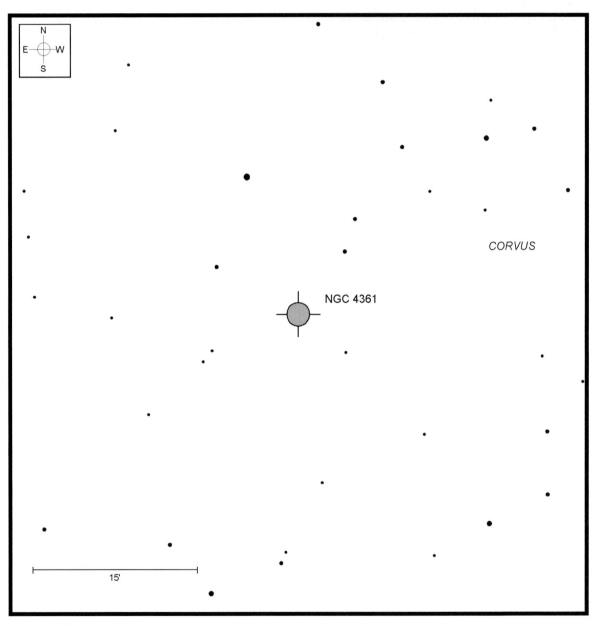

Chart 5.4 NGC 4361

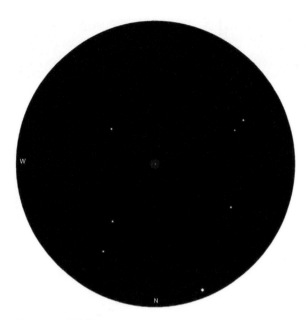

Figure 5.4 NGC 4361

Planetary nebula NGC 4361 is almost perfectly centered within Corvus's trapezoidal body. To find it, look through your finderscope about 1° south-southwest of Algorab [Delta (δ) Corvi] for a pair of 7th-magnitude suns. Drop another degree to the south and you'll find a small isosceles triangle of 7th-magnitude stars. NGC

4361 is less than a degree to their southwest. The lack of bright stars in its immediate area helps it to stand out more prominently than it might otherwise.

Most deep-sky observers are surprised by how large NGC 4361 appears in their telescopes. Its apparent diameter, some 2 minutes of arc, is unexpectedly large for the genre. At 54×, my 8-inch records a dull, round disk of grayish light surrounding a 13th-magnitude central star shown in Figure 5.4. But increase the magnification three-fold and add in a narrowband filter, and that unexciting disk is transformed into quite the sight. With averted vision, the disk shows a stippled texture that upon closer examination resolves into two dim tendrils of light extending away from the nebula's central disk and into its fainter outer shell. These two regions curve away in opposite directions, one toward the northeast, and the other toward southwest. The overall appearance, which is strongly reminiscent of a barred spiral galaxy seen face-on, makes NGC 4361 one of the most fascinating planetaries to study through amateur telescopes.

Unlike many other planetaries, however, which demand the highest magnification that sky conditions will tolerate, NGC 4361 puts on its best show at more moderate powers, between perhaps 150× and 200×. Anything higher and the faint structural details lose contrast against the outer gaseous shell.

95 NGC 5053

Target	Type	RA	Dec.	Constellation	Magnitude	Size	Chart
NGC 5053	Globular cluster	13 16.4	+17 41.9	Coma Berenices	9.0	10.0'	5.5

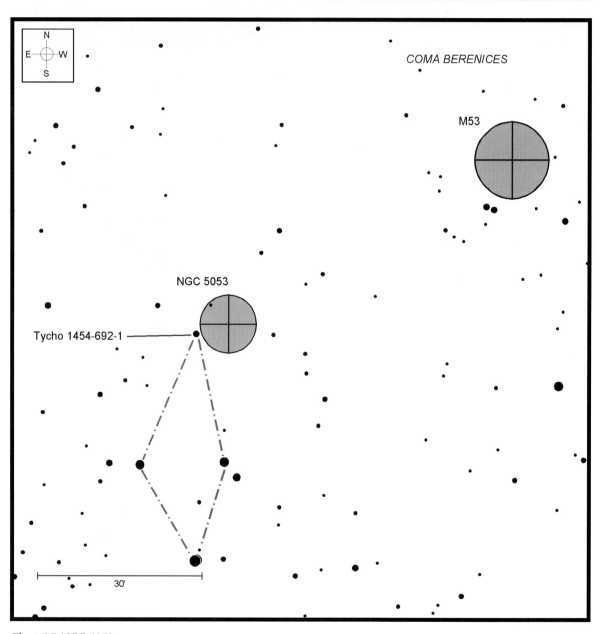

Chart 5.5 NGC 5053

Mention Coma Berenices to any devout deep-sky observer and immediately images of faint fuzzies will begin to dance in their heads. While Coma is indeed teeming with distant island universes, it also holds two much closer targets to test our mettle. Since both are located within 2° of the constellation's brightest star, Alpha (α) Comae, they are easy to pinpoint by telescope from suburban and urban skies. Actually seeing both may be another matter.

The first, M53, was discovered in 1775 by Johann Bode. Messier himself saw it for the first time on February 27, 1777. His description of a "nebula without stars; circular and conspicuous" closely matches the impression I had through 16×70 binoculars. My notes recall a "faint, circular patch of uniform light; unlike nearby M3, M53 does not possess a brighter core." Visible through 50-mm binoculars given light-free skies, M53 offers up a bonus challenge object for binocularists trolling the area.

While M53 should be an easy catch through 6-inch telescopes, seeing any of its 100,000 or so members takes some effort. Increasing aperture by just an inch or two helps. In my 8-inch reflector, I can spot several faint stars lined up in curving arcs that peel away from the core.

A true challenge for 6-inch telescopes lies 1° southeast of M53, and may just squeeze into the same low-power, wide-field eyepiece. NGC 5053 may have been missed by Bode and Messier during their explorations of M53, but its dim glow did not escape the eyes of William Herschel, who discovered it on March 14, 1784. Herschel added it to his growing list of deep-sky objects as "VI-7," which was his short-hand way of saying it was the 7th member of a class of objects described as "very compressed and rich clusters of stars."

Spotting those cluster stars is a tough test in a 6-inch scope, since like those in M53, none shines brighter than 13th magnitude. Indeed, seeing the cluster at all can prove nearly impossible because of its low surface brightness. Don't let its magnitude rating of 9.0 fool you. Any light pollution will conspire with the cluster's dim surface brightness, which is closer to magnitude 14, to fade it quickly into the background.

The cluster's low surface brightness is partly a result of its loose packing. The stars in NGC 5053 are scattered much more than M53, with little central concentration evident either visually or in photographs. Testifying to that weak construction is the cluster's

Figure 5.5 NGC 5053

rating of 11 on the Shapley–Sawyer scale of globular concentration. That, combined with a low population to begin with, adds to the challenge presented here. NGC 5053 contains perhaps 40,000 suns, as opposed to 100,000 or more for showpieces like M3 and M13.

Recent studies show that a number of so-called *blue stragglers* are scattered among the stars of NGC 5053. Blue stragglers are believed to be hotter and bluer than other stars in their respective cluster that have the same mass and luminosity. Why that is so remains a mystery more than half a century after American astronomer Allan Sandage discovered the first of the breed hidden in the globular cluster M3 in Canes Venatici. The leading theory at present dates to 1964 when astronomers Fred Hoyle and W. H. McCrea independently suggested that blue stragglers are formed when two passing stars capture each other to form a tight binary system, or possibly even merge. Computer models suggest that one star in the pair could tap the hydrogen of the other, resulting in a renewed and more vigorous rate of hydrogen fusion that could cause the star to appear hotter and bluer. Evidence in favor of this theory includes the fact that blue stragglers are most common in the cores of globular clusters, where stars are crammed together.

To locate NGC 5053, cast off from M53 toward the southeast. Just 1° away, look for the 9.5-magnitude star (Tycho catalog 1454-692-1) at the northern tip of a diamond-shaped asterism. NGC 5053 is just to the star's west, but don't be surprised if you can't find it. For the best chance of spotting its dim glow, begin your hunt with an intermediate magnification, between 75× and 90×. If you have no luck, then lower magnification to between 50× and 60×, and try again. Anything

lower, and I suspect the star will be swallowed up by the background sky; anything greater and the large size of the cluster will disperse it into the background sky. Be sure to wait for a special night when you can leave light pollution behind and travel out into the country.

96 M51's spiral arms

Target	Type	RA	Dec.	Constellation	Magnitude	Size	Chart
M51 spiral structure	Galaxy structure	13 29.9	+47 11.8	Canes Venatici	9.0b	10.3′×8.1′	5.6

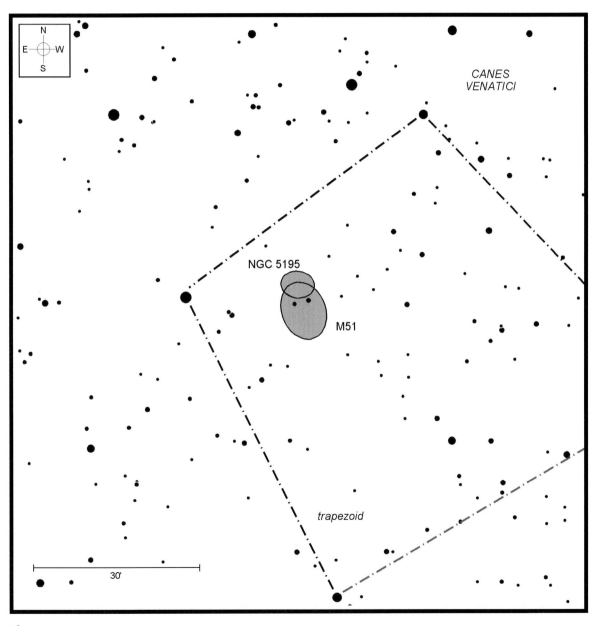

Chart 5.6 M51

Of the thousands of galaxies visible through intermediate apertures, one stands above the rest in terms of visual interest: M51, the famous Whirlpool Galaxy in Canes Venatici. Everything adds up in M51's favor. We are seeing it very nearly face-on, its spiral arm halo is bright and peppered with star clouds and vast regions of nebulosity, and it brings with it a friend in the form of a smaller companion galaxy that can even be seen through giant binoculars.

Charles Messier was first to lay eyes on the Whirlpool when he accidentally bumped into it on October 13, 1773. His notes recall a "very faint nebula without stars." The fact that he referred to it in the singular indicates that he saw only the bright core of M51 itself, and not its smaller companion, NGC 5195. The discovery of the latter is credited to Messier's friend and contemporary, Pierre Méchain, who noted a double core on March 21, 1781.

The first hint that there was more to see within M51 than just a pair of nebulous blobs came from an observation by John Herschel on April 26, 1830. Herschel recorded "a very bright round nucleus surrounded at a distance by a nebulous ring" through his 18.7-inch telescope. A later drawing by him recorded a large, bright core centered perfectly in a fainter surrounding ring. The companion, NGC 5195, is also shown as round, but smaller than M51's core and positioned outside of the mysterious ring. Herschel later mused, "Supposing it to consist of stars, the appearance it would present to a spectator placed on a planet attendant on one of them eccentrically situated towards the north preceding quarter of the central mass, would be exactly similar to that of our Milky Way."

Fifteen years later, in the spring of 1845, Herschel's puzzling nebulous ring was resolved into a pinwheel structure by Lord Rosse at Birr Castle in Ireland. Aiming toward it with his newly completed 72-inch "Leviathan" reflector, the largest telescope in the world at the time, Lord Rosse saw "spiral convolutions; with successive increase of optical power, the structure has become more complicated. The connection of the companion with the greater nebula is not to be doubted; the most conspicuous of the spiral class." Later, in 1861, Lord Rosse noted that "the outer nucleus unquestionably spiral with a twist to the left."

While discovering the spiral structure took a 72-inch aperture, knowing it's there gives you and me a distinct advantage. In fact, hints of M51's pinwheel construction have been reported through telescopes as

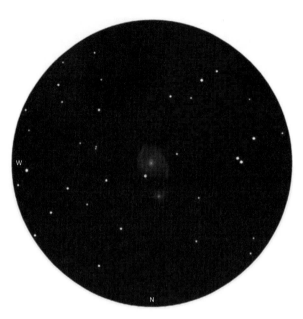

Figure 5.6 M51's spiral arms

small as 4-inches across, given extraordinary conditions and exceptionally sharp vision. I find those reports absolutely amazing, since spotting the spiral structure through my own 8-inch reflector, even given dark skies, is a rare treat. Figure 5.6 captures one of those moments.

Before we discuss strategy, let's zero in on M51. At 8th magnitude, M51 is bright enough to be visible through small binoculars even from suburbia. Start at Alkaid [Eta (η) Ursae Majoris], the end star in the handle of the Big Dipper. Hop to 24 Canum Venaticorum, a 4th-magnitude point just 2° to the west-southwest, then slide another 2° to the southwest to a trapezoid of faint stars. M51 lies inside the trapezoid's northeast corner.

M51 may be visible easily from suburban skies through 6- to 9.25-inch scopes, but sky darkness and transparency are the overriding factors when looking for its spiral structure. My first view of the spiral arms came in 1974 while observing through my 8-inch at the Stellafane amateur telescope makers' convention in Springfield, Vermont. The sky was especially dark that year, with M33 visible without optical aid (naked-eye Challenge 12). With the same equipment under lesser conditions, I see no hint whatsoever. Indeed, from my suburban backyard, it takes my 18-inch reflector to make out any suggestion of the arms.

If this is your first time looking for the spiral arms, strategy is everything. First, choose the right magnification. The best views seem to come with eyepieces producing an exit pupil of between 2 mm and 3 mm.[3] That narrow range seems to offer a good compromise between image size and contrast.

Next, you need to know how to look for the arms. Take a look at the glow surrounding the core of M51. It may look uniform at first, but careful study with averted vision will reveal some irregularities. One arm starts to the south of M51's core and hooks to the northeast, with the brightest portion lying halfway between the core and NGC 5195.

A second spiral arm begins just west of the core, curves to its south, and then spirals around toward the northeast. It fades from view as it extends toward NGC 5195. Using averted vision, can you also detect the faint nebulous bridge that reaches outward for the companion galaxy? Some observers report better success seeing the arms by focusing their attention on the dark gaps between them rather than looking for the bright arms themselves.

By waiting for an especially dark, clear spring night and letting your gaze sweep across the faint glow of the Whirlpool's halo, perhaps tapping the telescope gently to vibrate the image, Lord Rosse's elusive "spiral convolutions" should be discernible.

---

[3] From Chapter 1, exit pupil = aperture/magnification = (eyepiece focal length) / (telescope focal ratio).

97 NGC 6445

Target	Type	RA	Dec.	Constellation	Magnitude	Size	Chart
NGC 6445	Planetary nebula	17 49.3	−20 00.6	Sagittarius	13.2p	44″×30″	5.7

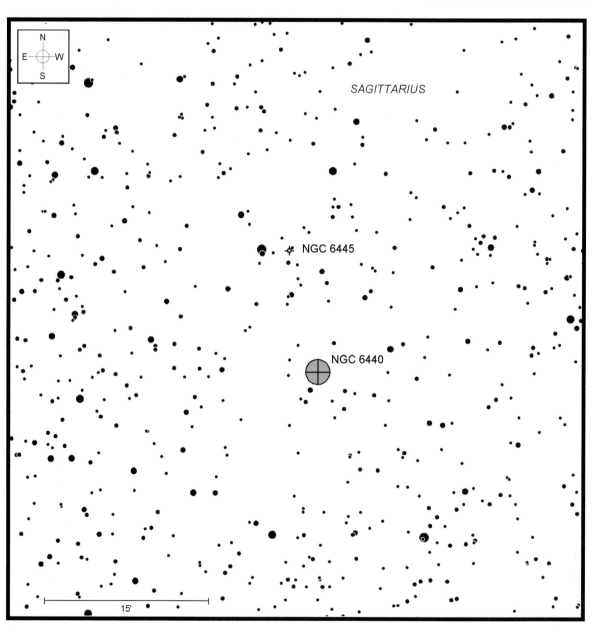

Chart 5.7 NGC 6445

The sky is full of weird sights. And, among planetary nebulae, NGC 6445 is one of the strangest. Although visible in smaller apertures, it takes a 6-inch telescope for NGC 6445's true, if bizarre, nature to shine through. The nebula's brighter central shell looks like a dented rectangle. Nature rarely creates an amorphous form with sharp edges, and indeed the peculiar appearance of NGC 6445 is due largely to our perspective as well as its age. But the look is very odd nonetheless. No wonder NGC 6445 has been nicknamed the Box Nebula.

As the material that will ultimately form a planetary nebula is expelled by its star, it takes on a cylindrical shape. If we look along its axis, we see the classic smoke-ring effect of the Ring Nebula, M57. But then, over time, gravitational influences from other sources, such as an outflowing stream of particles from the progenitor star, as well as from the gravity of companion stars or perhaps a family of planets, contort the shell into bizarre, distended shapes. Studies show that NGC 6445 is one of the sky's oldest planetary nebulae at an estimated 3,300 years, so there has been plenty of mixing time. From its full size, some $3' \times 1'$, and great distance, 4,500 light years, these same studies conclude that NGC 6445 is also one of the sky's largest, spanning perhaps 4 light years. Deep photos reveal its true asymmetrical bipolar structure, with a bright central ring surrounded by fainter nebulous tendrils. It is believed that most, if not all, planetary nebulae show bipolar tendencies owing to highly energetic streams of particles that flow from their progenitor stars. These streams, called *bipolar outflows*, are focused into cones of gas by the star's magnetic fields or perhaps by binary companions.

When examined visually under high magnification, the nebula's disk expands into a strange specter floating amid a very rich field of stars. Figure 5.7 shows the scene through my 8-inch reflector at 112×. I can readily see that the nebula is not only rectangular, but also has what appears to be a hollow center, akin to a withered version of M57. Closer scrutiny also discloses several brighter patches stitched within the nebula's outer edges, with the most prominent knots seen toward the eastern and southern limits.

Figure 5.7 NGC 6445

One of the challenges presented by NGC 6445 is simply finding it among the star-rich fields of the Sagittarius Milky Way. The easiest way to starhop there is to begin at the bright open cluster M23, itself a wonderful target at low power. Slide 1° to the southwest toward a yellowish 7th-magnitude sun, and then head due west for another degree, passing between two 8th-magnitude stars, to another 8th-magnitude point. Through the telescope, this last star is joined by two fainter companions that collectively form an isosceles triangle that points right at our quarry.

A bonus object, globular cluster NGC 6440, is nestled just a third of a degree south of our planetary challenge. Both will squeeze into a wide-field eyepiece's field, but high magnification is needed to enjoy the cluster, as it is with the planetary. Even at 284×, my 8-inch gives no hint of resolution within this packed swarm of ancient stars. Instead, NGC 6440 is simply a small, faint, diffuse ball of fuzz surrounding a somewhat brighter central core. The stars in the globular shine no greater than 16th magnitude, bringing them just within the range of giant backyard scopes.

98 NGC 6453

Target	Type	RA	Dec.	Constellation	Magnitude	Size	Chart
NGC 6453	Globular cluster	17 50.9	−34 35.9	Scorpius	10.2	7.6′	5.8

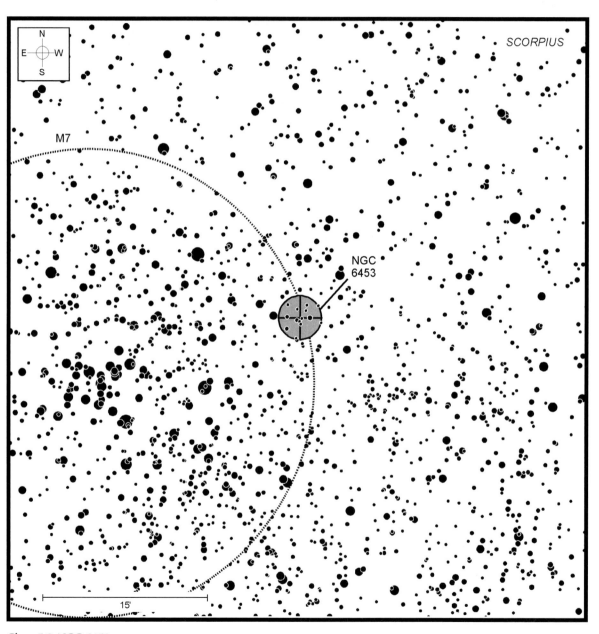

Chart 5.8 NGC 6453

The open cluster team of M6 and M7, floating just north of the Scorpion's stinger, creates one of the prettiest binocular sights in the summer sky. Even the smallest pocket binocular will show that the brightest stars in M6 form a rectangular pattern that inventive eyes can imagine as the outline of a butterfly. M7 is even larger and brighter than M6, and bursts into an exceptionally striking assortment of stars covering an area larger than the Full Moon. Several of its 80 stars show subtle hues of yellow and blue.

Although conventional wisdom says that M7 is far too large to enjoy through most telescopes because of their narrow fields of view, studying it through 6-inch or larger instruments reveals a buried treasure nestled near the cluster's western edge. John Herschel is credited with discovering this little gem, NGC 6453, on June 8, 1837. Herschel included it as h 3708 in his 1864 *Catalogue of Nebulae and Clusters of Stars*, which subsequently led John Dreyer to include it in his *New General Catalogue* as object number 6453. Echoing Herschel's original observation, Dreyer described the cluster as "considerably large, irregularly round, pretty much brighter (in the) middle, round."

NGC 6453 shines at 10th magnitude and measures nearly 8′ across. If it were alone in the sky, it would be a fairly routine catch in 8-inch and larger apertures. But NGC 6453 is not isolated by any means; instead, it is immersed in a heavily packed field of stars. Spotting it can be literally like finding a needle in a haystack!

To catch a glimpse, start with a low-power, wide-field eyepiece in your telescope and aim toward M7. From the jumble of bright stars in M7's center, look for HD 162391, a single 5th-magnitude point of pale yellowish light toward the cluster's northwestern edge. Switch to about 100× and then look immediately to the star's southwest for a slender triangle of 8th- and

Figure 5.8 NGC 6453

9th-magnitude points. By traveling away from HD 162391 along the triangle's long dimension, you'll bump right into NGC 6453.

At 112×, my 8-inch reflector displays an unresolved smudge of light buried in a stellar ocean, as shown in Figure 5.8. Don't mistake the few faint points scattered across the globular for object resolution. These are most likely foreground stars, since the cluster's brightest stellar citizens are no brighter than 14th magnitude. That's because the cluster, which is 31,300 light years from Earth, is positioned inward toward the galactic center. Its location, only 5,900 light years from the core of the Milky Way, confounds our view because of many intervening clouds of cosmic dust.

99   NGC 6517

Target	Type	RA	Dec.	Constellation	Magnitude	Size	Chart
NGC 6517	Globular cluster	18 01.8	−08 57.5	Ophiuchus	10.1	4.0′	5.9

100   NGC 6539

Target	Type	RA	Dec.	Constellation	Magnitude	Size	Chart
NGC 6539	Globular cluster	18 04.7	−07 35.2	Serpens	8.9	7.9′	5.9

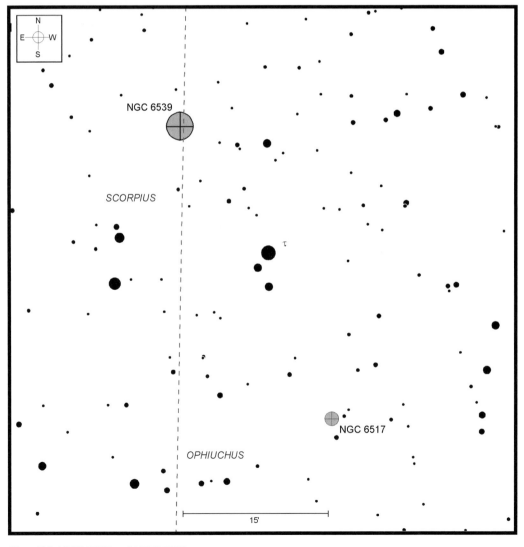

Chart 5.9 NGC 6517 and NGC 6539

Discovered on June 16, 1784, by William Herschel, NGC 6517 resides in a relatively star-poor region of eastern Ophiuchus. That's both a blessing and a curse. While the few surrounding stars means that the globular should stand out well, part of the reason for the lack of stars is that the area is polluted with clouds of intergalactic dust. Not only does this dust block the light from more distant stars, it also dampens our view of the globular.

NGC 6517 is found between 3.3-magnitude Nu (ν) Ophiuchi and 5th-magnitude Tau (τ) Ophiuchi. Nu marks the eastern point of a large isosceles triangle formed with the stars Cebalrai [Beta (β) Ophiuchi] and Sabik [Eta (η) Ophiuchi] along the eastern side of the constellation's large frame. Aim your finderscope halfway between Nu and Tau, and take a look through your telescope. NGC 6517 should lie very close to the center of the eyepiece field, just north of a 10th-magnitude star and east of an 11th-magnitude point.

At 100× in my 8-inch reflector under a country sky, NGC 6517 is seen as a faint smudge of light measuring no more than a few arc-minutes across and elongated slightly northeast–southwest. A brighter center surrounded by a dimmer halo is evident, but there is no hope of seeing any of its very faint constituent stars. Doubling magnification does little to improve the view. In fact, even my 18-inch at 220× fails to resolve any of the cluster's stars, although a slight graininess at the edges begins to hint at its true nature. Few of us will ever resolve this distant target, since the brightest cluster stars shine at only a pitiable 16th magnitude.

If NGC 6517 proved a little too tough, then NGC 6539 should come a little easier. To find it, switch back to a low-power eyepiece, center on Tau, and then move your telescope's aim slowly toward the northeast. As Tau Ophiuchi moves to the edge of the field, look along the opposite side for a dim smudge of light. That will be NGC 6539.

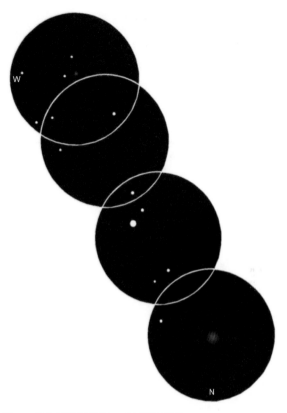

Figure 5.9 NGC 6517 and NGC 6539

Interestingly, even though NGC 6539 is brighter than its neighboring globular cluster, William Herschel failed to see it when he discovered NGC 6517. The Danish astronomer Theodor Brorsen was first to lay eyes on NGC 6539 in September 1856. Through modern-day 6- to 9.25-inch telescopes, it appears as an unresolved blur of starlight punctuated by a brighter, round core (Figure 5.9). Larger scopes add a few faint points scattered across. Appearances aside, these stars are not true members of NGC 6539, but rather just chance foreground objects. The brightest stars within this distant globular shine at 16th magnitude, just like those in NGC 6517.

101 Barnard 86

Target	Type	RA	Dec.	Constellation	Opacity	Size	Chart
Barnard 86	Dark nebula	18 03.0	−27 52.0	Sagittarius	5	5.0′	5.10

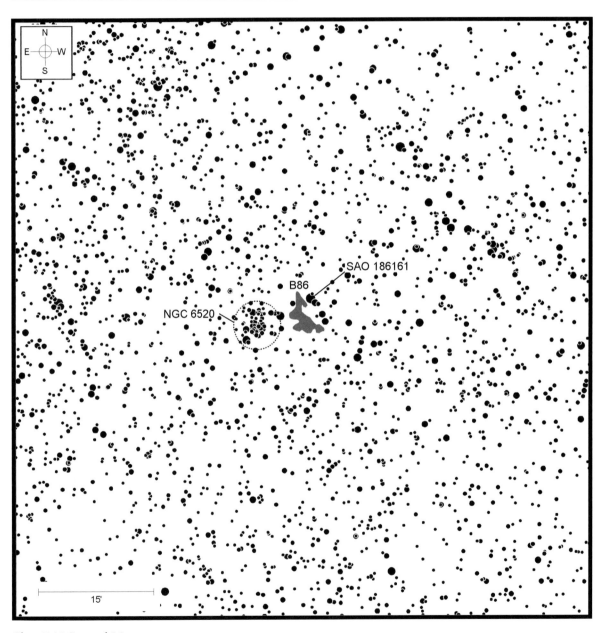

Chart 5.10 Barnard 86

Clouds of dust dampened NGC 6517 in our last challenge. If those clouds were dense enough, we couldn't see through them at all. Instead, we would see the region as a starless void, almost looking like a hole in the sky. The summer sky is full of these voids, but for much of the nineteenth and early twentieth centuries, astronomers were at a loss to explain what they were. Were they actually empty regions or were they some sort of dark material that simply blocked the light from stars that lay beyond?

The first astronomer to make an in-depth study of these mysterious regions was Edward Emerson Barnard. Barnard's 1919 paper in the *Astrophysical Journal*, "On the Dark Markings of the Sky with a Catalogue of 182 such Objects," outlined the results of his initial investigation: that these voids were anything but empty. Rather, they were vast clouds of opaque interstellar matter. Barnard's research was ongoing at the time of his death in 1923, but ultimately led to a second, expanded listing of *A Photographic Atlas of Selected Regions of the Milky Way* being released posthumously in 1927. This second paper expanded Barnard's catalog of dark nebulae to 349 total entries. His research still stands today as the benchmark for all subsequent investigations.

Whether it is due to light pollution, a lack of good charts, or perhaps because there are so many bright objects to observe, many amateurs tend to shy away from seeking out dark nebulae. In a way, that makes sense. No other deep-sky genre so easily falls victim to less-than-perfect sky conditions than dark nebulae. The slightest haze or light pollution is usually enough to increase the sky's ambient background brightness enough that it swallows up these dark, cold clouds of interstellar dust grains.

Barnard 86 is impossible to see in an urban sky since there is no light-pollution filter made to counter the destructive effects that our over-illuminated civilization has had on its visibility. Even a decent suburban setting could be bright enough to render Barnard 86 invisible. But head out to dark, rural skies and you can spot this inky black cloud easily with a 6-inch telescope (Figure 5.10).

Figure 5.10 Barnard 86

Barnard 86 enjoys several distinct advantages over some other dark nebulae when it comes to visibility. For one, it lies sandwiched between the open cluster NGC 6520 and the orangish 7th-magnitude star SAO 186161. Its position in between these two bright targets makes zeroing in on its exact location a fairly simple exercise. Another plus is that, unlike many dark nebulae that show only vague outlines, B86 has sharply defined borders. We can see exactly where it starts and where it ends. Finally, its rating of 5 on a 1-to-5 opacity scale indicates that this cloud is truly dark.

Since Barnard 86 measures only 5' across, a magnification of between 100× and perhaps 125× will serve up the best view of it and its two bright neighbors. B86 appears decidedly wedge-shaped through amateur telescopes, with its western border facing HD 164562 perhaps twice as long as the side closest to the cluster. NGC 6520 also spans about 5', with some two dozens stars visible through my 8-inch reflector. Together, they make a wonderful study in contrast.

102   NGC 6886

Target	Type	RA	Dec.	Constellation	Magnitude	Size	Chart
NGC 6886	Planetary nebula	20 12.7	+19 59.3	Sagitta	12.2p	6″	5.11

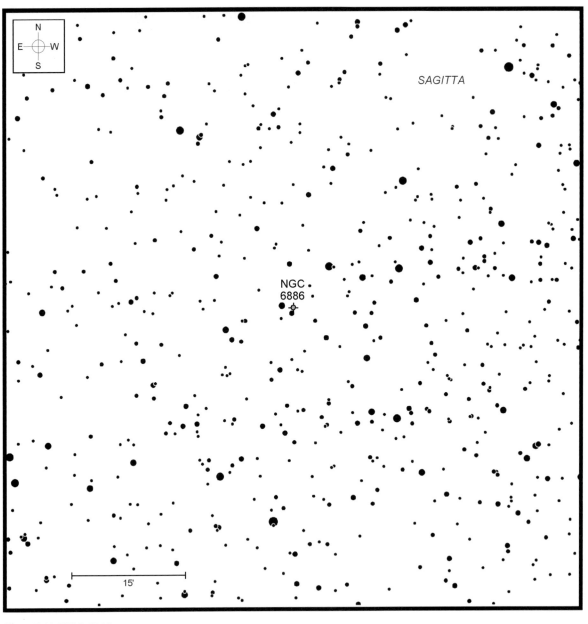

Chart 5.11 NGC 6886

Table 5.3 *Vorontsov-Velyaminov scale of planetary nebulae morphologies*

1	Stellar image
2	Smooth disk (a, brighter toward center; b, uniform brightness; c, traces of a ring structure)
3	Irregular disk (a, very irregular brightness distribution; b, traces of ring structure)
4	Ring structure
5	Irregular form, similar to a diffuse nebula
6	Anomalous form

Note: Planetary nebulae with more complex structures are characterized by combinations of classes. For instance, NGC 6886 is rated "2+3" for its complex disk morphology.

NGC 6886 is a relatively bright, but very small planetary nebula. To a casual eye glancing through a telescope, it will look just like another field star. And since it is poised along the eastern edge of the summer Milky Way in Sagitta the Arrow, there are plenty of imposters to weed out before the true challenge is uncovered.

"Now, wait a minute," you're thinking. "Look at those numbers in NGC 6886's data table." The nebula's apparent diameter is listed as 6 arc-seconds. That's small, but certainly resolvable as a disk with less than 100×. Unfortunately, numbers can be misleading, as we know from every automobile advertisement that we read. "Your mileage may vary." And it certainly will here.

NGC 6886 is classified as a 2+3 planetary nebula, a short-hand way for describing NGC 6886 as having a smooth disk surrounded by an irregular shell. This planetary-nebula rating system, called the Vorontsov-Velyaminov scale, was devised by the Russian astrophysicist Boris Vorontsov-Velyaminov (1904–1994). His 6-point system describing planetaries' morphologies is summarized in Table 5.3.

Photographs taken with the Hubble Space Telescope explain NGC 6886's multipart listing by revealing two "wings'' extending an additional 2 arc-seconds on either side of the nebula's circular inner shell. This

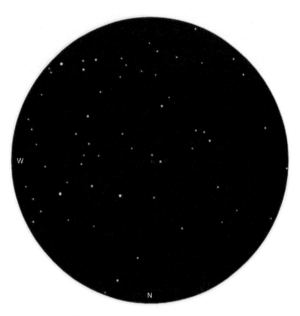

Figure 5.11 NGC 6886

brings its total diameter to 6 arc-seconds. Through our backyard telescopes, however, only the inner disk will be apparent, which measures only about 2″ across.

Begin your journey toward NGC 6886 from 5th-magnitude Eta (η) Sagittae, marking the pointy tip of the Sagitta Arrow. NGC 6886 lies 1.8° due east of Eta. If you're like me and prefer to let someone else do the work for you, aim at Eta, turn off your telescope's clock drive if it has one, and then sit back and relax for exactly 7 minutes 12 seconds. In that time, Earth's rotation will steer your telescope away from Eta and exactly toward the planetary. With a 50× eyepiece in place, look for a small isosceles triangle made up of three faint stars pointing toward the northeast. The "star" at the triangle's southwestern corner is actually NGC 6886 (Figure 5.11). Can't tell for sure? Increasing magnification four-fold may help, but to confirm which is the planetary, blink the field with a narrowband or, better still, an oxygen-III filter. (Refer to small-scope Challenge 58 in Chapter 4 for "blinking" details.) The nebula's central star, once four times more massive than our Sun, now shines at barely 18th magnitude.

103   IC 4997

Target	Type	RA	Dec.	Constellation	Magnitude	Size	Chart
IC 4997	Planetary nebula	20 20.1	+16 43.9	Sagitta	11.6p	2″	5.12

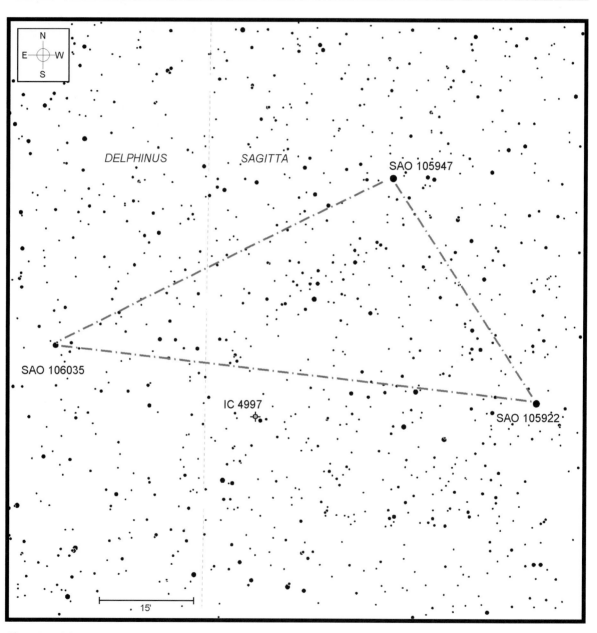

Chart 5.12 IC 4997

Like our last challenge, the planetary nebula IC 4997 lies within the borders of Sagitta, and is surprisingly bright, but extremely small. That combination makes this a great object for everyone, whether you are observing under the veil of light pollution or from a dark, rural location. Its intensity should shine through all but the most extreme situations.

Studies show that this little planetary varies in brightness. Long-term studies conducted by E. B. Kostyakova of Moscow University reveal that IC 4997 dimmed by as much as a half a magnitude between the years 1968 and 1985, only to slowly brighten again in the ensuing years. The cause of these variations is probably attributable to its misbehaving central star, whose temperature continues to fluctuate. These changes, along with the high density of its inner shell, point to IC 4997's extremely young age, possibly no more than 200 years old.

While NGC 6886 lets us sit around until it came to us, finding IC 4997 is going to require more active participation on your part. It lies in an empty void near the intersection of Sagitta and Delphinus, about halfway between 5th-magnitude Eta (η) Sagittae and 4th-magnitude Beta (β) Delphini. The bright stars in its immediate vicinity shine at only 7th magnitude. Three orangish suns – SAO 105922, SAO 105947, and SAO 106035 – form a right triangle just to its north and make a handy reference. IC 4997 is just south of the halfway point along the triangle's hypotenuse, marked by SAO 105922 and SAO 106035. There, you will see that IC 4997 forms a small obtuse triangle with a 10th-magnitude star situated just 1.1′ to its southwest and a 12th-magnitude sun 2′ to its west.

While it is easy to mistake those field stars for the planetary, they actually make handy comparison points

Figure 5.12 IC 4997

when blinking the field with an oxygen-III nebula filter. That's really the only way to confirm which is which, since IC 4997 remains perfectly stellar even at 300× (Figure 5.12). With a filter in place, however, the nebula will appear brighter than the 10th-magnitude star; with it removed, the star will outshine the planetary. The planetary's very subtle bluish tint may also help to identify it from among the field of white stars. After you confirm which "star" is, in fact, IC 4997, consider that what you have just found was never seen by William Herschel as he no doubt surveyed the same area. In fact, not only did he not see it; it didn't even exist in his lifetime.

104   NGC 6905

Target	Type	RA	Dec.	Constellation	Magnitude	Size	Chart
NGC 6905	Planetary nebula	20 22.4	+20 06.3	Delphinus	11.9p	72″×37″	5.13

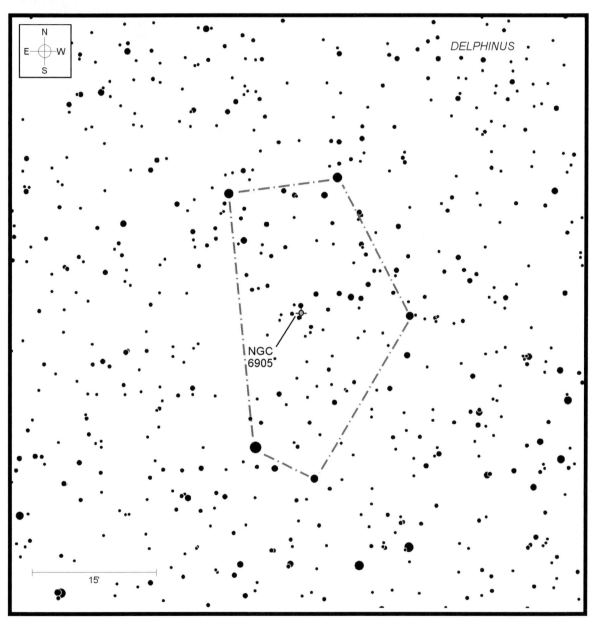

Chart 5.13 NGC 6905

Can you catch the Blue Flash? That's the nickname of NGC 6905, a 12th-magnitude planetary nebula in the constellation Delphinus. John Mallas, a prolific deep-sky observer in the 1950s–1970s, bestowed the epithet in an article he penned entitled "Visual Atlas of Planetary Nebulae" in the July/August 1963 issue of *Review of Popular Astronomy* magazine. That was 181 years after NGC 6905 was discovered by William Herschel.

Unlike the last two challenging planetary nebulae, the test posed by NGC 6905 does not stem from its small size. Quite the contrary. NGC 6905's inner shell measures $72'' \times 37''$, large as planetaries go; and its outermost edge is nearly double that diameter.

No, the challenge presented by the Blue Flash is in the hunt. NGC 6905 is in northwestern Delphinus, where nary a naked-eye star is found. Admittedly, that means little if you are using a GoTo telescope. Just punch "N-G-C-6-9-0-5" into the hand controller and off it goes in a whiz. If that's your preferred method, then that's fine. NGC 6905 should be easy to snag once aimed at the right field. But if you'd prefer the challenge of the quest, then follow me.

Begin again at Eta (η) Sagittae, which should now be an old friend after the last two planetaries. Better still, we can start our hunt from NGC 6886 itself, since it lies halfway between Eta and NGC 6905. Keep NGC 6886 centered as you switch to a low-power eyepiece, and then carefully head due east. As a reference, you will pass an 8th-magnitude star in about 40'. Keep going another $1\frac{1}{2}°$ until you spot a crooked pentagon of five 7th- and 8th-magnitude stars. Can you see a tight triangle of 11th- and 12th-magnitude stars just north of the pentagon's center? NGC 6905 will look like a bluish disk along the western side of that tiny triangle.

If you try and try again without success, use the same approach to find NGC 6905 as we did for NGC

Figure 5.13 NGC 6905

6886. That is, let Earth be your GoTo telescope mount. Center on Eta, turn off the tracking motor, and set a spell. As before, in 7 minutes 12 seconds, NGC 6886 will be passing through the field, but sit tight and let more time pass. Since NGC 6905 is also due east of Eta, the Earth will turn you toward it in due time – 16 minutes to be precise.

Once it is in view, the nebula's pale blue-green color should help to set it apart from the crowd of field stars. At $150\times$, my 8-inch exposes a slightly oval disk elongated north–south (Figure 5.13). The cloud's ellipticity is confirmed in photographs, which show two "wings" extending away from the bright central core. Spotting the 14th-magnitude central star is just possible through 8- and 9.25-inch scopes under ideal conditions.

105   IC 5217

Target	Type	RA	Dec.	Constellation	Magnitude	Size	Chart
IC 5217	Planetary nebula	22 23.9	+50 58.0	Lacerta	12.6p	7″	5.14

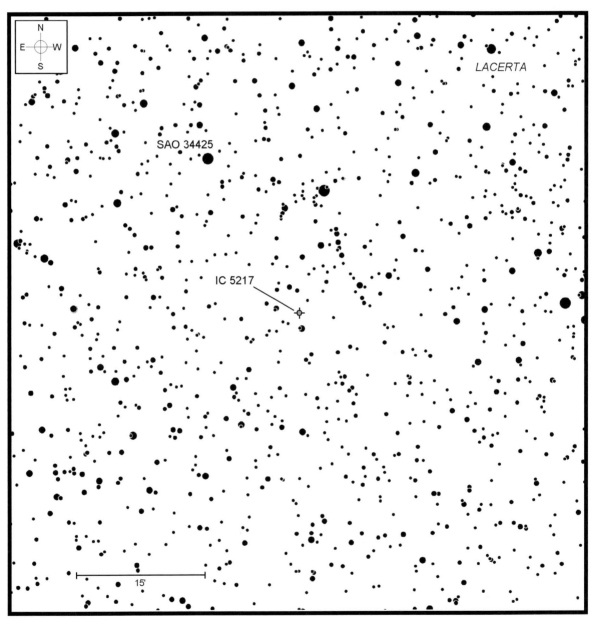

Chart 5.14 IC 5217

Figure 5.14 IC 5217

Our fall inventory kicks off with a taxing planetary nebula in an equally taxing constellation. IC 5217 lies among the faint stars of Lacerta the Lizard. Finding the nebula is a big part of the challenge because of its dim surroundings.

To spot IC 5217, use a technique offered several years ago by Walter Scott Houston in his Deep-Sky Wonders column. Begin by finding a four-star keystone asterism formed by Alpha (α), Beta (β), and 4 Lacertae, as well as 5.4-magnitude SAO 34143, about 18° east-northeast of Deneb [Alpha (α) Cygni]. The keystone measures about 3° across, small enough to fit into just about any finderscope. IC 5217 is located slightly northeast of the center of the keystone, in the same low-power field as the very red 6.5-magnitude star SAO 34425. The planetary is nestled just 3′ west of a slender isosceles triangle of 11th- and 12th-magnitude stars.

Since IC 5217 shines at fainter than 12th magnitude and measures just 7 arc-seconds across, an eyepiece switch will be needed to pick it out once its field is centered in view. Through my 6-inch Schmidt-Cassegrain telescope at 96×, the planetary was disguised as a starlike point. Increasing magnification to 240×, however, expanded the nebula's disk enough to identify it positively from the surrounding field. It appears grayish at this aperture (Figure 5.14), but will take on a faint bluish tint through 12-inch and larger telescopes. Try blinking the field with an O-III filter to enhance the nebula's disk. There is little hope of seeing the central star, since it shines at magnitude 15.5. In fact, the central star remains unseen through even the largest amateur telescopes because of the nebula's brightness. Those same instruments, however, may hint at this planetary's bipolar structure. Photographs reveal a bright equatorial ring and fainter bipolar lobes that extend north and south.

106 The Deer Lick Group

Target	Type	RA	Dec.	Constellation	Magnitude	Size	Chart
Deer Lick Group	Galaxy group	22 37.3	+34 26.1	Pegasus	–	~6′	5.15

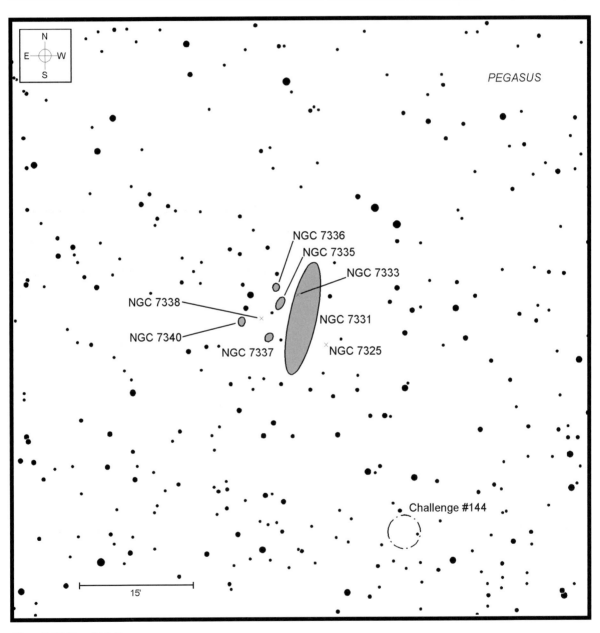

Chart 5.15 Deer Lick Group

Table 5.4 *Deer Lick Group*

	Target	Type	RA	Dec.	Constellation	Magnitude	Size
106a	NGC 7325	Double star	22 36.8	+34 22.0	Pegasus	14, 15	15″
106b	NGC 7331	Galaxy	22 37.1	+34 25.0	Pegasus	9.4	14.5′×3.7′
106c	NGC 7333	Single star	22 37.2	+34 26.0	Pegasus	15	–
106d	NGC 7335	Galaxy	22 37.3	+34 26.9	Pegasus	14.4b	1.7′×0.7′
106e	NGC 7336	Galaxy	22 37.4	+34 28.9	Pegasus	16.8	1.1′×0.4′
106f	NGC 7337	Galaxy	22 37.4	+34 22.5	Pegasus	15.2p	1.3′×0.9′
106g	NGC 7338	Double star	22 37.5	+34 24.8	Pegasus	14	–
106h	NGC 7340	Galaxy	22 37.7	+34 24.6	Pegasus	14.7p	1.1′×0.7′

The autumn sky abounds with little bundles of galaxies scattered throughout its stars. One of the best known is the group of seven galaxies that surround the magnificent spiral NGC 7331 in Pegasus, the Flying Horse. An observer could easily spend an hour or more just soaking in all that this small patch of sky has to offer.

NGC 7331 is reason enough to hunt down the group. Under dark skies, it makes a great challenge object for a pair of 50-mm binoculars. To find it, center on the star Matar [Eta (η) Pegasi]. Matar is often portrayed as one of the horse's front knees, just northwest of Sheat [Beta (β) Pegasi] in the Great Square. Look $5\frac{1}{2}°$, or about a binocular field of view, to its north-northwest for a pair of orangish 6th-magnitude stars. Drop a degree back and NGC 7331 should be centered in view.

This galaxy takes on a life of its own through backyard telescopes. At 84×, my 8-inch reflector shows an intense, perfectly circular galactic heart enveloped by the elongated whisper of the galaxy's spiral-arm halo, which is oriented north–south. Careful study through the same instrument at 119× reveals a subtle, mottled texture to the halo. As shown in Figure 5.15, a prominent dust lane along the western edge of the spiral-arm disk also comes into view and causes the core to appear slightly off-center. Can you make out any of the lane's blotchiness or hints of the spiral arms themselves that appear so prominently in photographs?

Four fainter galaxies are found just to the east of NGC 7331. Together, they collectively form the "Deer Lick Group." NGC 7331 represents the salt lick enticing the quartet of deer – NGCs 7335, 7336, 7337, and

Figure 5.15 *NGC 7331 and the Deer Lick Group*

7340. In other circles, these four are known as "the Fleas," perhaps in search of a deer to call home.

Fleas or deer notwithstanding, the most prominent of the four is NGC 7335, a tight-armed spiral spanning just 1.7′×0.7′. Look for its faint, oval blur $3\frac{1}{2}°$ to the northeast of NGC 7331. NGC 7336 is situated 2′ further north, on the other side of a 13th-magnitude field star. Spotting it through a telescope in the 6- to 9.25-inch range is difficult, if not impossible. Indeed, it puts up a valiant fight even through my 18-inch. Use averted vision to pick out its vague, elliptical form. If you have gifted vision, a top-notch telescope, and perfect conditions, you just might spy it.

Southeast of NGC 7331's core is an obtuse triangle of three 13th-magnitude field stars. If you can find them, then you have also found the home of NGC 7337, just west of the triangle's apex star. Although its spiral arms span about an arc-minute, only the galaxy's very compact core is likely to be detectable through amateur telescopes.

The easternmost deer, NGC 7340, strikes me as the second brightest. Its disk appears almost perfectly round, with just a hint of a brighter core. You'll find it just south of a second, brighter obtuse triangle of stars, this one oriented north–south.

NGC 7331 may look like the mother duck, and the others all ducklings following in line, but in reality they are at very different distances away. NGC 7331 is estimated to lie 50 million light years from the Milky Way, while NGCs 7335, 7337, and 7340 are some 300 million light years away. The last, NGC 7336, is another 100 million light years further still. But NGC 7331 may

not be a loner after all. Studies conducted in the mid 1990s indicate that NGC 7331 may be gravitationally associated with NGC 7320, a dim galaxy 5′ to the southwest. NGC 7320 is a member of a most peculiar group of five galaxies known as Stephan's Quintet. Stephan's Quintet is large-scope Challenge 144, discussed in Chapter 6.

Incidentally, if you check the original *New General Catalogue*, you will find five other entries in the immediate vicinity: NGCs 7325, 7326, 7327, 7333, and 7338. None is a distant galaxy, however. Rather, each is a misinterpreted close-set double star, a faint individual star, or perhaps a figment of a tired astronomer's imagination. Of these, NGC 7325 is the brightest of the "missing" deer, although it is nothing more than a very faint double set $4\frac{1}{2}′$ southwest of NGC 7331's core. Even its brighter component, shining at 14th magnitude, is beyond the grasp of many a backyard telescope.

107   NGC 7354

★
★

Target	Type	RA	Dec.	Constellation	Magnitude	Size	Chart
NGC 7354	Planetary nebula	22 40.3	+61 17.1	Cepheus	12.9p	36″	5.16

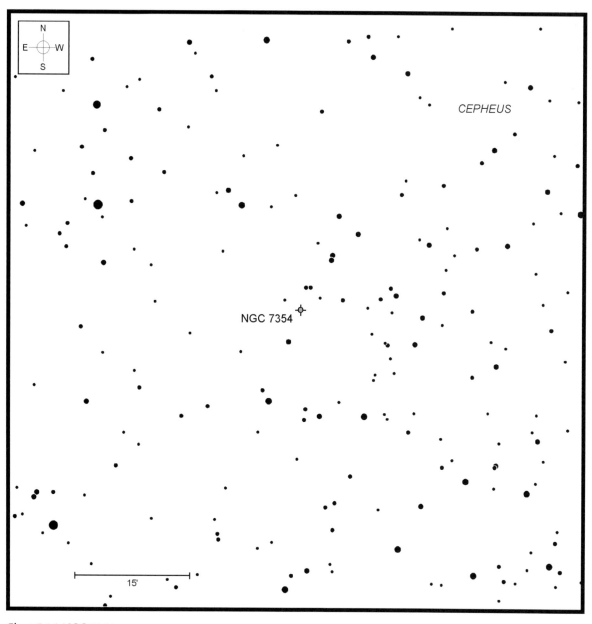

Chart 5.16 NGC 7354

Figure 5.16 NGC 7354

Of the constellations that line the autumn Milky Way, King Cepheus is trod upon by relatively few amateur astronomers. While this is most likely because the constellation's brightest stars are faint compared with his wife, Queen Cassiopeia, the King has many royal deep-sky subjects is his own right that merit a look, including this next challenge.

Discovered by William Herschel in 1787, NGC 7354 is a splendid little planetary nebula, even if it does not receive much attention. It lies among the faint stars

of southeastern Cepheus, not far from the Cassiopeia border. To find it, place the famous variable star, Delta (δ) Cephei near the southern edge of your finder's field. There, a 5th-magnitude star, 30 Cephei, should just pop into view along the field's northern edge. Center your aim about two-thirds of the way from Delta to 30, and then shift about half a degree east. NGC 7354 should be in your telescope's field, nestled between a pair of 11th-magnitude stars to its northwest and a lone 11th-magnitude sun to its southeast.

From my suburban backyard, my filterless 6-inch Schmidt-Cassegrain telescope shows NGC 7354 as a round, grayish disk at $127\times$ (Figure 5.16). The disk appears perfectly uniform and, with an apparent diameter of half an arc-minute, is clearly identifiable as nonstellar. Try the highest magnification that sky conditions and your telescope can bear for the best view. Its central star shines at only 16th magnitude.

Photographs of NGC 7354 reveal a brighter, oval inner shell encircled by a round, fainter outer shell. Flame-like FLIERs are also evident in some images. FLIERs, an acronym for "Fast, Low Ionization Emission Regions," are red in color and appear to shoot outward from planetary nebulae. Their exact cause and creation remain a mystery, but some suggest that FLIERs result from subsequent bursts of matter flung outward from the central star after the planetary itself formed. Given their high rates of speed, it seems certain that, whatever their cause, FLIERs are created independently from, and are formed after, the more slowly expanding planetary nebula.

108 NGC 7537 and NGC 7541

	Target	Type	RA	Dec.	Constellation	Magnitude	Size	Chart
108a	NGC 7537	Galaxy	23 14.6	+04 29.9	Pisces	13.9b	2.2′×0.5′	5.17
108b	NGC 7541	Galaxy	23 14.7	+04 32.0	Pisces	12.4b	3.5′×1.2′	5.17

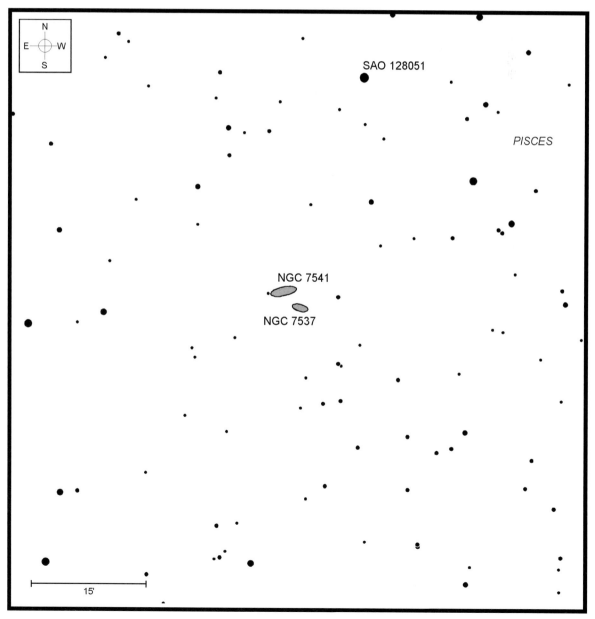

Chart 5.17 NGC 7537 and NGC 7541

NGC 7537 and NGC 7541 are two of the more visually interesting galaxies that hide among the faint stars of Pisces, the Fishes. Only 3 arc-minutes separate these nearly edge-on spirals, creating an attractive pair of faint fuzzies that float in a field of dim stars.

Both galaxies are nestled in the far southwestern corner of Pisces, not far from 4th-magnitude Gamma (γ) Piscium, the brightest star in the Circlet of Pisces asterism. The Circlet, a five-star pentagon marking the constellation's western fish, can be found south of the Great Square of Pegasus. Depending on your local light-pollution level, you might need some optical help to make out the Circlet; spanning 6° side-to-side, it should just cram into the field of most finderscopes.

Aim your telescope toward yellowish Gamma and then look about 2° to its northwest for a wide pair of 7th-magnitude stars. The western star of the pair shines with a golden glint, while the eastern star (SAO 128051) has a bluish tinge. Our challenge is found about three-quarters of the way between Gamma and the pair's eastern sun. Look for the galaxies just east of an 11th-magnitude field star.

NGC 7541 is the big kid on the block, appearing brighter and larger than its neighbor. My 8-inch reflector at 118× reveals a long, very thin spike of light oriented almost exactly east–west. A 12.5-magnitude Milky Way star shines just beyond the galaxy's eastern tip. Don't be fooled into thinking you've discovered a supernova if you spot it.

Back in 1998, however, a supernova was spotted in NGC 7541 by astronomers using the Katzman Automatic Imaging Telescope, or KAIT. The KAIT, part of the Lick Observatory complex on Mount Hamilton near San Jose, California, is a fully automated 30-inch Cassegrain reflector designed specifically to search for extragalactic supernovae and other transient phenomena. Since dedicated in 1996, KAIT has been the backbone of the Lick Observatory Supernova Search, discovering nearly 1,000 supernovae – indeed, when you read this, that remarkable number may have been exceeded. NGC 7541's supernova, designated SN1998dh, only reached 15th magnitude, keeping it well below the threshold of visibility for all but the largest amateur instruments.

Subsequent studies showed that SN1998dh was a Type Ia supernova. In very broad terms, supernovae can be divided into two categories, Type I and Type II, with

Figure 5.17 NGC 7537 and NGC 7541

the former further refined into Types Ia, Ib, and Ic. Type Ia supernovae occur in binary star systems where one of the system's members is a white dwarf star. White dwarfs are the end result of a star that has used up all of its fuel and has collapsed under the crushing influence of gravity. If a white dwarf lies close enough to a partner star in a binary system, mass may be transferred between the two. As material is shifted from the companion star to the white dwarf, the increase in pressure and density raises the temperature of its core. If enough mass is transferred, then the white dwarf may exceed its maximum mass limit, the *Chandrasekhar limit*, of about 1.4 times the current mass of our Sun. When that happens, the crushing effect is overwhelming, causing the white dwarf to detonate violently.

NGC 7537 lies in the same field as NGC 7541, but its smaller size and fainter nature presents us with a greater challenge. Indeed, when I first spotted NGC 7541 through my 8-inch reflector many years ago under only mediocre skies, NGC 7537 was nowhere to be found. Some time later under better viewing conditions, the same telescope detected the missing galaxy as an exceedingly faint oval glow. NGC 7537 is also oriented east–west, tilted slightly askew to its more prominent neighbor.

109 STF 3057

Target	Type	RA	Dec.	Constellation	Magnitude	Separation	Chart
STF 3057	Binary star	00 04.9	+58 32.0	Cassiopeia	6.7/9.3	3.9″	5.18

110 STF 3062

Target	Type	RA	Dec.	Constellation	Magnitude	Separation	Chart
STF 3062	Binary star	00 06.3	+58 26.2	Cassiopeia	6.4/7.3	1.3″	5.18

111 Lambda Cassiopeiae

Target	Type	RA	Dec.	Constellation	Magnitude	Separation	Chart
Lambda Cas	Binary star	00 31.8	+54 31.3	Cassiopeia	5.3/5.6	0.3″	5.18

The Rayleigh Criterion and Dawes' Limit were formulated to predict just how close a pair of stars could be resolved, at least partially, through a given aperture. Using the formulae discussed in Chapter 1, a 6-inch aperture should be able to discern the duality of a binary star with components separated by 0.91″ (Rayleigh) and 0.76″ (Dawes). At the other end of this chapter's aperture range, a 9.25-inch instrument should be able to resolve, at least partially, a pair of stars separated by 0.59″ and 0.49″, respectively. These are ideal numbers based a pair of 6th-magnitude yellow stars. In the case of Dawes' Limit, these values predict how close those stars can be to each other, and appear elongated, but not necessarily resolved separately.

Can your telescope meet the Rayleigh and Dawes challenge? Here are three pairs of stars that will prove worthy adversaries for 6- to 9.25-inch telescopes. All lie in fairly close proximity to each other and offer a broad range of separation distances. Let's see how well you do.

First up are Struve 3057 (often abbreviated as Σ 3057 or STF 3057) and Struve 3062 (STF 3062), both

from Friedrich Georg Wilhelm von Struve's monumental catalog of binary stars published in 1827.[4] Both lie 50′ southwest of Caph [Beta (β) Cassiopeiae], the westernmost star in the constellation's familiar five-star W pattern. Together, Caph and the two doubles form a prominent isosceles triangle that's easy to find through finderscopes.

STF 3057, at the triangle's northwestern corner, presents a challenge not because of its close-set stars – they are separated by nearly 4 arc-seconds – but rather because of the disparity in their magnitudes. The primary sun shines at magnitude 6.7, but its companion shines at only magnitude 9.3.

Marking the triangle's southern corner, STF 3062 presents a different sort of test, one more in line with Dawes' original concept. Here, we find two stars

---

[4] Friedrich Georg Wilhelm von Struve was the first astronomer to search for and study binary stars. He compiled those studies into a catalog published in 1827 entitled *Catalogus Novus Stellarum Duplicium*, or simply the *Dorpat Catalogue*.

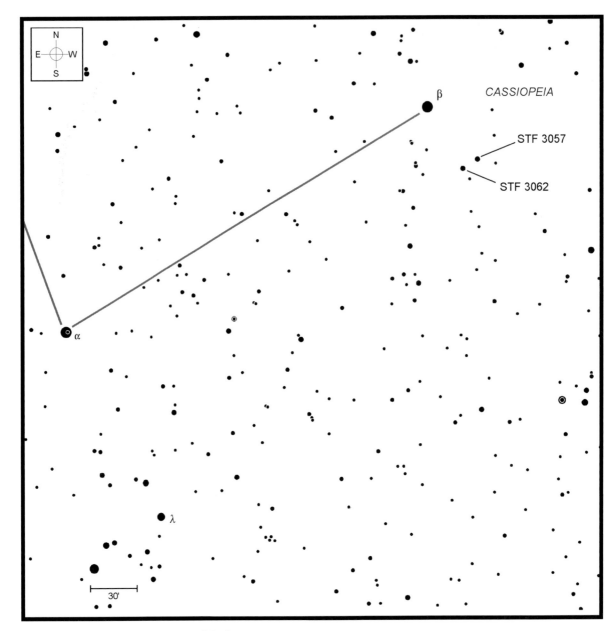

Chart 5.18 STF 3057, STF 3062, Lambda Cas

shining at magnitudes 6.4 and 7.3, and separated by 1.3″. Given reasonably good seeing conditions and optical quality, a 6-inch should be able to resolve STF 3062 fairly handily (Figure 5.18); indeed, STF 3057 might prove more difficult.

Looking for tougher game? Take aim at Lambda (λ) Cassiopeiae (STT 12, for its listing in Otto Struve's

catalog[5]). Lambda's components shine at magnitudes 5.3 and 5.6, a little brighter than Dawes' ideal test star. The resulting glare makes this test all the harder. These

---

[5] Otto Struve was the great-grandson of Friedrich Georg Wilhelm von Struve, and created the *Pulkowo Catalog* of binary stars as an expansion of Friedrich's earlier work.

Figure 5.18 STF 3062

two blue-white main-sequence stars orbit a common center of mass once every 640 years and are currently separated by only 0.3 arc-seconds. If you own an 8-inch or larger scope, Lambda Cas is a formidable opponent indeed.

112   IC 10

Target	Type	RA	Dec.	Constellation	Magnitude	Size	Chart
IC 10	Galaxy	00 20.3	+59 18.1	Cassiopeia	11.8b	6.3′×5.1′	5.19

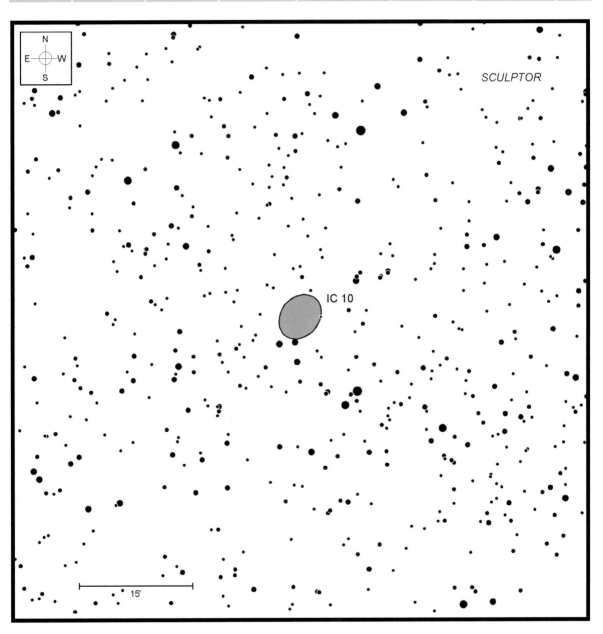

Chart 5.19 IC 10

I debated where to include this next challenge object. Should it be in this chapter or perhaps in the next chapter, among the challenges for 10- to 14-inch telescopes? I finally decided to keep it here, but with a disclaimer. Spotting IC 10, an outlying member of our Local Group of galaxies, takes an especially dark, transparent sky regardless of aperture. If you can't see M33 naked eye (at least a "Bortle 4" sky on John Bortle's scale detailed in Chapter 1), then I suspect this galaxy will jump an aperture class. But let's not give up so easily – at least, not just yet.

Despite its prominent position within the constellation Cassiopeia, IC 10 was missed by Messier, Méchain, and the Herschels. Instead, American astronomer Lewis Swift, famous more for his comet observations than uncovering new deep-sky objects, discovered IC 10 from Warner Observatory in Rochester, New York, in 1889. That was a year too late to be added to John Dreyer's *New General Catalogue*, but he subsequently included it in the first supplement to the NGC, volume 1 of the *Index Catalogue*. However, its portrayal in the *Index Catalogue* was all wrong. IC 10 was described there as a "faint star involved in extremely faint and very large nebula." Forty years would pass before IC 10's suspected extragalactic nature came to light. Edwin Hubble, who once described IC 10 as "one of the most curious objects in the sky," later proposed that it might be a member of the Local Group, along with the Milky Way and the Andromeda Galaxy. It would be another three decades before Hubble's suspicions could be confirmed.

We now know that IC 10 is a dwarf irregular galaxy, resembling in many ways the Milky Way's Large Magellanic Cloud. Unlike the Large Magellanic Cloud, however, IC 10 is gravitationally bound to the Andromeda Galaxy. Closer studies also show that IC 10 has more Wolf-Rayet stars than all other dwarf galaxies in the Local Group combined. Wolf-Rayet stars are extremely hot blue stars that are losing mass at a furious rate. Why IC 10 contains such a disproportional number of these rare stars is a question that remains unanswered.

IC 10 lies 1.4° due east of star Caph [Beta (β) Cassiopeiae], the westernmost point in the Cassiopeia W and the same starting point as the last challenge. Scanning there, look for a small right triangle of three

Figure 5.19 IC 10

10th-magnitude stars. IC 10 is just north of the triangle's right angle.

Once aimed its way, be sure to use a moderately low power eyepiece, say, between 75× and 100×. One problem with spotting IC 10 is its surroundings. Remember, we're looking through the plane of the Milky Way when we strain to see IC 10. There are an awful lot of stars to push through before IC 10 comes into view.

The galaxy's low surface brightness only compounds the problem. After viewing it from under very dark skies through my 8-inch at 84×, I described the IC 10 as "extremely faint, visible weakly only with averted vision; surprisingly large, but with soft edges that make it difficult to tell where the galaxy ends and the background sky takes over; slightly oval, oriented approximately east–west." Figure 5.19 reflects that observation. Again, sky conditions are critical. Notes made through my 18-inch reflector under 5th-magnitude suburban skies recall simply "a dim undefined glow just above the background" at 121×.

Can you see IC 10 through your 9.25-inch telescope? An 8-inch telescope? Smaller?! If so, you have great vision, a great telescope, and pretty good skies to boot.

113   Sculptor Dwarf Galaxy

Target	Type	RA	Dec.	Constellation	Magnitude	Size	Chart
Sculptor Dwarf Galaxy	Galaxy	01 00.2	−33 43.3	Sculptor	10.5b	39.8′×30.8′	5.20

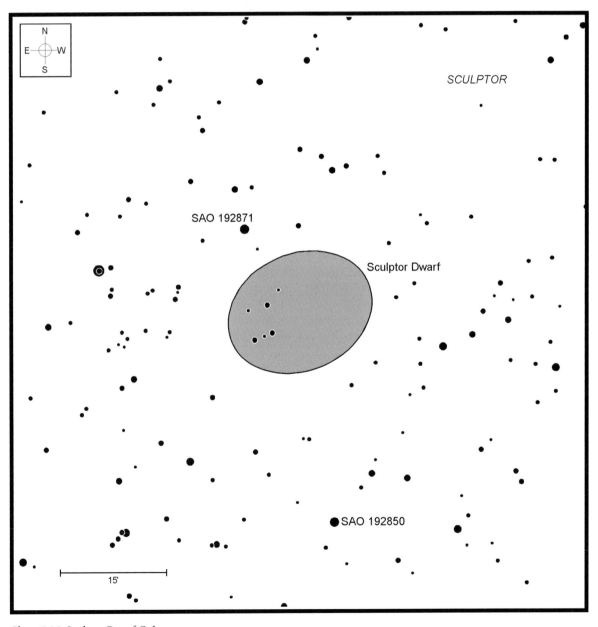

Chart 5.20 Sculptor Dwarf Galaxy

It's probably reasonable to expect that a galaxy comparatively close to the Milky Way will appear big and bright in our sky. In fact, this "rule" is violated by most "nearby" galaxies in our Local Group. While the Andromeda Galaxy, the Triangulum Spiral, and the Magellanic Clouds meet this expectation, other members of the gang do not, as we just saw when looking for the last challenge, IC 10.

As tough as IC 10 is, this next test is even tougher. The Sculptor Dwarf Galaxy spans a huge area measuring $40' \times 31'$. It also has an integrated magnitude of 10.5, which sounds reasonably bright. You've seen much fainter objects through your telescope, believe me. So, from those statistics, we might expect that this galaxy would be simple enough to spot through a 6-inch telescope, or even smaller instruments equipped with wide-field eyepieces. But in fact, it is not. Indeed, the Sculptor Dwarf Galaxy, cataloged officially as ESO 351-30, is one of the toughest challenges in this book regardless of aperture. If you're able to spot this one, you are in the big leagues among observational astronomers!

The Sculptor Dwarf was not discovered until March 1938,[6] when American astronomer Harlow Shapley noticed its dim glow on photographic plates taken at Harvard's Boyden Observatory in South Africa. In his discovery announcement, Shapley accurately said "nothing quite like it is now known," a true statement since he had just discovered the first dwarf spheroidal (dSph) galaxy. Dwarf spheroidal galaxies are similar in appearance to the more common dwarf elliptical galaxies in that they possess little evidence of interstellar gas and dust and show no signs of recent star formation.

An even-dozen dwarf spheroidal galaxies are now known to be gravitationally associated with the Milky Way, including two challenges in Chapter 7, Leo I and Leo II (monster-scope Challenges 158 and 159, respectively). Once interpreted as large, low-density globular clusters, dwarf spheroidal galaxies are now known to contain a more diverse stellar population. Like globulars, we find an abundance of old stars. But, unlike globular clusters, we also find stars of more intermediate ages.

Another distinct difference is their mass-to-luminosity ratios. By studying the motion of individual stars within a galaxy, astronomers can

Figure 5.20 Sculptor Dwarf Galaxy

calculate the total mass of that galaxy. Doing so with the Sculptor Dwarf and others of the type reveals that these galaxies are far more massive than can be accounted for by their stars alone. There has to be hidden mass – invisible mass – lurking somewhere inside. The presence of this invisible mass, known popularly as dark matter, is often cited as one of the primary differences between dwarf spheroidals and globular clusters.

Finding the dark matter in the Sculptor Dwarf is beyond present-day technology. Indeed, finding the galaxy itself is tough enough. Not only does Sculptor lie low in the southern sky for most us at mid-northern latitudes, but the constellation's brightest star, Alpha (α) Sculptoris, shines at only 4th magnitude. To find it, extend the eastern side of the Great Square of Pegasus southward for 34° to find 2nd-magnitude Deneb Kaitos [Beta (β) Ceti], and then another 12° south-southeast to Alpha Sculptoris. Once there, scan another 2° southward to 6th-magnitude Sigma (σ) Sculptoris. The Sculptor Dwarf is another 2° south, set between two 8th-magnitude stars, SAO 192871 half a degree to its northeast and SAO 192850 twice the distance to its south.

With an apparent diameter larger than the Moon's, the Sculptor Dwarf Galaxy hardly lives up to its tiny name. To squeeze it into view, you will need a low power and a wide field, but that also creates a problem.

[6] "A Stellar System of a New Type," *Harvard College Observatory Bulletin*, No. 908, March 1938.

Owing to the galaxy's tenuous structure, its surface brightness is much lower than its integrated magnitude. If the sky background at your observing site is at all impacted by light pollution, the galaxy's surface brightness is probably so close to the surroundings that it will be invisible. Only with very dark skies will its subtle glow be brighter than the sky itself.

My sole view of the Sculptor Dwarf came under naked-eye limiting magnitude 6.0+ skies one special autumn night a few years ago. Peering through my 8-inch reflector at 56×, I could make out its faint glow barely just to the west of a triangle of 10th-magnitude stars. Figure 5.20 is based on that observation, but image contrast has been necessarily enhanced for the galaxy to show in print. The word "subtle" doesn't begin to describe how understated this galaxy appeared. Its dim glimmer finally registered on my eye's retina only after combining averted vision with very gently tapping the side of the telescope, a technique discussed in Chapter 1. Try it with this object; it works!

114   NGC 1343

Target	Type	RA	Dec.	Constellation	Magnitude	Size	Chart
NGC 1343	Galaxy	03 37.8	+72 34.3	Cassiopeia	13.5p	2.5′×1.5′	5.21

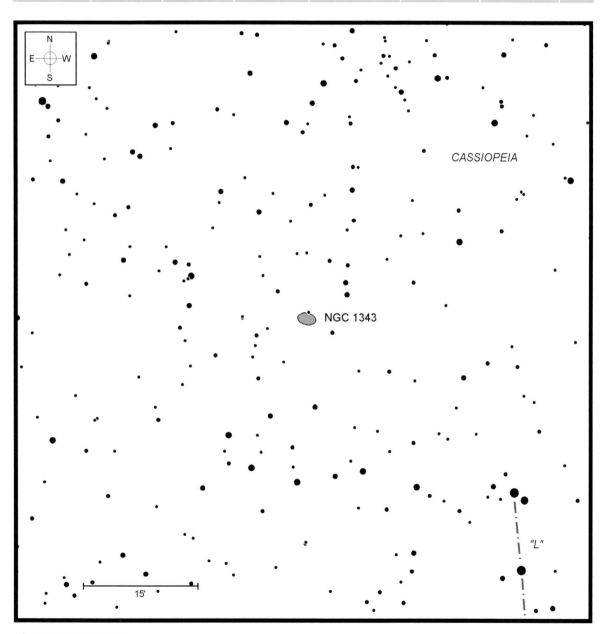

Chart 5.21 NGC 1343

Cassiopeia is most famous among deep-sky observers for the many open star clusters scattered within its borders. It is hardly known as a Mecca of extragalactic treasures, yet concealed among its rich star clouds are several galaxies to test our powers of observation. One is NGC 1343, a little known and little observed galaxy just tucked inside the constellation's extreme northeastern corner.

To find NGC 1343, extend a line from Ruchbah [Delta (δ) Cassiopeiae] to Segin [Epsilon (δ) Cassiopeiae], the two easternmost stars in the Queen's W pattern and continue twice that distance toward the northeast. Aim there to find a pair of 5th-magnitude stars, Gamma (γ) Camelopardalis and SAO 5000. From Gamma Cam, head northwest back into Cassiopeia, and keep your eyes peeled for an asterism of 8th-magnitude stars shaped like a capital "L." Switch to your telescope and scan northeastward of the star at the top of the L for 36′ to NGC 1343.

My first meeting with this odd little galaxy took place more than two decades ago. Notes made through my 8-inch reflector recall a better than average night, with a naked-eye limiting magnitude of 6.0. At 142×, NGC 1343 displayed a small, diffuse glow with a faint stellar nucleus buried within (Figure 5.21). A faint double star, with 13.0- and 14.3-magnitude components separated by 15′, could be seen just off its north-northwest.

Long-exposure photographs reveal that NGC 1343 has a faint ring of material surrounding the galactic core. Many suggest that this odd feature resulted from a galactic collision, with a smaller galaxy striking nearly perpendicular to the plane of the larger galaxy. The collision caused gravitational tidal forces to generate a

Figure 5.21 NGC 1343

ring around the galaxy's nucleus. To illustrate this, imagine dropping a small stone into a pond. As the stone strikes the water, concentric waves are sent rippling away from the point of impact. Others believe that high-resolution photographs with the Hubble Space Telescope show that the ring may actually be spiral arms wrapped very tightly around the core. This faint, oval-shaped halo is beyond the grasp of my 8-inch scope, and even evades my 18-inch from my suburban backyard. Readers with access to large apertures under dark skies, however, should be able to catch a hint of it, elongated roughly 2-to-1 and oriented east–west.

115 Abell Galaxy Cluster 373

Target	Type	RA	Dec.	Constellation	Magnitude	Size	Chart
AGCS 373	Galaxy cluster	03 38.5	−35 27.0	Fornax	–	180′	5.22

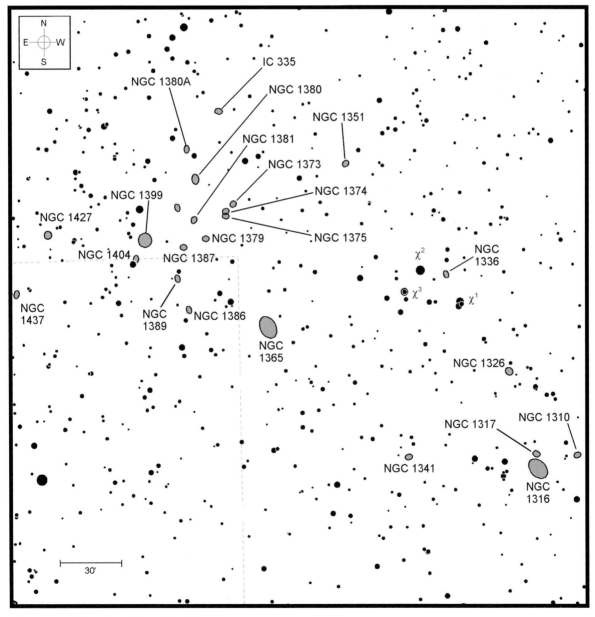

Chart 5.22 Abell Galaxy Cluster (South) 373

Nestled in the southeast corner of the dim early winter constellation Fornax, adjacent to the distinctive triangle formed by 6th-magnitude Chi-1 (χ 1), Chi-2 (χ 2), and Chi-3 (χ 3) Fornacis, is an attractive cluster of galaxies known as Abell Galaxy Cluster – Southern Supplement (AGCS) 373. In addition to his research that led to the discovery of more than 80 new planetary nebulae in the 1950s, George Abell also examined the overall structure of the universe. He did so by studying and cataloging 2,712 galaxy clusters that had been captured on the then-new National Geographic Society–Palomar Observatory Sky Survey taken with the 48-inch Samuel Oschin Schmidt camera at Palomar Observatory. In 1958, he published the results of his study as a paper entitled "The Distribution of Rich Clusters of Galaxies" in the *Astrophysical Journal Supplement*.[7] Although Abell died in 1983, his catalog was expanded in 1989 by Harold Corwin and Ronald Olowin with the publication of their article "A Catalog of Rich Clusters of Galaxies," again in the *Astrophysical Journal Supplement*.[8] The Corwin/Olowin addendum to Abell's original catalog encompasses a total of 4,073 rich galaxy clusters.

Also known informally as the Fornax Galaxy Cluster, AGCS 373 lies nearby, as galaxy clusters go, at an estimated distance of 62 million light years. At least 18 of its members are within range of 6- to 9.25-inch telescopes.

Let's start with two of the most interesting. In 1966, American astronomer Halton Arp published his monumental *Atlas of Peculiar Galaxies*, a photographic survey of oddball galaxies that he made with the 200-inch Hale reflecting telescope and the 48-inch Schmidt telescope at Palomar between 1961 and 1966. The Arp catalog's 338 entries comprise a fascinating collection of interacting and merging galaxies. Most of the Arps are in the realm of large and giant backyard scopes, although a dozen entries involve Messier objects.[9]

Arp 154 involves two of the galaxies within the Fornax cluster. NGC 1316, at 9th magnitude the brightest galaxy in the bunch, is set in the group's

western suburbs. Deep photographs reveal that NGC 1316 contains many dust clouds and is surrounded by a complex envelope of faint material, several loops of which appear to engulf a smaller galaxy, NGC 1317, 6′ to the north. Astronomers consider this to be a case of galactic cannibalism, with the larger NGC 1316 devouring its smaller companion. The merger is further signaled by strong radio emissions being telegraphed from the scene.

In my 8-inch reflector, NGC 1316 appears as a bright, slightly oval disk with a distinctly brighter nucleus. NGC 1317, about 12th magnitude and 2′ across, is visible in a 6-inch scope, although averted vision may be needed to pick it out. Try about 150× for the best view.

With NGC 1317 centered in your field, turn off your telescope drive and wait 5 minutes. The Earth will turn your view eastward to NGC 1341, a challenging 12th-magnitude barred spiral. I could only see it fleetingly with my 8-inch reflector from a dark site on the south shore of Long Island, New York. Its featureless disk, only $1\frac{1}{2}′$ long, is just north of a faint star.

There is another barred spiral, NGC 1326, about halfway between NGC 1316 and Chi-l. In the 8-inch, it appeared as an 11th-magnitude oval smudge visually measuring about 2′ long and half as wide. It also has a stellar nucleus centered within.

The heart of the Fornax cluster lies at right ascension 03h 38.5m, declination −35° 27.0′, halfway between the Chi Fornacis triangle to the west and Sigma (σ) Eridani to the east. A telescope with a 1° field aimed toward this position will embrace eight galaxies brighter than 14th magnitude, with 10th-magnitude elliptical NGC 1399 lying dead center (Figure 5.22). This galaxy, set 15′ south of a 7th-magnitude field star, appears as a perfectly round glow 2′ in diameter with a brighter nucleus.

NGC 1404, another 10th-magnitude elliptical, is just 10′ south and slightly east of NGC 1399. I found it slightly oval with its long axis stretching 2′. Like many of the galaxies in this cluster, NGC 1404 has a brighter nucleus. Another 20′ west-southwest of NGC 1399 lies NGC 1387. I was surprised at how bright this spiral appeared in my 8-inch, considering it is listed as magnitude 11.7 in blue light.

NGC 1389 is 14′ south and a bit east of NGC 1387, just over the border into Eridanus. Averted vision is a must for any scope smaller than 10 inches. Most references label this 2′-long elliptical as 12th

[7] George O. Abell, "The Distribution of Rich Clusters of Galaxies," *Astrophysical Journal Supplement*, Vol. 3 (1958), p. 211.
[8] George O. Abell, Harold G. Corwin Jr., and Ronald P. Olowin, "A Catalog of Rich Clusters of Galaxies," *Astrophysical Journal Supplement Series*, Vol. 70 (1989), pp. 1–138.
[9] M32 (Arp 168), M49 (Arp 134), M51 (Arp 85), M60 (Arp 116), M65 (Arp 317), M66 (Arp 16 and Arp 317), M77 (Arp 37), M82 (Arp 337), M87 (Arp 152), M90 (Arp 76), and M101 (Arp 26).

Figure 5.22 NGC 1399

Table 5.5 *Members of AGCS 373 (bold entries are discussed in the text)*

Object	RA	Dec.	Magnitude	Size
NGC 1310	03 21.1	−37 06.1	12.6b	1.9′×1.5′
NGC 1316	**03 22.6**	**−37 12.8**	**9.4b**	**11.1′×7.2′**
NGC 1317	**03 22.7**	**−37 06.2**	**11.9b**	**2.5′×2.2′**
NGC 1326	**03 23.9**	**−36 27.9**	**11.4b**	**3.9′×2.8′**
NGC 1336	03 26.5	−35 42.8	13.1b	2.1′×1.4′
NGC 1341	**03 28.0**	**−37 09.0**	**12.3**	**1.5′×1.2′**
NGC 1351	03 30.6	−34 51.2	11.5	2.8′×1.7′
NGC 1350	03 31.1	−33 37.7	11.2b	5.8′×2.7′
NGC 1365	**03 33.6**	**−36 08.3**	**10.3b**	**11.3′×6.6′**
NGC 1373	**03 35.0**	**−35 10.3**	**13.3**	**1.1′×1.0′**
NGC 1374	**03 35.3**	**−35 13.6**	**11.1**	**2.5′×2.4′**
NGC 1375	**03 35.3**	**−35 16.0**	**12.4**	**2.2′×0.9′**
IC 335	03 35.5	−34 26.8	12.9p	2.5′×0.6′
NGC 1379	**03 36.1**	**−35 26.5**	**11.8b**	**2.3′×2.3′**
NGC 1380	**03 36.4**	**−34 58.6**	**10.9b**	**4.8′×2.7′**
NGC 1381	**03 36.5**	**−35 17.7**	**11.5**	**2.3′×0.7′**
NGC 1380A	03 36.8	−34 44.4	12.4	2.6′×0.8′
NGC 1386	**03 36.8**	**−36 00.0**	**12.1b**	**3.4′×1.3′**
NGC 1387	**03 37.0**	**−35 30.4**	**11.7b**	**2.8′×2.6′**
NGC 1389	**03 37.2**	**−35 44.8**	**12.4b**	**2.2′×1.3′**
NGC 1399	**03 38.5**	**−35 27.0**	**10.6b**	**6.9′×6.4′**
NGC 1404	**03 38.9**	**−35 35.6**	**10.0**	**3.4′×3.0′**
NGC 1427	03 42.3	−35 23.6	11.8b	3.6′×2.4′
NGC 1437	03 43.6	−35 51.2	12.4b	2.9′×1.9′

magnitude, but I estimate it to be a half magnitude fainter. Somewhat brighter and larger is the elliptical NGC 1386, found about 15′ south-southwest of NGC 1389 and set a bit deeper into Eridanus. At 120×, I saw its brighter nucleus.

Moving back to NGC 1387, you can sweep 14′ north-northwest to NGC 1381, a faint, 12th-magnitude, cigar-shaped elliptical 2′ long. About the same distance west-northwest of NGC 1387 is another 12th-magnitude galaxy, NGC 1379. It appears as a circular glow about 1′ across.

NGC 1374 measures just 2 in diameter, but should be visible in a good 4-inch scope. Can you see NGC 1375 located just 2′ to its south? It's a magnitude fainter, so a 6-inch might be required. A third, fainter smudge is an equal distance to the north of NGC 1374. That's NGC 1373, a tough target in a 6-inch.

A 4-inch, however, should show the long, thin disk of NGC 1380 to the northeast of the NGC 1374 trio. Because of its distinctive lenticular shape, this 11th-magnitude galaxy is an intriguing target for astrophotographers.

Another object with a pronounced shape is the barred spiral NGC 1365, which, at 10th magnitude, is

the third brightest member of the Fornax cluster. Visually it appears as an oval nebulous patch that grows steadily brighter toward its center. Photographically it is one of the most impressive examples of a barred spiral south of the celestial equator. It has long curving arms that extend north and south from a pronounced central bar running east and west.

Table 5.5 lists these as well as several other galaxies in the area that are within range of amateur telescopes. Be sure to pay homage to each.

Target	Type	RA	Dec.	Constellation	Magnitude	Size	Chart
Jonckheere 320	Planetary nebula	05 05.6	+10 42.4	Orion	12.9p	26″×14″	5.23

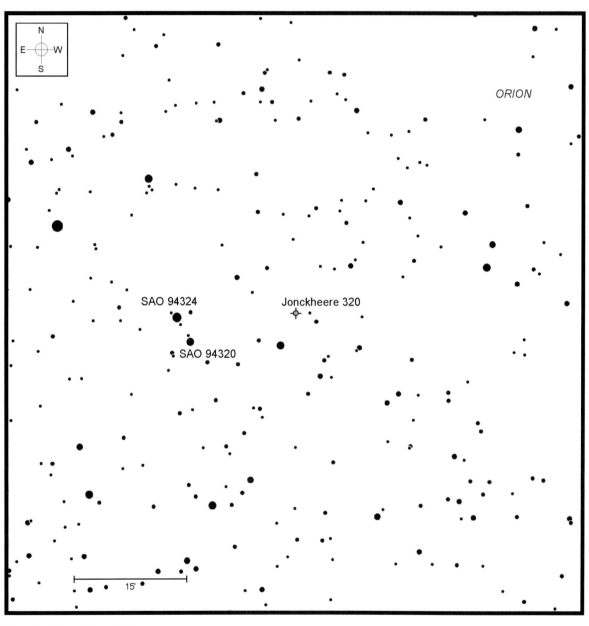

Chart 5.23 Jonckheere 320

On January 22, 1916, while revisiting some of the stars in his catalog of double stars,[10] French astronomer Robert Jonckheere returned to a vague double in Orion, which he had previously designated as entry number 320. Jonckheere was later to write of this encounter through the 28-inch refractor at the Royal Greenwich Observatory: "I noticed that the object I have catalogued as J 320 is not a double star, but . . . it appears with the larger instrument to be an extremely small bright elongated nebula. As is the case with J 900 [small-scope Challenge 73], this object also appears to be new as a nebula."[11]

Cross-listed as PK 190-17.1 in Perek and Kohoutek's comprehensive catalog of planetary nebulae, J 320 is in northern Orion, 7° northwest of Bellatrix [Gamma (γ) Orionis], the Hunter's western shoulder. To get there, first hop 6° due west from Bellatrix to 5th-magnitude 16 Orionis, and then continue another 2° further northwest to a pair of 8th-magnitude suns, SAO 94320 and 94324. You'll find the nebula lying just 15′ to their west and just 5′ northwest of a 9th-magnitude field star.

That is, you *should* see it there. That's the rub. J 320 shines at about 12th magnitude, which is bright enough to be seen in an 8-inch telescope trapped under the veil of suburban light pollution. But there are so many stars in the same field that picking out which one is the planetary is a tough job. J 320 only measures 26″ × 14″ across, and indeed is easy to mistake for a close-set double star if viewed at low power, as Jonckheere probably did during his initial discovery. Again, you may need to flash the planetary, if you'll pardon the phrase, by holding a narrowband or oxygen-III filter between your eye and eyepiece. Doing so will suppress the field stars, but not the planetary. The culprit will have no choice but to surrender.

My notes made through my 8-inch reflector at 56× recall a small, extended object that indeed looked like a pair of close-set stars on the brink of resolution. Switching to 203×, however, quickly dispelled that notion. The planetary's disk, though still fuzzy, was easy

Figure 5.23 Jonckheere 320

to tell apart from a double star, with two egg-shaped spheres just touching each other tangentially (Figure 5.23). Its central star shines at magnitude 14.4, but evades detection even through my 18-inch, perhaps masked by the nebula's high surface brightness.

Photographs reveal that J 320's ellipticity reflects its lobed structure, which resembles a butterfly in flight. A study[12] conducted with the Hubble Space Telescope's Wide Field Planetary Camera (WFPC2) in 2003 revealed a more complex internal structure than the typical bipolar planetary. Two pairs of bipolar lobes are clearly visible extending from the nebula's core. One is approximately aligned north–south and the other southeast–northwest. In addition, the Hubble images uncover two pairs of faint knots just outside the center of the nebula. As a result of this complex morphology, J 320 has been classified as not just a bipolar planetary, but rather as an example of a much less common breed known as a polypolar planetary. Try to say that three times fast!

[10] *Catalogue and Measures of Double Stars discovered visually from 1905 to 1916 within 105° of the North Pole and under 5″ Separation,* (Royal Astronomical Society, 1917)
[11] R. Jonckheere, "A New Stellar Nebula," *The Observatory,* Vol. 39 (March 1916), pp. 134–5.

[12] D. J. Harman, M. Bryce, J. A. López, J. Meaburn, and A. J. Holloway, "J320 (PN G190.3–17.7) as a poly-polar planetary nebula surrounded by point-symmetric knots," *Monthly Notices of the Royal Astronomical Society,* Vol. 348, No.3, pp. 1047–54.

117   IC 418

Target	Type	RA	Dec.	Constellation	Magnitude	Size	Chart
IC 418	Planetary nebula	05 27.5	−12 41.8	Lepus	10.7p	12″	5.24

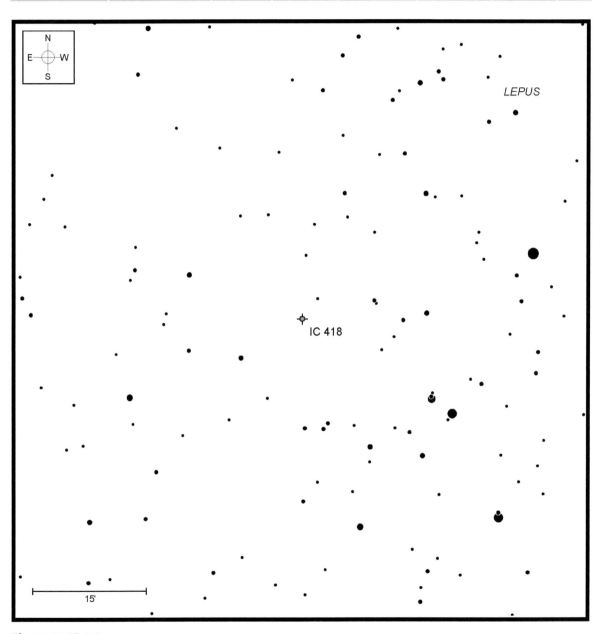

Chart 5.24 IC 418

If you have ever glanced at a compendium of images taken with the Hubble Space Telescope, then chances are you have seen this next challenge. You may not know it by its catalog number, IC 418, but instead may recognize it by its nickname, the Spirograph Nebula. Those Hubble images show an amazingly complex cloud of entangled filaments that create a strange, oval cloud that looks like it could have been drawn using a child's Spirograph toy. Remember those? You would trace intertwining arcs by rolling a color pen in a circle along the inside or outside of another circle.

Despite its disk shining at magnitude 10.7, IC 418 remains an underappreciated target among amateur astronomers. Why, I don't know. It might be that there is just so much to look at in the winter sky that nobody pays much attention to a planetary nebula that even the Herschels missed during their sky surveys. That could lead some to think that any planetary listed in the *Index Catalogue* is probably so difficult to see that they don't even try. Too bad, because they are missing a nice catch.

To find IC 418, drop 4° southward from Rigel [Beta (β) Orionis] to a keystone of four 4th- and 5th-magnitude stars – Iota (ι), Kappa (κ), Lambda (λ), and Nu (ν) Leporis. By tracing a line from Iota southeastward through Nu, and continuing that line an equal distance farther beyond, you'll come right to the field of IC 418. Look for a close-set pair of 12th-magnitude stars lying just 10′ to its west-northwest.

Once you have the nebula in view, switch to as high a magnification that seeing conditions will allow for the best view. Notes jotted down several years ago at the eyepiece of my 8-inch reflector at 203× evoke memories of a "small, bright disk, perhaps a greenish-gray, surrounding an obvious central star. Although clearly not a ring, averted vision suggests a darker central area adjacent to the central star, just to its north and south." The sketch in Figure 5.24 was also made at the time.

Figure 5.24 IC 418

What color is IC 418? In its most famous Hubble rendition, IC 418 shows off a burnt orange edge fading into a purplish disk. A blue inner disk surrounds the white hot progenitor star buried within. That color, however, is false, induced to accent subtle contrasts in the Spirograph-like structure.

How about you? What color do you see when you look at IC 418? Observers seem to disagree. Some, like me, see a gray disk with just a hint of a greenish tinge. Others recall a pinkish or reddish tint. The issue appears to boil down to aperture. The larger the instrument, the more distinct IC 418's ruddy hue. Magnification also plays a role. While higher magnifications are needed to see the planetary's disk, they tend to dilute any coloring. To see the reddish or pinkish effect that has led to IC 418's alternate nickname, the Raspberry Planetary, stick to magnifications below about 175×.

118   NGC 1924

Target	Type	RA	Dec.	Constellation	Magnitude	Size	Chart
NGC 1924	Galaxy	05 28.0	−05 18.6	Orion	13.3b	1.5′×1.1′	5.25

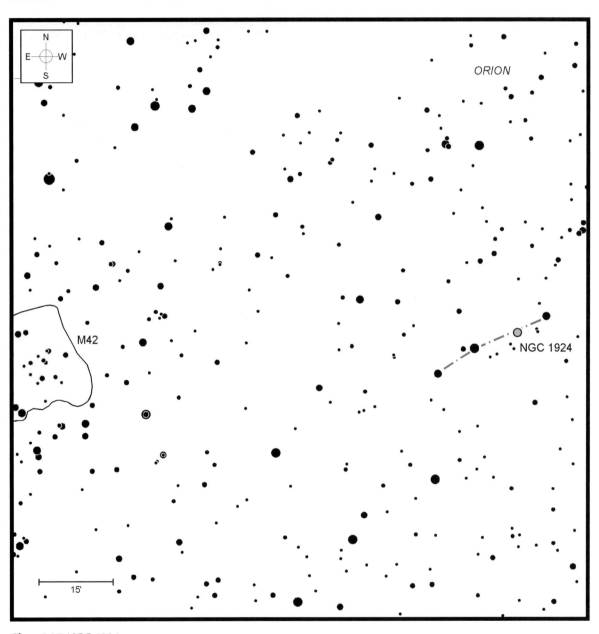

Chart 5.25 NGC 1924

You've heard of the Andromeda Galaxy and the Orion Nebula, but what about the Orion Galaxy? Probably not. But would you believe the *New General Catalogue* lists 21 galaxies in Orion, and the *Index Catalogue* adds another 9? That's a pretty respectable tally. Of those 30 Orion galaxies, I find our next challenge particularly intriguing because it lies so close to everyone's favorite winter deep-sky object, M42. Yet, I am sure that very few observers have seen it.

NGC 1924, a barred spiral galaxy, was discovered in 1785 by William Herschel using his 18.7-inch reflector, undoubtedly on an evening when he too was admiring the Orion Nebula. And why not? After all, M42 lies less than 2° to the east. That's pretty good company to keep, but, at the same time, it can also be a curse, since the Orion Nebula can be fairly distracting.

Finding NGC 1924 is easy enough by starting at M42 and scanning due west. Some $1\frac{1}{2}°$ into your scan, you will come to a diagonal path of three 8th- and 9th-magnitude field stars oriented northwest–southeast from one another. NGC 1924 lies along the path, like a distant galactic steppingstone equally spaced between two of those Milky Way suns.

When summarizing its appearance for his *New General Catalogue*, John Dreyer described it as "very faint, pretty large, irregularly round, stars nearby." My 8-inch shows NGC 1924 as a faint, oval disk accented by a stellar nucleus. It lies between two 8th-magnitude stars set amidst a sparkling field of fainter stardust (Figure 5.25).

Figure 5.25 NGC 1924

Larger apertures bring out additional subtle details. Through my 18-inch, the galaxy reveals a brighter outer edge and a starlike central core, closely mimicking the look of a planetary nebula. Add in the spectacular surroundings and how they add a faux-3D effect, and the beauty of this little treasure really comes through. It's a challenge that you are sure to return to time and again as you further explore its more affluent neighbor.

119 PK 189+7.1

Target	Type	RA	Dec.	Constellation	Magnitude	Size	Chart
PK 189+7.1	Planetary nebula	06 37.3	+24 00.6	Gemini	13.4p	32″×15″	5.26

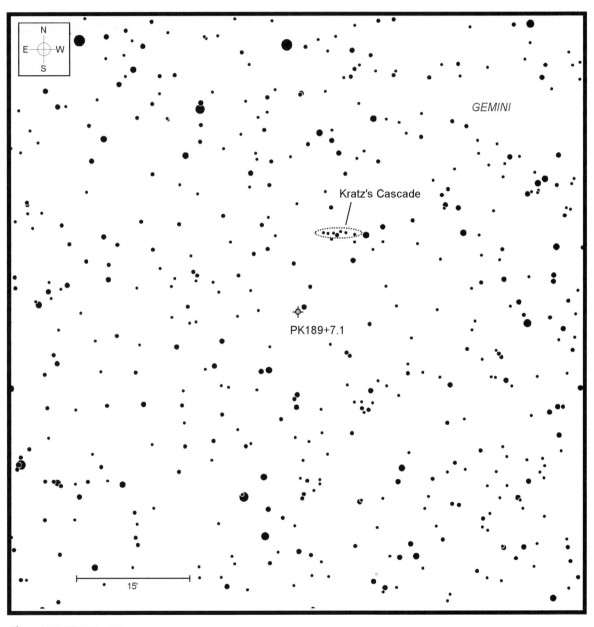

Chart 5.26 PK 189+7.1

With their feet standing on the winter Milky Way, the Gemini twins play host to a wide variety of deep-sky targets. Some, like M35, are well known among amateur astronomers, while others, like this next planetary nebula, go unappreciated by all but the most devout followers.

PK 189+7.1 lies along the leg of the twin brother Castor, a third of the way from Epsilon (ε) Geminorum to Mu (μ) Geminorum. Aim about $1\frac{1}{2}°$ southwest of Epsilon for an obtuse triangle of three 6th-magnitude stars, its apex pointing southward. With a low-power eyepiece in place, scan 25′ southwestward from the apex star for a curious line of 11 faint stars meandering east-to-west. Although these stars are apparently not physically associated with each other, they form an intriguing little asterism that is bound to catch your attention. I call this Kratz's Cascade, after Dave Kratz, a veteran deep-sky observer from Poquoson, Virginia, who first made note of the group several years ago while observing PK 189+7.1.

PK 189+7.1 is in the same field as Kratz's Cascade, just 11′ to its southeast and next to a 10th-magnitude field star. It looks like a faint star itself at magnifications below about 100×, but above, it is unmistakably nonstellar. Through a 6-inch at 232×, its faint, gray disk appears perfectly round and evenly illuminated, with no hint of the central star that begat the planetary (Figure 5.26). Since that star glows weakly at 19th-magnitude, it's a good bet that few amateurs have ever seen it.

Many references mention PK 189+7.1 by its alter ego, Minkowski 1-7, or simply Mink 1-7. Rudolph Leo B. Minkowski (1895–1976) was a German-born astronomer who immigrated to the United States in 1935 to join the staff of Mount Wilson Observatory. During his career, Minkowski specialized in the search for, and study of, planetary nebulae, and in the process,

Figure 5.26 PK 189+7.1

doubled the number known. He also led the National Geographic Society–Palomar Observatory Sky Survey, a monumental task of photographing the entire northern sky, in the 1950s.

Minkowski compiled his work with planetary nebulae into four listings, abbreviated Mink 1-xx to Mink 4-xx, with entries in each listed in order of increasing right ascension. Therefore, Mink 1-7 was the seventh item in his first listing, which formed the basis for a 1946 paper entitled "New Emission Nebulae."[13] Most of the 80 objects in Minkowski's first catalog were confirmed as planetary nebulae using the historic 60- or 100-inch telescopes at Mount Wilson.

---

[13] *Publications of the Astronomical Society of the Pacific*, Vol. 58, No. 344, p. 305.

120 **Sirius and the Pup**

Target	Type	RA	Dec.	Constellation	Magnitude	Size	Chart
Sirius and the Pup	Binary star	06 45.1	−16 42.8	Canis Major	−1.46/8.30	varies; see below	5.27

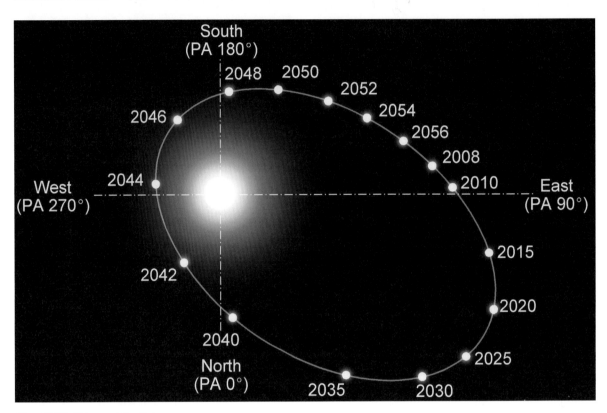

Chart 5.27 Sirius and the Pup

Ask an amateur astronomer to name binary stars that are difficult to resolve and one of the most common responses will probably be "Sirius." While there are more difficult targets, Sirius is always a perennial favorite. The challenge comes not from the close separation of the two stars in the system, however. Rather, the challenge here is from the extreme difference in the two stars' magnitudes. Sirius A, a type-A1V star, is about twice the mass of our Sun, while Sirius B is a white dwarf, the last stage of a star that was once approximately five to six times our Sun's mass. Sirius B, also known as "The Pup," is believed to contain about 0.5 solar mass at present, the rest having exhausted into space over 100 million years ago. Today,

all we see left from that once mighty star is the incredibly hot remnant core, a white dwarf.

The Pup was thought to exist long before it was actually discovered. In 1844, German astronomer Friedrich Bessel noticed that Sirius wobbled ever so slightly against the background sky. From this, he reasoned that the gravity from an unseen companion must be causing the star's odd behavior. Nearly two decades later, Bessel was proven correct when American telescope-maker and astronomer Alvan G. Clark spotted the little companion on January 18, 1862, while testing an $18\frac{1}{2}$-inch refractor, then the world's largest, that he had made for the University of Mississippi. Clark had selected Sirius for testing the level of chromatic aberration in the telescope; it is unlikely that he was specifically looking for the companion or that he even knew of Bessel's conclusions.

Sirius B measures no more than twice the Earth in diameter, yet its mass is nearly equal to the Sun's. Were we able to scoop up a teaspoon of Sirius B, transport it back to Earth and place it on a scale, we would find that it weighs several tons. Such is the stuff of white dwarfs. Eventually, Sirius B's energy emissions will ebb and cool, leaving just a cinder of superdense carbon. What do you get when you compress carbon under extreme conditions for eons of time? You get a diamond. Yes, eventually, Sirius B will become a diamond roughly the diameter of Earth!

Spotting Sirius B takes just the right combination of excellent optics, both in the telescope as well as in the sky. If the sky's optical behavior – that is, seeing – is uncooperative, then even the finest telescope optics will fail to resolve the Pup. Fortunately, things are getting easier. That's because Sirius B is heading toward the apastron point in its 50-year orbit around Sirius A. Over the course of half a century, the separation of these two stars varies from 3 arc-seconds to 11.5 arc-seconds. The pair appeared closest in 2000 and their distance apart has been widening ever since. Both will continue to grow apart until 2023, after which they will close on each other. Table 5.6 lists the separation and position angle of Sirius B over the next 20 years.

Even when the separation is widest, seeing the Pup takes strategy. Beyond steady skies and high magnification, determine where the companion should be in the view relative to Sirius itself, and then move Sirius just out of the field. Keep in mind, however, that, depending on your eyepiece, edge distortions could distort the Pup out of existence. Therefore, many use an occulting bar across the center of the field to hide the glare of Sirius. Rotate the eyepiece until it matches the Pup's position angle, which is currently toward the east-northeast. Incidentally, if you are using a Newtonian or Cassegrain reflector that has a spider mount holding the secondary mirror in place, double

Table 5.6 *Sirius B 2018–2037*

Year	Position Angle (°)	Separation (″)
2018.0	72.2	10.91
2019.0	70.3	11.04
2020.0	68.3	11.15
2021.0	66.5	11.22
2022.0	64.6	11.27
2023.0	62.7	11.28
2024.0	60.9	11.25
2025.0	59.0	11.19
2026.0	57.1	11.09
2027.0	55.1	10.96
2028.0	53.1	10.78
2029.0	51.1	10.57
2030.0	48.9	10.31
2031.0	46.6	10.00
2032.0	44.2	9.64
2033.0	41.5	9.23
2034.0	38.6	8.77
2035.0	35.3	8.24
2036.0	31.6	7.65
2037.0	27.2	7.00

check that a diffraction spike from the spider does not inadvertently cover the star.

Give the Pup a try when Sirius is highest in the southern sky, but, again, wait for those nights when there is exceptional seeing. Be sure to begin your hunt before the sky fully darkens, since a twilight sky will help absorb some of Sirius's glare. Sirius B is magnitude 8.3, so the sky does not have to be darkened fully to see it. Set your telescope up before the Sun goes below the horizon, let the optics cool to ambient temperature, and then, as you are waiting for twilight to wane, focus on Sirius and see what you discover.

121   NGC 2298

Target	Type	RA	Dec.	Constellation	Magnitude	Size	Chart
NGC 2298	Globular cluster	06 49.0	−36 00.3	Puppis	9.3	5.0′	5.28

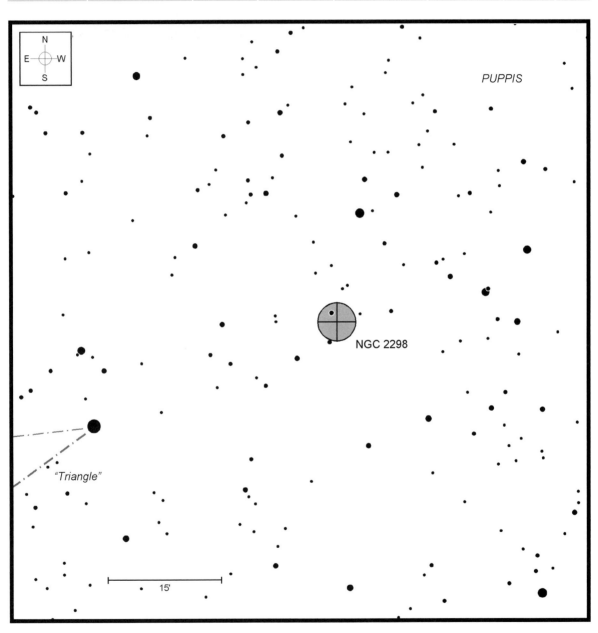

Chart 5.28 NGC 2298

Although most globular clusters line the summer sky as they huddle around the core of our galaxy, there are a few renegades that have stepped out on their own to occupy regions far beyond the rest. One such globular, nestled behind the rich Milky Way star fields of Puppis, is NGC 2298. Discovered on May 30, 1826, by James Dunlop, NGC 2298 is a loner, some 35,000 light years from Earth. It orbits the Milky Way along a very broad, elliptical path that carries it as close as 6,500 light years from the galactic core and as far as 49,000 light years away across 304 million years.

A study published in 2008 based on data collected by the Hubble Space Telescope suggests that NGC 2298 is losing mass at a higher rate than might be expected from its remote location. As the authors discuss, "observations over the past years have revealed a growing number of globular clusters severely depleted of low-mass stars."[14] A remote globular like NGC 2298 should show a greater abundance of low-mass stars than globulars that are closer to the Milky Way's core. But it does not. The study concludes from this that NGC 2298 is losing mass more rapidly than expected. In fact, as the title of the study surmises, NGC 2298 is on its way to disruption.

Not to worry, however; NGC 2298 will be around for while longer. You can take your time finding it. Begin your quest at Aludra [Eta (η) Canis Majoris], the tail star of the Large Dog. Slipping 3° southward brings a right triangle of three bright stars into view, accompanied by many fainter points that collectively form the little-known star cluster Collinder 140. Its large size and sparseness masked the cluster's true nature until 1931, when Swedish astronomer Per Collinder included it as number 140 in his catalog.

Drop another 5° southwest to the golden 3rd-magnitude star, Pi (π) Puppis, and another underappreciated open cluster, Collinder 135.

---
[14] G. de Marchi and L. Pulone, "NGC 2298: A Globular Cluster on its Way to Disruption," *Astronomy & Astrophysics*, Vol. 467, No. 1 (2007), pp.107–15.

Figure 5.27 NGC 2298

Collinder 135 also includes the double star v1 and v2 Puppis, just north of Pi, and a solitary 5th-magnitude star to the west that collectively give the group a distinctive arrowhead shape. Your finderscope will probably offer a better view of both clusters than your telescope. Finally, turn toward the northwest and look 4°, or about half a finder field, for a small triangle of 6th- and 7th-magnitude stars. NGC 2298 is $1\frac{3}{4}°$ away to the west-southwest of the triangle and 15' due south of an isolated 8th-magnitude field star.

NGC 2298 will give most 6-inch telescopes a run for their money, but seeing it through 8- and 9.25-inch scopes should pose little problem as long as the atmosphere is clear of any horizon-hugging haze and clouds. My notes made through my 8-inch at 56× recall a dim ball of starlight punctuated by a vague central condensation. As Figure 5.27 shows, a few faint points were just visible around the periphery by using averted vision and raising magnification to 142×. None of the cluster's stars shines brighter than magnitude 13.5.

122 NGC 2371–72

Target	Type	RA	Dec.	Constellation	Magnitude	Size	Chart
NGC 2371–72	Planetary nebula	07 25.6	+29 29.4	Gemini	13.0p	55″	5.29

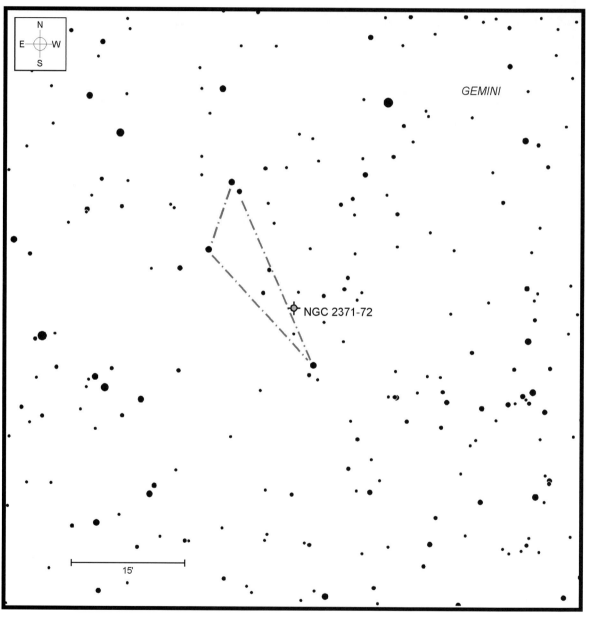

Chart 5.29 NGC 2371–72

Figure 5.28 NGC 2371–72

Although I have read reports of its visibility through telescopes half the size, NGC 2371–72 always proves troublesome through my 6-inch scope from my suburban backyard. It is, therefore, presented as a challenge here, although observers with 3- to 5-inch telescopes might want to give it a go from dark skies.

The dim disk of this elusive planetary is located about 3° southwest of Castor [Alpha (α) Geminorum] and about $4\frac{1}{2}°$ west-northwest of Pollux [Beta (β) Geminorum]. To find it, proceed due west from Pollux for $3\frac{1}{2}°$ to 64 and 65 Geminorum, a pair of 5th- and 6th-magnitude stars. Turn 90° to the northwest, and move your telescope $1\frac{1}{2}°$ to three 8th- and 9th-magnitude stars that form a gentle arc measuring $\frac{1}{4}°$

long. Our target is about 20 arc-minutes farther to the northwest, adjacent to a pointy triangle of 9th-magnitude points. If you have trouble picking it out, switch to 100× or so and slowly scan the region. Narrowband and oxygen-III filters will also help to show its tiny disk.

Once spotted, NGC 2371–72 is best seen at higher magnifications. By cranking up the magnification to at least 150×, and by using averted vision, the nebula's disk appears to have been severed in two (Figure 5.28). The two halves are oriented northeast–southwest, with the southwestern lobe (cataloged separately as NGC 2371) impressing me as slightly brighter than the other (NGC 2372). Together, they resemble a cosmic peanut. Others are reminded of M76, the "Little Dumbbell," a similar bipolar planetary in Perseus. Maybe we should christen NGC 2371–72 the "Littlest Dumbbell."

Many early observers mistakenly interpreted NGC 2371–72 as two separate objects. William Herschel, who discovered this two-faced object in 1785, described it as "two, faint of equal size, both small, within a minute [of arc] of each other; each has a seeming nucleus, and their apparent atmospheres run into each other." This led John Dreyer to list this planetary twice in his *New General Catalogue*. Dreyer also listed M76, the Little Dumbbell, in Perseus twice (NGC 650 and NGC 651) for much the same reason.

Today, we understand that NGC 2371–72 is a single object with bipolar tendencies. The angled view that we have of the planetary from Earth results in the unusual two-lobed structure that we see. In the case of NGC 2371–72, its central star, the incredibly hot remnant of a red giant, has a surface temperature estimated at 240,000° Fahrenheit. The central star shines at 15th magnitude and probably requires at least a 12-inch telescope to be seen.

123   NGC 2438

Target	Type	RA	Dec.	Constellation	Magnitude	Size	Chart
NGC 2438	Planetary nebula	07 41.8	−14 44.1	Puppis	10.1p	64″	5.30

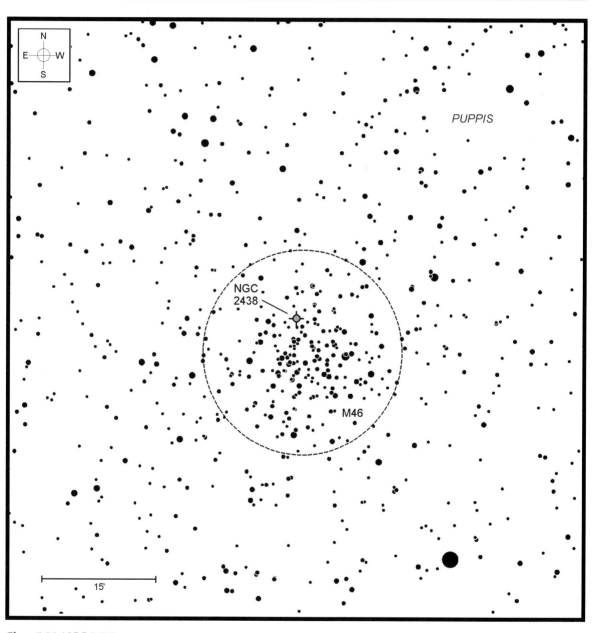

Chart 5.30 NGC 2438

M46 in Puppis is one of my favorite open clusters and a striking sight through almost any telescope. More than 500 stars are crammed into an area just a Moon's diameter across, creating one of the most jam-packed throngs in the winter sky.

M46 was discovered by Charles Messier on February 19, 1771, only three days after he had published the first edition of his catalog covering Ms 1 through 45. Of his new catch, Messier wrote "A cluster of very small stars, between the head of the Great Dog and the two hind feet of the Unicorn; one cannot see these stars but with a good refractor." But when William Herschel gazed upon M46 through his 18.7-inch reflector on March 19, 1786, he saw something else, something Messier apparently had missed or overlooked, floating among the cluster stars north of the group's center. He probably thought to himself, "That's not a star at all. That's a tiny disk of light." Herschel included his find as H-IV-39, the 39th planetary nebula in his catalog, but today, we know it best as NGC 2438.

While NGC 2438 may look like it belongs to M46, in reality it is much closer to Earth. Modern reckoning places it "only" about 2,500 light years away. M46 is another 1,500 light years farther still. How do astronomers know? One way is by comparing the spectra of the planetary with that of the stars in M46. These show that both M46 and NGC 2438 are moving away from the Solar System, but at two different speeds. If the planetary and cluster were physically connected, they would be moving through space at the same speed. A second, even more obvious, basis for this conclusion is that stars in open clusters are usually quite young, in contrast to planetary nebulae, which form from stars that are comparatively old.

M46 and NGC 2438 are easiest to find by dropping 5° due south from 4th-magnitude Alpha (α) Monocerotis, the brightest star in Monoceros. Trying to find *that* star is its own challenge, especially with less-than-perfect sky conditions. Fortunately, a line extended from Sirius [Alpha (α) Canis Majoris] through Gamma (γ) Canis Majoris for 11° to the east points right at Alpha Mon. Use your binoculars or finderscope to trace the line, and then shift southward to find M46

Figure 5.29 NGC 2438

within a slender stellar triangle. Incidentally, you will also find another open cluster, M47, at the triangle's western tip, just $1\frac{1}{2}°$ west of M46. They make a spectacular couple in rich-field telescopes.

As striking as that low-power view is, NGC 2438 will take at least 150× to tell it apart from just another cluster star. Focus your attention on the stars in the northern part of the cluster, keeping an eye out for a tiny, softly glowing disk of greenish light. That will be the 10th-magnitude planetary. Through my 8-inch reflector at 203× and with an oxygen-III filter in place, the nebula's ring shape (Figure 5.29) is clearly evident and appears very slightly oval. Removing the filter and using averted vision adds a 13th-magnitude star within the ring, just slightly offset to the northwest of center. Don't be fooled into thinking that you are seeing the nebula's forbearer, however. NGC 2438's actual central star barely cracks 18th magnitude. The dim sun we are seeing is most likely a distant member of M46. Another of M46's stars, an 11th-magnitude point, appears to just brush the nebula's southeastern edge.

124   Moon: Alpine Valley rille

Target	Type	Best lunar phases (days after New Moon)	Chart
Moon: Alpine Valley rille	Lunar rille	Days 7 and 20–21	5.31

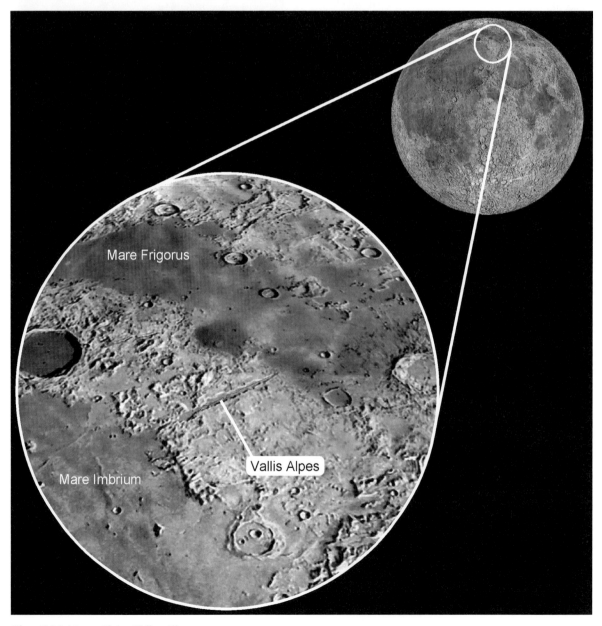

Chart 5.31 Moon: Alpine Valley rille

One of the most striking trios of lunar features lies along the north shore of Mare Imbrium, the Sea of Rains. Farthest west of the three is Sinus Iridum, the Bay of Rainbows. Originally, Sinus Iridum was a complete crater, but lava from Mare Imbrium breeched its southern wall, flooding the floor to the point that we only see half of the original impact site. Move eastward and you'll come to the dark-floored crater Plato, which we visit will later in this chapter as medium-scope Challenge 128.

For now, let's survey a gash in the lunar Alps that looks like a canal connecting Mare Imbrium to Mare Frigorus. Welcome to Vallis Alpes, or the Alpine Valley. Discovered in 1727 by Francesco Bianchini, the Alpine Valley is not a canal, but a *graben*, a sunken stretch of lunar crust. This almost unnaturally straight wound spans 100 miles (166 km) in length, but is no more than 6 miles (10 km) across in the middle. The Valley tapers noticeably at either end.

Under the right lighting conditions, the Alpine Valley can be seen through steadily held 12× binoculars, but that is not the challenge presented here. Instead, let's try to spot the narrow rille that slices along the valley floor. Rilles are common features on the lunar surface and can be divided into three basic types. Sinuous rilles meander randomly across the surface, arcuate rilles form sweeping arcs, while straight rilles, such as that flowing through the Alpine Valley, cut long, linear paths. Although the three types of rilles probably have three different origins, all appear to be the result of lava- or volcanic-induced events.

The Alpine Valley Rille measures more than 87 miles (140 km) in length, but even at its widest, it is never more than 0.6 mile (1 km) across. Through our telescopes, it looks like a dark, pencil-thin line in the middle of the Alpine Valley. You'll need at least 150× to see it at all, and then, only with steady seeing. Atmospheric turbulence can smudge the rille even when the valley appears sharp.

Your best chance of spotting the Alpine Valley rille comes the night after First Quarter, when the valley sees sunrise. By the time 8 days have passed since New Moon, the Sun has climbed high enough in the lunar sky for light to fill the floor of the valley and illuminate the rille. But don't wait too long to see it. By Day 12, shadow relief will be fading, causing the rille to blend into the surroundings.

★★★★
★★★

125 Moon: Craters Armstrong, Aldrin, and Collins

Target	Type	Best lunar phases (days after New Moon)	Chart
Craters Armstrong, Aldrin, & Collins	Lunar crater trio	Days 5 and 19	5.32

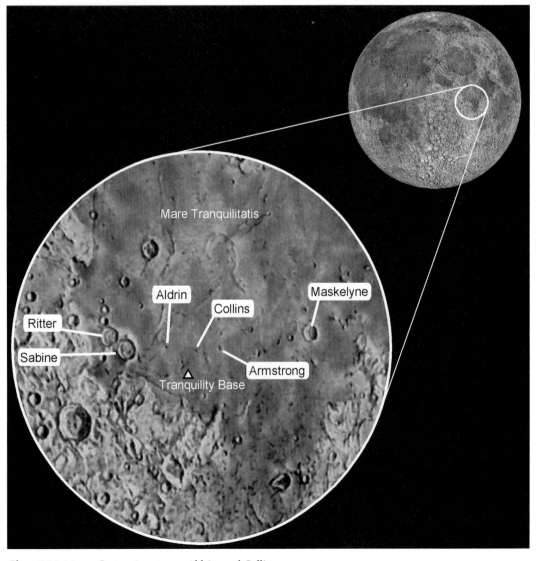

Chart 5.32 Moon: Craters Armstrong, Aldrin, and Collins

Binocular Challenge 46 dared you to find all six of the Apollo landing sites. Here, we return to Mare Tranquilitatis, the scene of Apollo 11, to find three small craters that bear the names of that historic mission's crew members: Neil Armstrong, Edwin Aldrin, and Michael Collins.

The reason you are reading this book right now is because I am a product of the 1960s space race, when the United States and the former Soviet Union battled each other for supremacy in space. The goal, set by President John F. Kennedy in 1961, was "to land a man on the Moon and return him safely to the Earth" by the end of that decade. It seemed impossible, given that we had just barely scraped outer space at the time with Alan Shepard's suborbital flight. But eight years later, Kennedy's dream was realized, if posthumously, when Neil Armstrong took that "one small step for (a) man, one giant leap for mankind." It was one of those moments in history that, if you were alive at the time, you can recall exactly where you were. Me? I was transfixed to the television screen (we even had color TV, although the Apollo camera was only black-and-white) in the living room of my childhood home in Connecticut. I was hooked for life.

The names of the three Apollo 11 astronauts are now permanently commemorated on the surface of the Moon by a trio of small craters that lie near the mission's landing site along the southwestern shore of Mare Tranquilitatis. Since the largest, named for mission commander and first man on the Moon, Neil Armstrong, only measures 3 miles across, however, high magnification and steady seeing is needed to see them.

The best opportunity for finding the three Apollo astronaut craters comes six days after New Moon, when the Sun has risen high enough in their sky for light to fall on a pair of twin craters to their west. Those two craters, Ritter (19 miles, or 30.5 km, diameter) and Sabine (18 miles, or 29 km, diameter) ride the Tranquility coastline, tucked just inside the mare's southwestern corner. Once both are in view, switch to about 200× and scan to the east-northeast of Sabine. Look some nine crater diameters to the west for the lone crater Maskelyne. Armstrong crater lies almost exactly between Sabine and Maskelyne. Scanning further back toward Sabine, a second, slightly smaller, crater can be found. That's Collins, 2 miles (3.2 km) in diameter. Finally, the least distinct of the three, Aldrin, is also 2 miles across and found about halfway between Collins and Sabine. All three Apollo craters are nearly equally spaced, with Armstrong and Collins slightly closer to each other than Collins to Aldrin.

As you look their way, play back that famous line in your mind and realize the history this area has seen. "Houston, Tranquility Base here. The Eagle has landed." Tranquility Base lies some 14 miles (22.5 km) to the southwest of Collins.

★
★ ★
★

126 Moon: Hortensius dome field

Target	Type	Best lunar phases (days after New Moon)	Chart
Hortensius dome field	Lunar volcanic domes	Days 9 and 22	5.33

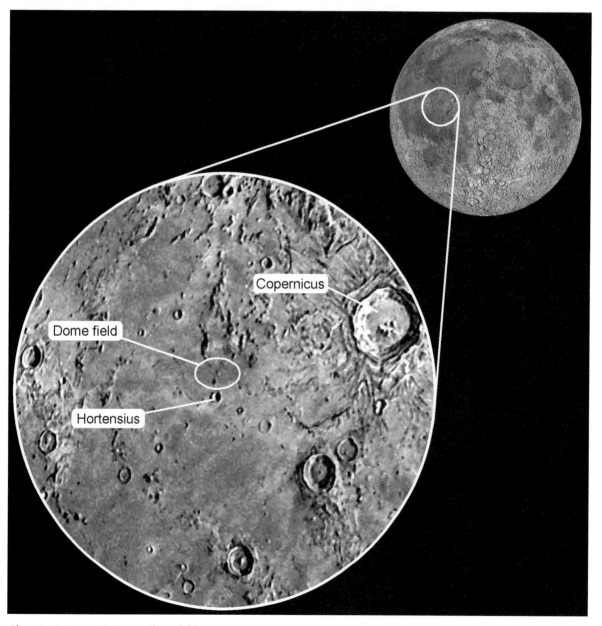

Chart 5.33 Moon: Hortensius dome field

Thanks to the lunar samples brought back by Armstrong, Aldrin, Collins, and the other Apollo astronauts, it is well established that the vast majority of craters are impact craters formed when leftover debris from the formation of the Solar System slammed into the Moon. But if we look carefully, scattered among all of those impact scars is direct evidence that the early Moon was also a hotbed of volcanic activity.

Some of the most intriguing evidence of that activity is the so-called lunar domes. Experts tell us that the lava that flooded the giant impact basins to form the maria we see today had a low viscosity. In other words, it flowed quickly due to its high temperature. As the lunar core cooled over time, however, the erupting lava decreased in flow rate as well as temperature, which increased its viscosity. As it continued to percolate through vents, the lunar lava's ability to flow away decreased, creating in the process low shield-like volcanoes, the lunar domes.

Most lunar domes are clustered together in small groups and lie in, or are immediately adjacent to, maria. A typical dome measures between 3 and 12 miles (4.8–19 km) in diameter, appears round or elliptical in shape, and has an average slope of only 2° to 5°. Owing to this gentle grade, volcanic domes are only readily visible right after sun-up. All quickly fade from view after the Sun moves higher in their sky. But when sunlight just grazes their tops, domes can appear quite striking.

At first pass, the crater Hortensius appears to be of no special significance. Just another impact crater among the myriad, right? Closer inspection shows that,

although Hortensius itself is rather bland, its immediate surroundings are anything but. Visit here just as the light from the rising Sun grazes the adjacent plain of Mare Insularum (the Sea of Isles) and as many as six unusual mounds, or bumps, can be seen to the crater's northeast. The Hortensius dome field, as it is most often called, forms the Moon's best-known region of lava domes.

Although it only spans 9 miles (14.4 km) from edge to edge, Hortensius is easy to pinpoint thanks to two prominent craters on either side, Copernicus to its (lunar) east and Kepler to its (lunar) west. Since Kepler still lies in darkness during the dome field's prime viewing time, we need to rely on Copernicus as our guide. Hortensius resides just beyond the southwestern edge of Copernicus' pronounced ray pattern, or ejecta field, and forms a not-quite-equilateral triangle with the larger crater Reinhold, also to Copernicus' southwest.

Once you've identified Hortensius, switch to between 150× and 200×, or higher if conditions permit, and focus your attention just to its northeast. Can you spot five or six "bumps" in the otherwise flat plain? If so, look carefully and you should also see that five of the domes are punctuated by tiny craterlets – volcanic vents – centered on their summits. The vent of the sixth dome must have been concealed by lava.

Remember, timing is critical. The 10-day-old Moon, smack dab in the middle of the waxing gibbous phases, is best for spotting the Hortensius dome field, as is the 25-day-old waning crescent phase. Both position the Sun low in the dome field's sky, maximizing the visibility of their gentle topography.

127 Moon: Mons Hadley and Rima Hadley

Target	Type	Best lunar phases (days after New Moon)	Chart
Mons Hadley and Rima Hadley	Lunar mountain peak and rille	Days 7 and 20–21	5.34

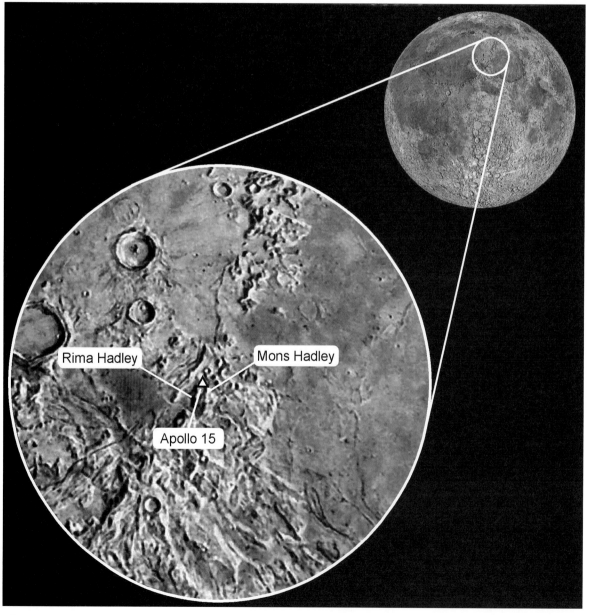

Chart 5.34 Moon: Mons Hadley and Rima Hadley

On July 30, 1971, Commander David Scott and Pilot James Irwin navigated their Apollo 15 lunar module, nicknamed Falcon, to land among the lunar Apennine mountains, while Alfred Worden remained in orbit aboard the command module, Endeavor. Scott and Irwin guided Falcon to a soft landing between the edge of a deep precipice and the base of a tall mountain.

The remains of Hadley Base, as the landing site became known, may be invisible through our telescopes, but its spectacular surroundings are a must-see stop on any tour of the Moon. Follow the curve of the Apennines toward their northern end, where, to the east, lies a flat area known as Pales Putredinis, the Marsh of Decay. Just west of the Marsh, in the lunar highlands, is the prominent peak Mons Hadley. Hadley rises an impressive 14,500 feet (4,420 m) above its surroundings, about half the height of Mount Everest, the tallest mountain on Earth.

Look carefully just to the east-southeast of Mons Hadley for a thin, sinuous channel that snakes through the adjacent flat plane. That's Hadley Rille, or more accurately, Rima Hadley. Rima Hadley meanders for 48 miles (77 km), severed almost perfectly in the middle by the 4-mile (6.4-km)-diameter crater Hadley C. The rille, cut by flowing lava 3.8 billion years ago, is 2 miles (3.2 km) wide at its broadest and drops a $\frac{1}{4}$ mile (400 m) to its floor below.

Rima Hadley is a textbook example of a sinuous rille, one of the three types of rilles found on the Moon. Sinuous rilles probably show the flow path that lava followed from the mouth of a lunar volcano back when the Moon's core was still molten. Schröter's Valley (small-scope Challenge 84) in Oceanus Procellarum is the largest sinuous rille on the Moon.

The best time to visit the area of Hadley Base is at First Quarter. That night, the Sun has broken over the mountains and poured into the flat surrounding plain. With steady seeing conditions, magnifications in excess of 300× add a three-dimensional beauty to the scene, with the towering mountain rising dramatically above the steep gorge. From the looks of it, lava flowed from a broad gash in the hills at the southern end of the rille and migrated northward, making several sharp turns along the way, before emptying into the mare. Apollo 15 landed just north of the rille's northernmost crook, where it veers suddenly to the west.

128 Moon: Plato's craterlets

Target	Type	Best lunar phases (days after New Moon)	Chart
Plato's craterlets	Walled plain and interior craterlets	Days 8 and 21	5.35

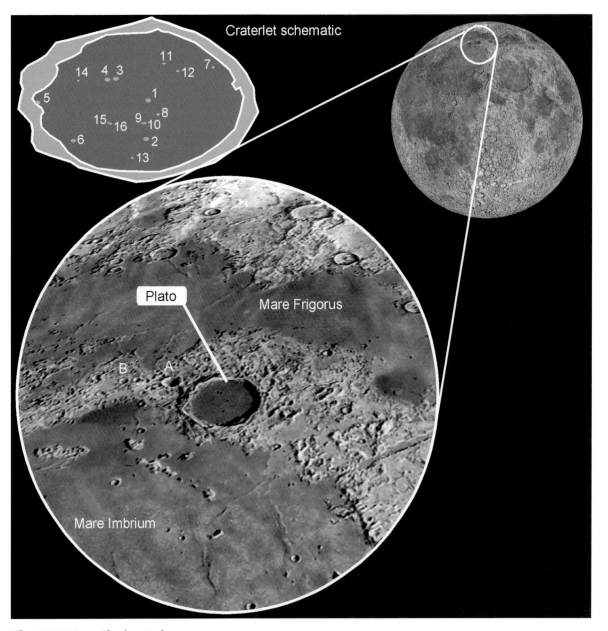

Craterlet schematic

Mare Frigorus

Plato

B   A

Mare Imbrium

Chart 5.35 Moon: Plato's craterlets

One of the great challenges facing even the most devoted lunar observers is trying to see the many small craters that dot the lava-covered floor of the walled plain Plato. Plato itself is a prominent impact scar measuring 62 miles across. It takes no more than 10× binoculars to see Plato after the Sun rises in its sky one day past First Quarter.

Unlike typical impact craters, Plato shows no sign of a central peak or a chaotic floor. Instead, it appears perfectly smooth and, at first blush, featureless. That's because shortly after the blast had excavated the bowl of Plato, lava welled up to flood its interior, hiding any central remnants of the original impact. As a result, we see an unusually dark floor that stands in stark contrast against the more lightly shaded Mare Imbrium to its south and Mare Frigorus to its north.

While viewing Plato through his telescope in 1824, German astronomer Franz von Paula Gruithuisen noticed something peculiar. He spotted a tiny speck on the floor of Plato, the first observation of a Plato craterlet. While Gruithuisen's place in history was tarnished by a second discovery he made, that of an expansive "lunar city" to the north of the crater Schröter, his sighting of a Plato craterlet prompted others to take a closer look.

Over the next several decades, interest in Plato increased dramatically, as did the number of craterlets. In a report published in the March 1883 issue of *The Observatory* (Vol. 6, pp. 85−91), British observer A. Stanley Williams described how, over the years 1879 to 1883, he and a number of other observers had observed and accounted for more than 40 "spots" on Plato's floor, along with several crisscrossing "streaks."

In a sense, Williams was to Plato what Percival Lowell was to Mars. Lowell believed that Mars was inhabited by intelligent beings who constructed an elaborate array of crisscrossing canals that spanned the globe of the Red Planet. Although Williams didn't see canals within Plato, he did believe that the surface of that distant crater was home to a variety of so-called transient lunar phenomena. He would periodically see streaks come and go across the floor, or perhaps a new crater suddenly pop forth where none had existed before. While today we understand that what he perceived as changes across the floor of Plato was most likely caused by changes in our own turbulent atmosphere, alternately masking and revealing fine detail just at the edge of visibility, Williams was

Table 5.7 *Plato craterlets*

Craterlet designation	Diameter (km)	Diameter (miles)
1	2.4	1.5
2	2.1	1.3
3	2.2	1.4
4	2.0	1.2
5	2.6	1.6
6	1.8	1.1
7	1.4	0.9
8	1.3	0.8
9	1.2	0.7
10	1.0	0.6
11	1.2	0.7
12	1.1	0.7
13	1.1	0.7
14	1.0	0.6
15	1.0	0.6
16	0.9	0.6

steadfast in his belief that these phenomena were related to lunar volcanic activity.

Although we shouldn't expect any stray volcanic emissions spewing across Plato, spotting the tenuous detail that so intrigued Williams is a challenge that many enjoy to this day. Several small craterlets are indeed strewn across Plato's otherwise smooth floor.

First, we need to take care of some bookkeeping. Since none of the Plato craterlets have official International Astronomical Union (IAU) designations, we are left to our own devices to create some. Although references sort them by letter, that system may be easily confused with the IAU's convention of naming so-called satellite features after a major landmark. Plato has more than a dozen satellite features associated with it, labeled Plato A, Plato B, and so on.

Instead, let's go with numbers, as Williams and other classic lunar observers did more than a century ago. Using this system, the most obvious craterlet is denoted on Chart 5.35 as "1." When sunlight strikes it just so, its steep walls appear brightly lit in contrast to the dark surrounding floor of Plato. Telescopes as small as 4-inchers can resolve it as a circular pit.

Three other craterlets, plotted here as 2, 3, and 4, are also visible with difficulty through a 6-inch telescope. Craterlets 3 and 4 are especially close-set; can you resolve them both? The best chances to see all four as true depressions occur when the Sun is either rising or setting in their sky, on the nights immediately after the quarter phases. Closer to Full Moon, the stark lighting turns them into bright, almost starlike points against Plato's floor.

Craterlet 5 is removed from the rest, embedded in Plato's eastern wall. Although it's the largest craterlet, the location against Plato's wall tends to mask it more than those centrally located on the floor. The best chance for spotting it comes during the waxing phases, since it remains hidden by the wall's shadow during the waning phases.

If you have success with the Plato quintet, try your luck with five smaller understudies. Craterlets 6, 7, 8, 9, and 10 will probably need at least an 8-inch aperture and 200× to be seen, and even then only with great difficulty. Yearning for even more? Try your luck with all sixteen listed in Table 5.7.

129 Mercury: Surface markings

Target	Type	Best time
Mercury	Surface markings	Autumn mornings when Mercury is near greatest western elongation.

For many, just seeing the planet Mercury against the bright twilit sky is challenge enough. For others, however, the real test comes not from just seeing the planet, but from seeing the *real* Mercury. By that, I mean the challenge enjoyed by some devout planetary observers is detecting subtle features on the barren surface of the Solar System's smallest planet.

The problems with viewing Mercury have already been discussed in naked-eye Challenge 18. Rather than repeat that advice here, let's instead discuss strategy for viewing Mercury telescopically in the hope of seeing evidence of some of the magnificently rugged terrain photographed by NASA's Mariner 10 and, more recently, Mercury MESSENGER spacecraft. Both of those unmanned missions revealed a crater-strewn world that has been punished for 4.5 billion years by meteoric bombardment and searing heat from the Sun, only 36 million miles (58 million km) away.

Just as seeing Mercury by eye is confounded by its nearness to the Sun in our sky, so too is seeing any evidence of surface terrain. The planet always hugs the horizon after sunset or before sunrise, where haze, clouds, smog, and turbulence in Earth's atmosphere are constant problems. Add to that the fact that Mercury isn't much larger than our Moon, and is never closer to us than 57 million miles (92 million km), and it is easy to see why most observers look elsewhere.

Frank Melillo, coordinator of the Association of Lunar and Planetary Observers' (ALPO's) Mercury section, disagrees:

Much popular astronomy literature contends that surface detail cannot be seen on Mercury. Under excellent seeing conditions, however, I have seen unquestionable detail on the surface. With practice, clean optics, and appropriate filters and magnification, you should also see features on the planet.

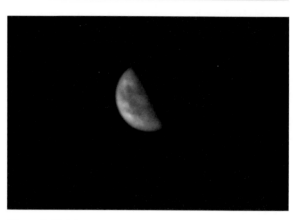

Figure 5.30 *Mercury*

What is needed to see these sketchy surface markings? Telescopes as small as 4 inches across have been able to uncover some hints but, as with spotting other small-scale structural details, larger apertures will increase resolution. A 6- to 9.25-inch aperture is ideal. Anything larger will only amplify air turbulence and blur the image. No matter the aperture, seeing is absolutely critical since Mercury will be so low in the sky. Review some of the hints in Chapter 1 for ways to tone down atmospheric turbulence.

Mercury's small diameter, only 3,030 miles (4,880 km) across, combined with its distance from us means that the apparent diameter of the planet's disk ranges between only $4\frac{1}{2}$ and 13 arc-seconds. As a result, it will take at least 200× to see any surface mottling. A clock-driven mount will prove immensely helpful, since it will let you to concentrate on viewing the planet rather than keeping up with it.

Melillo, who took the photo in Figure 5.30, also recommends using an orange or red filter to counteract some of the damaging effects of our atmosphere and to reduce glare. "I prefer a Wratten #23A light red filter.

Others find #21 orange to be best. A #25A red is useful, but only with larger apertures because of its density. Blue filters, #80A and #38, are also worth trying."

Morning observations are preferred to evening, since there is no need to rush before the planet sets. That's the key: relax, take your time. In fact, as Mercury gets higher in the predawn sky, not only is it rising above our atmosphere's densest layers, but the increasing morning twilight will also reduce glare that might otherwise blind the observer's eye.

The first thing you will notice about Mercury is its phase, but look more closely and low-contrast shadings might pop in and out of view. All will be extremely vague – far more so than surface features on Mars – so a careful and considerate eye is needed to distinguish what is there and what is an illusion. Sketching what you see can go a long way to improving your observing skills as well as in spotting delicate features that come and go as seeing conditions vary.

130 Mars: Phobos and Deimos

★
★★
★★
★

Target	Type	Best time
Phobos and Deimos	Satellites orbiting Mars	Near Martian opposition

In his classic book *Gulliver's Travels*, Jonathan Swift created a fanciful world called Laputa. Laputa could fly or float thanks to its inhabitants discovering the power of magnetic levitation. The Laputians made many other scientific discoveries, including two moons orbiting the Red Planet, Mars.

They have likewise discovered two lesser stars, or satellites, which revolve about Mars; whereof the innermost is distant from the centre of the primary planet exactly three of his diameters, and the outermost, five; the former revolves in the space of ten hours, and the latter in twenty-one and a half.

When *Gulliver's Travels* was published in 1726, two moons orbiting Mars was just fancy. But 151 years later, Swift's premonition was proven fact by Asaph Hall. Viewing through the historic 26-inch Alvan Clark refractor at the US Naval Observatory in Washington, DC, on August 11, 1877, Hall discovered "a faint object on the following side and a little north of the planet." Four cloudy nights intervened but, by the 16th, Hall returned to Mars to confirm that his object seen on the 11th had moved with Mars against the background star field. To his surprise, a second faint point was also seen near Mars the following night, as well. The following day, news of Hall discovering two tiny moons orbiting Mars was announced. The moon that he discovered on the 11th went on to be named Deimos, while that of the 17th is now known as Phobos.

The Martian moons are tiny, with Phobos measuring only 13 miles across and Deimos spanning just 7.5 miles (12 km). Neither is massive enough to have solidified into a spherical globe when it was formed 4.5 billion years ago. Instead, both appear irregular in form, similar to asteroids and shaped a bit like potatoes. Phobos, the innermost of the pair, takes 7 hours, 39.2 minutes to orbit the Red Planet, while

Deimos takes 30 hours, 17.9 minutes. Isn't it curious how Swift's fictional description of the moons is close to scientific fact?

It took a 26-inch telescope for Hall to discover Phobos and Deimos, but both are visible in much smaller instruments, provided certain conditions are met. First, as with seeing tenuous surface details on Mars itself, the best chance for spotting its attendant satellites comes at or near opposition. It's on these occasions that the Earth–Mars distance is at a minimum, which translates to Mars (and Phobos and Deimos) appearing the brightest and largest in our sky. Revisit Table 4.4 for a list of upcoming Martian oppositions.

As you can see from that table, some oppositions are much better than others. Those occurring in 2010, 2012, and the latter half of the 2020s are especially poor, since, even at its closest, Mars will remain more than 60 million miles away. But at others, notably the oppositions of 2018, 2020, 2033, and 2035, the distance between our worlds will be less than 40 million miles. It is at those instances when Phobos and Deimos stand their best chance of being seen through backyard telescopes, since they will shine as brightly as 11th and 12th magnitudes, respectively. From the sounds of those numbers, it seems that both should be visible in a 4-inch telescope. But that's if we were somehow able to view them in a dark sky, which we cannot, because of the planet's glare. Neither could Asaph Hall. In fact, contrary to expectations, larger and brighter Phobos is more difficult to spot since it also lies much closer to the planet.

When he discovered Phobos and Deimos, Hall understood that, to see any faint companions, he had to keep the planet's disk just out of the field of view. We need to do the same. If that will be your tack, be sure to use a comparatively narrow field eyepiece, such as an orthoscopic. Wider-field eyepieces will make spotting

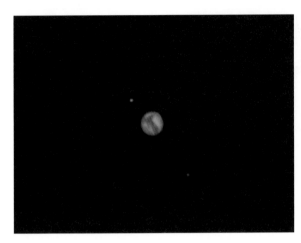

Figure 5.31 Mars with Phobos and Deimos

the moons more difficult, since they will be so far over to one side. Better still, use an eyepiece outfitted with an occulting bar. While no company sells such a thing commercially, one is easy enough to make at home. Refer back to the discussion in Chapter 1 under the heading "Your eyepieces."

Even armed with an occulting eyepiece and waiting for Mars to be near opposition, spotting the Martian moons (Figure 5.31) takes precise timing since both moons are always within 1 arc-minute of the planet. To be seen clearly, Phobos and Deimos must be as far from the planet as possible, at greatest eastern or western elongation. How will you know when that will next occur? Several planetarium-style computer programs, such as *Guide* by Project Pluto, plot the real-time positions of both moons. Simply plug in the time and date, and the positions will be plotted against the background sky. For positional data, you can also consult the current year's *Astronomical Almanac* published by the US Naval Observatory.

131 Saturn: Encke's Division

Target	Type	Best time
Encke's Division	Division in Saturn's rings	Rings near maximum tilt

There are many features hidden within the Saturnian rings that test our powers of observation, the acuity of our vision, and the sharpness of our telescope's optics. The most obvious is Cassini's Division (small-scope Challenge 89), the dark gap that separates the broad inner B ring, from the dimmer outer A ring. Seeing Cassini's Division is a fun challenge for a 2-inch aperture, but should be clear and distinct in telescopes covered in this chapter.

Cassini's Division was discovered in 1676, but it would be more than a century and a half before additional structural details in the rings were detected. In 1837, German astronomer Johann F. Encke recorded a low-contrast band in the A ring, sometimes referring to it as a line, while at others calling it a stripe. Although he did not give a dimension to this band, most historians today agree that what Encke saw was a grayish zone along the outer half of the A ring. The A ring is brightest along its interior circumference adjacent to Cassini's Division. About halfway out, its brightness drops off, continuing at that lower level to the outer edge. That is the effect that Encke first detected, and indeed is today referred to as the Encke Minimum. The Encke Minimum is not the same as the so-called Encke Division, as many believe. The Encke Division is a thin, but distinct gap near the outer edge of the A ring.

Who discovered the Encke Division? The question sounds too silly to ask, like who is buried in Grant's Tomb and how long did the Hundred-Years War last? But it turns out that Encke never saw the division that bears his name. Instead, it was discovered half a century later by the American astronomer James E. Keeler through the then-new 36-inch refractor at Lick Observatory in California. On January 7, 1888, Keeler recorded a thin gap within the Encke Minimum that he described as "a mere spider's thread." The gap went unseen again until March 2, 1889, when Keeler, along with observatory director Edward S. Holden and staff astronomers Edward E. Barnard and John M. Schaeberle, spotted its feeble presence.

In the years that followed, however, Keeler's discovery went on to become known as Encke's Division, a name sanctified by the International Astronomical Union. For reasons that are lost to history, the IAU subsequently decided to honor Keeler by naming a much smaller gap near the outer edge of the A ring the Keeler Gap. Unfortunately, Keeler couldn't possibly have seen his namesake feature, since it is visible only to passing spacecraft. Let's hope that one day the IAU will revisit and correct their error for much the same reason as they chose to reclassify Pluto as a dwarf planet in 2006. Admittedly it will not earn them the same press coverage.

Seeing the Encke Minimum as well as Keeler's Division – excuse me, Encke's Division – requires exceptional seeing as well as good geometry. As already mentioned in the last chapter when discussing the visibility of Cassini's Division, Saturn's rings appear edge-on twice during its $29\frac{1}{2}$-year orbit, effectively causing the rings to disappear. The Encke Minimum and Encke's Division are tricky at best, so maximize your odds by waiting until the rings are fully open, that is, when they are tilted near maximum angle. That next occurrence is in late 2016 and 2017, when the rings' north face will be open toward Earth. In the years that follow, the tilt shrinks, as will your chances of seeing this challenge. The odds improve once again in the years 2032–3 when the rings' southern face is fully tilted earthward.

Even under the best of circumstances, neither feature will be seen to encircle Saturn. Instead, focus your attention near the rings' *ansae*, the "bend" at either outer extreme. There, as shown in Figure 5.32, the

Figure 5.32 Encke's Division

foreshortening effect is least, giving the best opportunity for spotting these subtle features. The Encke Minimum is certainly the easier of the two, but will still require patience, high magnification, and steady seeing. Remember, you are looking for a shaded zone – not a line – around the outer half of the A ring. Unless you can convincingly hold Saturn in sharp view at 300× or more, the seeing is not steady enough. Wait for another night.

Keeler's "spider's thread" gap measures only about 0.2″ wide at maximum ring tilt, which from Dawes' Limit sounds much too small to resolve through just about any amateur telescope. Yet, we can. How come? Remember that Dawes' Limit applies to two point sources of light. Here, however, we are looking for a linear feature, which increases potential resolvability by a factor of about 3. As a result, an 8-inch telescope, with a calculated Dawes' Limit of 0.57″ can potentially resolve a linear feature measuring only 0.19″ wide. But to do so, you will need even higher magnification. By raising magnification to 400× or more, you just might be able to see a thread-thin line near the rings' ansae, and repeat history. Incidentally, Dawes himself saw Encke's Division in November 1850 using a 6.3-inch refractor at 460×.

Oh, and the Hundred-Years War? It lasted from 1337 until 1453.

6

Large-scope challenges

10- to 14-inch telescopes

We now enter the realm of double-digit apertures and step into the full depths of the deep-sky ocean. It wasn't that long ago that a 10-inch telescope was considered large by amateur standards. Bringing one to a club star party was a sure way to be the center of everyone's attention, as it stood alone amidst an ocean of 6-inch and smaller telescopes. A 12- or 14-inch telescope practically deified its owner.

Times have changed. Today, these same scopes are no longer the big kids on the block. But that doesn't change the fact that each can still bring its owner a lifetime of enjoyment. Indeed, there are so many challenging objects for these telescopes that trying to settle on the best-in-class is difficult to do. The targets included here represent some of my favorites, but you may well have others that you find equally demanding.

132	Abell 33

Target	Type	RA	Dec.	Constellation	Magnitude	Size	Chart
Abell 33	Planetary nebula	09 39.2	−02 48.5	Hydra	13.4p	270″	6.1

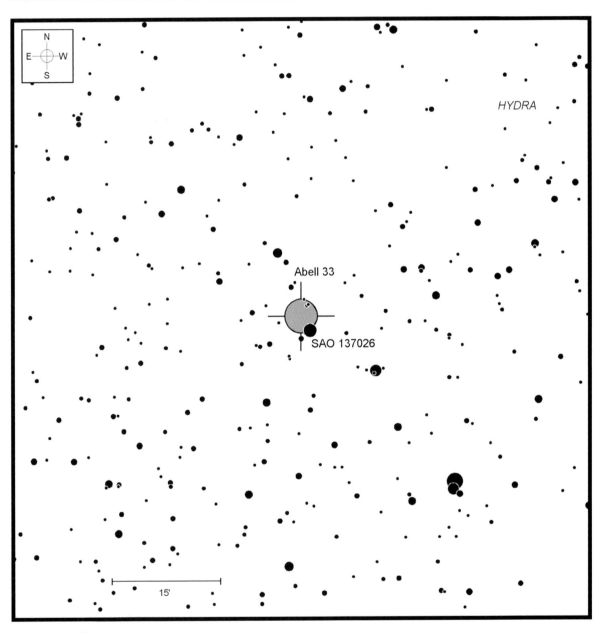

Chart 6.1 Abell 33

Several of the challenges in the last chapter involved hunting down tiny planetary nebulae. Many planetaries appear very small as seen from Earth, which can make them difficult to tell apart from surrounding stars. This also works in our favor, however, since their small size focuses all of the available light into small disks with high surface brightnesses. Their existence is also accented nicely by using narrowband and oxygen-III filters, which help suppress light pollution. That's why planetary nebulae are far better targets for urban observers than some other types of objects.

There are exceptions to every rule, however, and this challenge is one. Abell 33 was first spotted by George O. Abell in the 1950s as he and colleagues A. G. Wilson, Robert G. Harrington,[1] and Rudolph Minkowski pored over the then-new Palomar Observatory Sky Survey photographic plates. Abell announced their collaborative discoveries as part of a paper he published in the April 1966 issue of *Astrophysical Journal* entitled "Properties of Some Old Planetary Nebulae."[2] In the article's introduction, Abell noted that "these objects are large and faint and are probably at an advanced stage in their evolution as planetary nebulae; all 86 objects . . . are described as 'planetary nebulae,' with the cognizance of the fact that one or two of them may be improperly identified." Ultimately, four of his list's entries proved to be other types of object.

Many of the Abell planetary nebulae stand today as some of the ultimate tests for deep-sky observers using the largest backyard scopes. Abell 33 is one of the easier members of this elite list. It can be detected through a 10-inch telescope from under suburban skies – with a little help, that is.

First, let's zero in on its location within its home constellation of Hydra, the Water Snake. That's actually quite easy to do. From the serpent's pentagonal head, slither southeastward along its winding body, past Theta (θ) Hydrae to Iota (ι) Hydrae. Abell 33 is 1.6° due south of Iota, right next to 7th-magnitude SAO 137026. That star is the predicament. Abell 33's perfectly round

Figure 6.1 Abell 33

gaseous shell, which spans almost 5 arc-minutes in diameter, just overlaps that star. As a result, glare easily overwhelms the nebula's faint glow, especially if a telescope's or eyepiece's optics are dirty or contaminated.

Even with clean optics, seeing Abell 33 can be tricky. To improve the odds, add an occulting "bar" to an eyepiece that produces about 100× in your telescope by placing a piece of opaque black photographic tape across half of that eyepiece's field stop. Review the discussion under "Your eyepieces" in Chapter 1 for further information. Then, when you aim toward the planetary, hide the star behind the occulting tape before looking for the nebula's faint glow. As a reference, a faint double star is superimposed on the nebula's northwestern edge, while several even-fainter points litter the disk itself. With an oxygen-III filter and occulting tape in place, I have been able to spot Abell 33 with direct vision through a 12-inch telescope. Remove the tape and filter, and the planetary quickly disappears, even with averted vision. Spotting the central star, which only rates 15th magnitude, is a difficult chore through 10- to 14-inch scopes.

[1] No relation to your author!
[2] G. O. Abell, "Properties of Some Old Planetary Nebulae," *Astrophysical Journal*, Vol. 144 (1966), p. 259.

133 Hickson Compact Galaxy Group 44

Target	Type	RA	Dec.	Constellation	Magnitude	Size	Chart
Hickson 44	Galaxy group	10 18.0	+21 49.3	Leo	–	16′	6.2

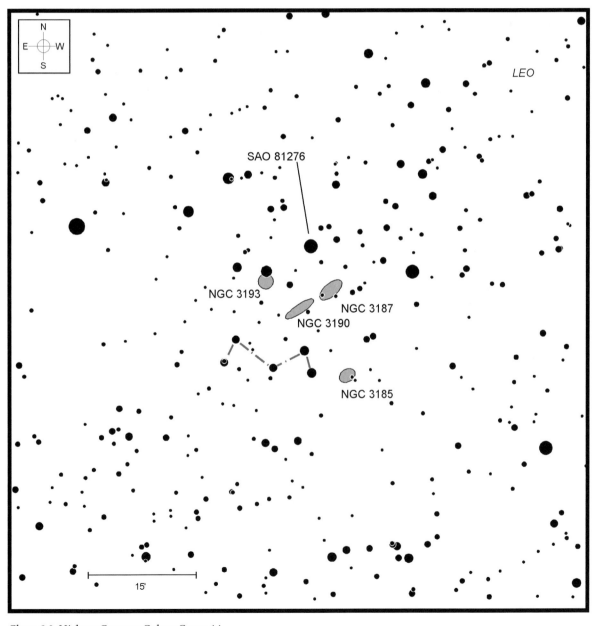

Chart 6.2 Hickson Compact Galaxy Group 44

Table 6.1 *Members of Hickson Compact Galaxy Group 44*

Object	RA	Dec.	Magnitude	Size
NGC 3185	10 17.6	+21 41.3	13.0b	2.3′ × 1.5′
NGC 3187	10 17.8	+21 52.4	13.4	3.6′ × 1.6′
NGC 3190	10 18.1	+21 50.0	12.1b	4.4′ × 1.2′
NGC 3193	10 18.3	+21 53.6	11.8b	2.0′ × 2.0′

In 1982, astronomer Paul Hickson, professor of astronomy at the University of British Columbia, published a study of 100 compact galaxy groups scattered throughout the sky. In his paper "Systematic Properties of Compact Groups of Galaxies",[3] Hickson defined a compact galaxy group as a small, relatively isolated collection of four or five individual systems that are set in close proximity to one another, and that differ in brightness by no more than 3 magnitudes. Further, so as to avoid including the central regions of dense galaxy clusters, Hickson stipulated an "isolation factor" requiring that there not be a nonmember galaxy of similar magnitude within three radii of the group's center.

Surveying the Palomar Observatory Sky Survey, Hickson created an inventory of 100 such groupings. The entries in the Hickson Compact Galaxy, or HCG, catalog are ordered numerically according to increasing right ascension. The classic example of a compact galaxy group, and the first of the genre to be discovered, is Stephan's Quintet in Pegasus (large-scope Challenge 144), which Hickson cross-referenced as HCG 92.

Stephan's Quintet is probably the best-known example of a compact galaxy group, but it is by no means the brightest. That honor falls to HCG 44, nicknamed the Leo Quartet. HCG 44, itemized in Table 6.1, is easily pinpointed a little less than halfway between the stars Adhafera [Zeta (ζ) Leonis] and Algieba [Gamma (γ) Leonis] along the sickle of Leo the Lion. If you center your finderscope on Adhafera (Zeta), you will see 6th-magnitude 39 Leonis just 20′ to its south-southeast. Without moving the aim, look for two 7.6-magnitude stars closer to the southern edge of the finder field. See them? Good.

Recenter your telescope on the western star in that pair, listed at SAO 81276, insert a medium-power

Figure 6.2 *Hickson Compact Galaxy Group 44*

eyepiece, and take a look just to its south. Two faint blurs should be immediately visible. The smaller and easternmost of the two is NGC 3193, found just 1′ to the south of a 9th-magnitude star. My 10-inch scope at 106× reveals this E2 elliptical as a bright, round glow highlighted by a brighter core.

NGC 3193 plays second fiddle to the group's brightest member, the edge-on spiral NGC 3190. Its cigar-shaped disk appears elongated northwest-to-southeast and surrounds a faint stellar nucleus. Photographs reveal that, like many edge-on spirals, NGC 3190 is bisected by a lane of opaque dust that runs along its galactic plane south of the central core. I have failed to see the lane through my 10-inch under suburban skies, but others report success under darker conditions with the same aperture. Larger instruments show that the dark lane's western tip appears to curl northward due to the tug of an external gravitation source.

That source of gravity is most likely NGC 3187, a dim barred spiral just 5′ to the northwest. Shining at magnitude 13.4 and spanning some 4′ × 2′, NGC 3187 is the faintest of the group. Its inherent low surface brightness is easily overwhelmed by the light of nearby SAO 81276, our finder star 6′ to the northeast. Move the star out of the field and the galaxy's faint glow should pop into view with averted vision. Look for a

[3] P. Hickson, "Systematic Properties of Compact Groups of Galaxies," *Astrophysical Journal*, Part 1, Vol. 255 (1982), pp. 382–91.

dim, featureless smudge of grayish light that is oriented perpendicular to NGC 3190 and just 1′ north of two 14th-magnitude field stars.

Finally, NGC 3185 is found 11′ southwest of NGC 3190 and 5′ due west of a five-star asterism shaped like a capital "M." Its diffuse glow draws to a slightly brighter core and appears extended slightly northwest–southeast. With averted vision, the southwestern edge of the galaxy appears to just kiss a 14th-magnitude field star. That star is well within the Milky Way, however, so be sure not to mistake it for an extragalactic supernova.

134 Abell Galaxy Cluster 1060

★
★★
★★
★

Target	Type	RA	Dec.	Constellation	Magnitude	Size	Chart
Abell Galaxy Cluster 1060	Galaxy cluster	10 36.9	−27 31.0	Hydra	–	168′	6.3

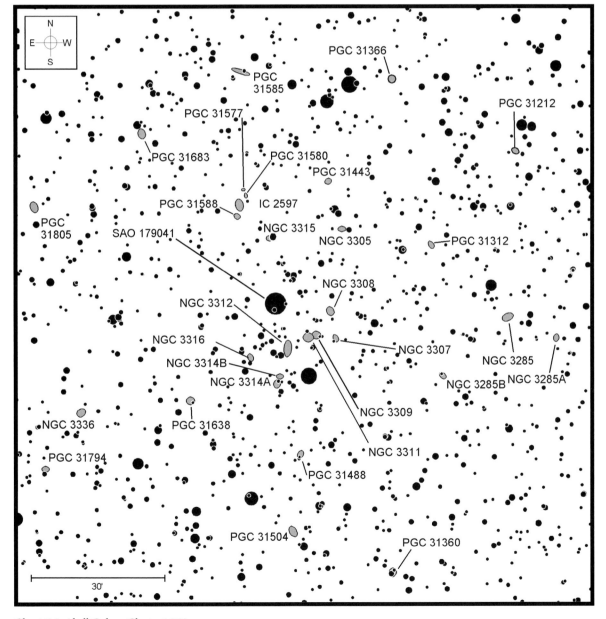

Chart 6.3 Abell Galaxy Cluster 1060

Entry number 1060 in the Abell Galaxy Cluster (AGC) listing forms our large-scope Challenge 134. Also known as Hydra 1, AGC 1060 covers nearly 3° of our southern spring sky and includes more than 100 individual galaxies. Fourteen of those have assignments in the *New General Catalogue*, while another belongs to the NGC's supplemental *Index Catalogue*. Six of those NGC galaxies shine brighter than 14th magnitude and lie within a 20′ field. The dozens of other cluster members are all too faint to have been included in those catalogs, but instead have individual entries in the *European Southern Observatory Galaxy Catalogue* (ESO), *Millennium Galaxy Catalogue* (MGC), and *Principal Galaxies Catalog* (PGC).

Finding AGC 1060 is a simple task as long as you can spot 4.5-magnitude SAO 179041. This red giant sun overlaps the center of the cluster and lies $4\frac{1}{4}°$ north of Alpha (α) Antliae. Of course, finding Alpha Antliae presents its own challenge, since it shines at only magnitude 4.2 and lies far from any handy reference stars. My best suggestion is use binoculars and start off from the trapezoidal body of Corvus the Crow. Extending a diagonal line southwestward from Algorab [Delta (δ) Corvi] through Epsilon (ε) Corvi and continuing on for another 24°, you will come to Alpha Antliae and, a degree to its northeast, Delta (δ) Antliae. SAO 179041 is now about two-thirds of a finder field to their north. Center SAO 179041 in your telescope field, switch to about 100×, and take a moment to orient yourself. In particular, note the 7th-magnitude sun, SAO 179027, 16′ to its south-southwest.

An arc of no fewer than six NGC galaxies curves south and west from SAO 179041. How many can you see? The brightest pair is formed by NGC 3309 and NGC 3311. Both lie just west of the midpoint between those two SAO reference stars. My 10-inch scope shows NGC 3309 as a small, circular disk with a concentrated core at its heart. Even though its integrated magnitude is greater, NGC 3311's lower surface brightness makes it less apparent than NGC 3309. It too appears circular, but evenly illuminated.

Another dim galactic ember glows to the east of the halfway point between the two SAO stars and just west of a pair of 12th-magnitude points. NGC 3312 appears quite elongated, oriented almost exactly north–south, and is highlighted by a bright stellar nucleus. Photos of the cluster reveal this to be a classic spiral galaxy tilted partially to our perspective. Monster backyard scopes

Figure 6.3 Abell Galaxy Cluster 1060

may also hint at the dark lane that runs along the edge of the spiral-arm halo.

NGC 3316 is nestled 8′ due east of NGC 3312. This little barred spiral is one of the toughest of the NGC galaxies in AGC 1060, since its tiny disk appears no more than 1′ across visually and shines at only 13th magnitude.

Veering off the arc for a moment, try your luck with an even dimmer pair, NGC 3314A and 3314B. Neither breaks the 13th-magnitude barrier, with the latter only managing a weak 14th-magnitude effort. A 14th-magnitude star is superimposed on the northern tip of NGC 3314A, while an even dimmer field star abuts NGC 3314B to the east.

Okay, back to SAO 179041. Scan 11′ due west for the faint patch of NGC 3308. A good eye will show that its dim disk is slightly oval, oriented south-southwest/north-northeast, and draws to a brighter central nucleus.

Now, look 20′ northwest of SAO 179041 for the faint glow of NGC 3305, just tucked inside the edge of the cluster. Its small, round disk lies just to the east of a 12th-magnitude star. The 10-inch offers no hint of a centralized core, but rather shows only a dim, evenly illuminated blur.

NGC 3315 is found some 14′ north of SAO 179041 and just east of an 11th-magnitude sun. At 106×, my 10-inch reveals a small, faint glow elongated

Table 6.2 *Central region of AGC 1060 (galaxies in **bold** are discussed in the text)*

Target	ESO	RA	Dec.	Magnitude	Size
NGC 3285A		10 32.8	−27 31.4	**14.5p**	1.1′×0.9′
PGC 31212	501–13	10 33.5	−26 53.8	13.9p	1.3′×0.5′
NGC 3285		10 33.6	−27 27.3	13.1b	2.5′×1.4′
NGC 3285B		10 34.6	−27 39.1	13.9p	1.5′×1.1′
PGC 31312	501–20	10 34.8	−27 12.8	14.5b	0.9′×0.6′
PGC 31310	436–44	10 34.8	−28 29.8	13.9p	1.1′×0.7′
PGC 31316	436–46	10 34.8	−28 35.0	13.4p	2.7′×2.0′
PGC 31360	437–4	10 35.4	−28 18.9	13.9p	1.8′×1.1′
PGC 31366	501–25	10 35.4	−26 39.8	14.1p	1.7′×0.8′
NGC 3305		10 36.2	−27 09.7	13.8b	1.1′
NGC 3307		10 36.3	−27 31.8	14.5v	0.8′×0.4′
PGC 31443	501–35	10 36.4	−27 00.0	14.2p	1.5′×0.5′
NGC 3308		10 36.4	−27 26.3	12.9b	1.7′×1.2′
NGC 3309		10 36.6	−27 31.1	12.6b	1.8′×1.5′
NGC 3311		10 36.7	−27 31.6	11.6v	2.1′×1.9′
PGC 31488	437–11	10 36.8	−27 55.2	14.3b	1.1′×0.5′
PGC 31504	437–15	10 37.0	−28 10.7	13.5p	2.4′×0.4′
NGC 3312		10 37.0	−27 33.8	12.7p	3.3′×1.2′
NGC 3314B		10 37.2	−27 39.5	13.5	0.3′×0.2′
NGC 3314A		10 37.2	−27 41.0	14	1.6′×0.7′
NGC 3315		10 37.3	−27 11.5	14.4b	1.1′×0.9′
NGC 3316		10 37.6	−27 35.6	13.6p	1.2′×1.0′
PGC 31577		10 37.7	−27 03.5	16.4b	0.6′×0.3′
PGC 31580		10 37.7	−27 02.6	17.0	0.3′
IC 2597		10 37.8	−27 04.9	12.8b	2.5′×1.7′
PGC 31588		10 37.8	−27 07.3	15.0b	0.9′×0.7′
PGC 31585	501–56	10 37.8	−26 37.8	13.8p	2.0′×0.4′
PGC 31638	501–65	10 38.6	−27 44.3	13.7p	1.7′×1.0′
PGC 31683	501–68	10 39.3	−26 50.4	14.3p	2.0′×0.6′
NGC 3336		10 40.3	−27 46.6	13.0p	1.9′×1.5′
PGC 31794	437–38	10 40.8	−27 58.0	14.4p	1.3′×0.6′
PGC 31805	501–75	10 41.0	−27 05.0	13.5p	2.2′×0.8′

northwest–southeast. Averted vision will probably be needed to see this little 14th-magnitude S0 system.

A tight group of four faint galaxies, cross-listed in Paul Hickson's Galaxy Group catalog as Hickson 48, is found another 9′ farther northeast still. Brightest of that bunch is IC 2597, a small, 13th-magnitude blur set just east of a faint star. Gathered round it are galaxies PGC 31588 (magnitude 15.0), PGC 31577 (magnitude 16.4), and PGC 31580 (magnitude 17.0). At those magnitudes, I'm afraid that we will have to leave that trio to the big guns.

Three NGC members guard the western flank of AGC 1060. To find them, move $\frac{3}{4}°$ due west to 7th-magnitude SAO 178978. NGC 3285 lies just 7′ to the star's southwest and reveals a small, oval glow tilted west-northwest to east-southeast. A faint stellar core is centered within. The other two NGC galaxies here, NGC 3285A, 12′ to the west-southwest of NGC 3285, and NGC 3285B, 18′ to its southeast, are each a paltry 14th magnitude.

Finally, the barred spiral NGC 3336 lies near the eastern edge of the cluster, about $\frac{3}{4}°$ east-southeast of center. This is a tough catch. Look for a dim, slightly elongated smudge of uniform grayish light.

Completing this challenge will give you 14 more notches in your galaxy belt, but there are still many fainter galaxies lying in wait. Table 6.2 includes the central region of AGC 1060, listing galaxies shining at magnitude 14.5 or brighter, while Chart 6.3 plots them amongst stars to magnitude 15. See how many of these other galactic denizens you can find.

135 NGC 3172

★
★
★
★

Target	Type	RA	Dec.	Constellation	Magnitude	Size	Chart
NGC 3172	Galaxy	11 47.2	+89 05.6	Ursa Minor	14.8	1.2′×1.1′	6.4

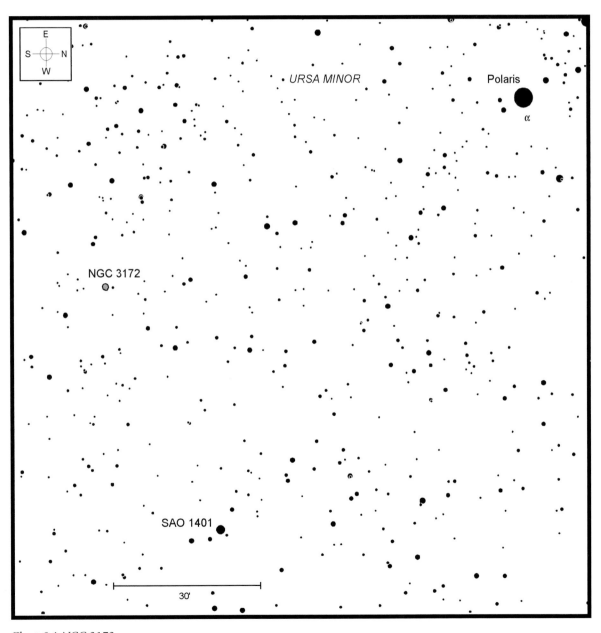

Chart 6.4 NGC 3172

Figure 6.4 NGC 3172

In real estate, as in astronomy, we have the three Ls: location, location, location. NGC 3172 is a challenge for all seasons – literally – as its location keeps it above the horizon throughout the year no matter what time of night you are looking. That's because NGC 3172 lies within 1° of the north celestial pole, closer than any other NGC object.

John Herschel discovered NGC 3172 during a deep-sky sweep with his 18.7-inch telescope in 1831. He later christened his new "nebula" Polarissima Borealis, or simply Polarissima, for its distinctive location.

Polarissima is easiest to find by casting off from Polaris, so take aim at the pole star through your finder. If you look carefully, you should see that it belongs to a circlet of faint stars that remind many of a heavenly engagement ring. Polaris serves as the diamond, while fainter stars fill out the rest of the ring. Two of the ring stars closest to Polaris, 6th-magnitude Lambda (λ) Ursae Minoris and 7th-magnitude SAO 1401, make handy reference stars, as NGC 3172 is located almost exactly halfway between the two.

Apart from its location near the top of the celestial sphere, NGC 3172 holds no distinguishing characteristics that set it apart from the throng of similar galaxies to its south. Classified as a spiral system, Polarissima shines at 14th magnitude and measures just 1′ in diameter. While it has been seen through apertures as small as 6 inches, it still presents a formidable test for 10-inchers, especially under less than ideal conditions. Regardless of aperture, NGC 3172 is an object that requires high magnification. Through my 10-inch at 181×, Polarissima shows off a subtle, round disk peppered with a brighter core and is accompanied by a 13th-magnitude field star just 2′ away.

136   Abell 36

Target	Type	RA	Dec.	Constellation	Magnitude	Size	Chart
Abell 36	Planetary nebula	13 40.7	−19 53.0	Virgo	13.0p	480″×300″	6.5

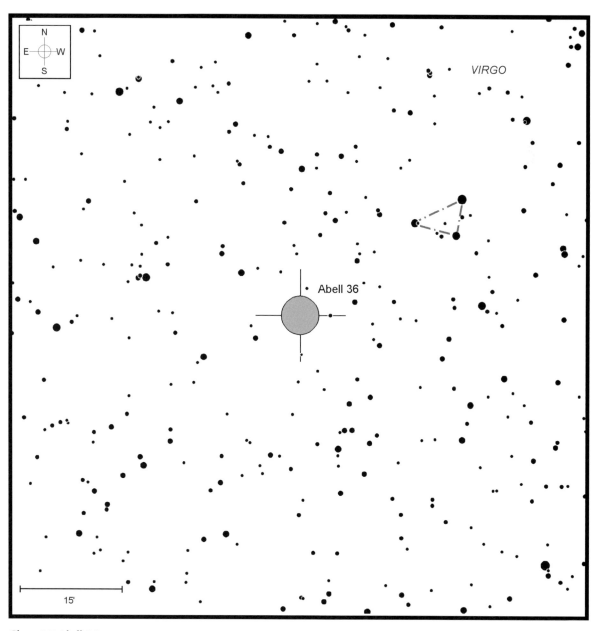

Chart 6.5 Abell 36

No wonder the Herschels missed this next planetary nebula. It lies right in the middle of galaxy country in the constellation Virgo. Mix in all those distractions with its large apparent size, very low surface brightness, and a bright central star, and you have an object that can confound the best of us. You have Abell 36.

Abell 36, another of the planetary nebulae discovered by American astronomer George Abell and colleagues as they scrutinized the Palomar Sky Survey photographic plates in the 1950s, lies south-southeast of Spica [Alpha (α) Virginis]. To find it, hop about 8° south-southeast of Spica to 6th-magnitude 73 Virginis. Once at 73 Vir, shift 40′ southeast to an 8th-magnitude field star, and another $1\frac{1}{4}$° farther southeast to a small right triangle of 9th-magnitude stars. Abell 36 is just 23′ southeast of that triangle. The nebula's central star shines at 11th magnitude, nearly twin in appearance to a field star set just 10′ to its northeast. A 13th-magnitude sun is also seen just 4′ to the nebula's west.

Unlike so many other planetaries, high magnification is not needed to pick out Abell 36; in fact, too much could prove counterproductive owing to the large size of the disk. Instead, start at around 100× to make an initial sighting, and then experiment to see which magnification works best. Regardless, a nebula filter will probably be needed in order to see the planetary's disk, even from the darkest observing sites. Some report better success with a narrowband filter than an oxygen-III filter, but try both.

Figure 6.5 Abell 36

Through my old 13.1-inch telescope at 68× with a narrowband filter in place, Abell 36 appeared as a very faint, semicircular glow offset to the central star's south. Averted vision revealed some faint extensions north of the star but only fleetingly. The star's brightness makes spotting these faint wisps very difficult even under ideal conditions.

137 Dissecting M101

Target	Type	RA	Dec.	Constellation	Magnitude	Size	Chart
Dissecting M101	Emission nebulae	14 03.2	+54 20.9	Ursa Major	–	–	6.6

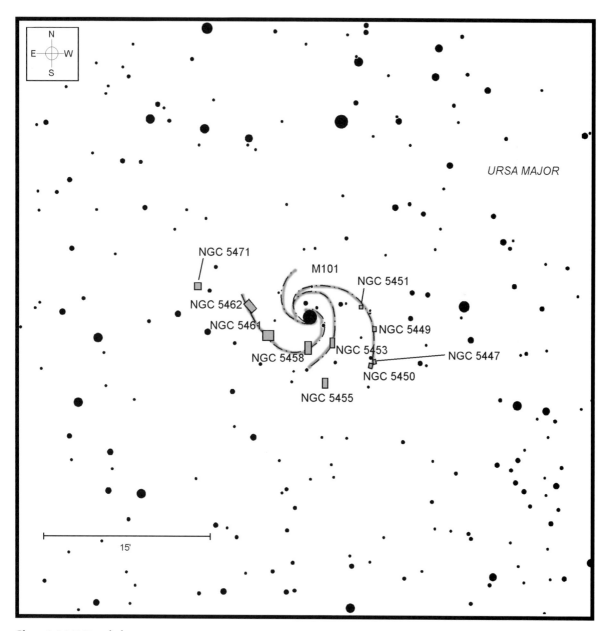

Chart 6.6 M101 nebulae

Just spotting the gigantic Pinwheel Galaxy, M101, can sometimes be challenging enough. Its low surface brightness can drive suburban observers crazy, especially when we see photographs that show it as very big and bright, or we see that it is listed as 8th magnitude. It all comes down to surface brightness, or more accurately lack of surface brightness. Spotting the dim glow of the galaxy's small core, or the even dimmer glimmer of the surrounding spiral arms, can take a concerted effort. But, with time and patience, M101 is visible, with difficulty, through 50-mm binoculars even given a sky with a naked-eye limiting magnitude of perhaps 4.5. You'll find those details in Chapter 2, binocular Challenge 27.

For double-digit apertures, the test presented by M101 is not only seeing the galaxy, but also finding latent structural details hidden within. Pierre Méchain may have discovered the Pinwheel in 1781, but it took the trained eye of William Herschel gazing through his 18.5-inch telescope to begin to crack M101's galactic vault and find the first hidden gems, three of the galaxy's interior clouds.

That was a good beginning, but was by no means the end of the story. The next chapter opened in 1845 when William Parsons, the third Earl of Rosse, first examined the galaxy through his monstrous 72-inch reflector at Birr Castle in Parsonstown, Ireland. Lord Rosse was the first to spot nine knots scattered throughout the galaxy's amazing spiral arms.

Rosse's discoveries were added to John Herschel's *General Catalogue* in 1864 and subsequently to John Dreyer's 1888 *New General Catalogue*. Today, the M101 family of hydrogen-II regions holds eleven entries in the NGC, more than any other single object. Each of these clouds is a huge expanse of ionized hydrogen surrounding embedded stars, like the Orion Nebula (M42) and the Lagoon Nebula (M8), among others.

Use the instructions found under binocular Challenge 27 and a wide-field eyepiece producing no more than 100× to find M101 initially and to trace out the full breadth of its spiral-arm disk. Can you repeat Lord Rosse's historic observation by spotting the subtle serpentine arms curving away from the galactic core? One arm branches off the southern tip of the core, curling around the core and twisting toward the west and south. The second major arm curves away from the northern edge of the core, hooks to the west, and then comes around the other side, where it divides.

Figure 6.6 M101 nebulae

There has been long-standing confusion over the exact locations of many of the NGC targets within M101 dating back to a drawing that Lord Rosse made in 1861. John Herschel subsequently used that drawing to determine the positions of those objects for inclusion in his *General Catalogue*, which ultimately resulted in errors that have been carried over to the present day. After more than a century, these galactic boo-boos were finally corrected by Harold G. Corwin, Jr., of the California Institute of Technology. The positions as well as the locations of each object listed in Table 6.3 and plotted on Chart 6.6 are based on Corwin's research.

Our first stop is NGC 5471 at the far end of the eastern arm, 11.5′ northeast of the core. Heinrich Louis d'Arrest was the first to spot it in 1863. Its isolation so far from the heart of M101 led many twentieth-century observers to conclude that NGC 5471 was actually a separate galaxy, and in fact its appearance through my 10-inch at 254× mimics a small elliptical galaxy perfectly, with an amorphous glow surrounding a brighter central core. Today, there is no longer any question as to its true nature. Photos taken with the Hubble Space Telescope reveal a glowing area some 200 times as vast as the Orion Nebula with several brighter regions embedded within. Detection of extremely strong X-ray emissions emanating from it

Table 6.3 *Nebulae within M101*

Target	Type	RA	Dec.	Magnitude	Size
NGC 5450	Emission nebula	14 02.5	+54 16.2	–	20″×6″
NGC 5447	Emission nebula	14 02.5	+54 16.8	–	8″
NGC 5449	Emission nebula	14 02.5	+54 19.8	–	~15″
NGC 5451	Emission nebula	14 02.6	+54 21.8	–	~10″
NGC 5453	Emission nebula	14 02.9	+54 18.5	–	<10″
NGC 5455	Emission nebula	14 03.0	+54 14.5	–	15″
NGC 5458	Emission nebula	14 03.2	+54 17.9	–	~20″
NGC 5461	Emission nebula	14 03.7	+54 19.1	–	25″×15″
NGC 5462	Emission nebula	14 03.9	+54 21.9	–	60″×18″
NGC 5471	Emission nebula	14 04.5	+54 23.8	–	25″

have led researchers to conclude that NGC 5471 is home to no fewer than three supernova remnants.

Traveling inward along the same spiral arm, we next come to NGC 5462, one of Herschel's trio of discoveries. Unlike NGC 5471, which appears nearly circular, NGC 5462 looks quite distended, oriented northeast–southwest. It is slightly dimmer than NGC 5471, but should still be apparent in a 10-inch telescope. NGC 5462 shows little improvement with a narrowband or oxygen-III nebula filter.

Closer in along the same spiral arm, we next come to NGC 5461, another of Herschel's finds. NGC 5461 is about 5′ south-southeast of the galaxy's nucleus and looks like a faint, slightly fuzzy star through my 10-inch. My 18-inch at 411× begins to hint at some of the cloud's subtle structure, including what appears to be a stellar brightening at its northeastern edge. Again, a narrowband filter offers only a modicum of help.

Finally, NGC 5458 is situated along the same spiral arm, just prior to where it wraps into M101's nucleus. Look for a very small, very faint glow measuring less than 30″ across set 5′ directly south of the core.

M101's western arm also offers a variety of H II regions. Working out from the galactic nucleus, we first come to NGC 5451, found about 5′ to its west. This is a tough catch. Unless your skies and optics are close to perfect, the low surface brightness of this nebulous tuft

will probably escape unnoticed. A pair of faint field stars is only 1′ to the cloud's west, so use them as a guide. But unless you can see those stars *and* the nebula, odds are good that you are only seeing the stars. NGC 5449, about 2′ further south along the arm, is also a difficult target. Use high power for both.

A close-set pair of nebulous knots, NGC 5447 and NGC 5450, is found toward the southern tip of the western arm. Less than ideal seeing conditions will merge these into a single, elongated blur but, under steady skies, each can be resolved as a separate glow just south of a 14th-magnitude Milky Way star. NGC 5447 is a huge association of hot O- and B-type stars, while NGC 5450 is an H II region that may eventually evolve to resemble its neighbor.

Following a fork in the western arm that hooks back toward the galactic center brings us to NGC 5453. Look for its tiny presence about 2′ west-northwest of NGC 5458.

NGC 5455 is found nearly half a degree south of M101's core, near the outskirts of the galaxy's vast spiral-arm halo. Curiously, some computer software programs plot NGC 5455 as just another field star, failing to recognize its true extragalactic nature. It forms the southern point in an equilateral triangle with two 14th-magnitude field stars, one to its northeast and the other to its northwest.

138 Dissecting M13

Target	Type	RA	Dec.	Constellation	Magnitude	Size	Chart
M13 propellers	Globular cluster	16 41.7	+36 27.6	Hercules	5.8	20.0′	6.7

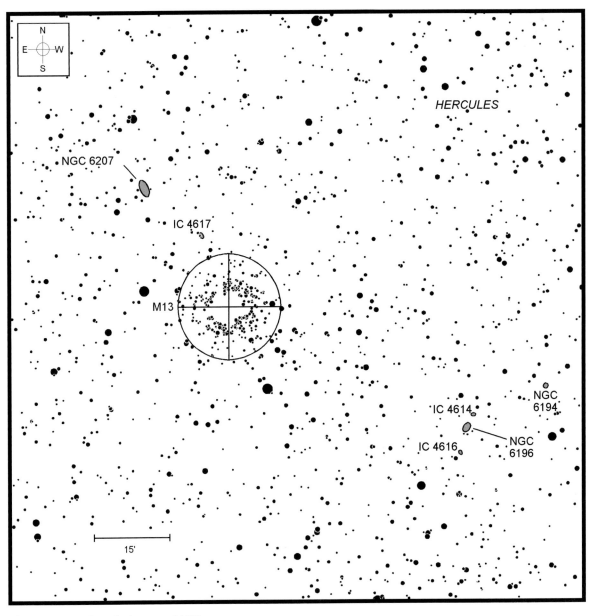

Chart 6.7 M13 plus galaxies around M13

There are few deep-sky objects more spectacular through a large telescope than globular clusters. Each globular cluster may contain hundreds of thousands to several million stars, all swarming together around a core so densely packed that seeing individual points may defy resolution. But the view is spectacular nonetheless.

To the casual eye, all globular clusters may look the same. Just a big ball of stars, right? Nothing could be further from the truth. Each has a personality all its own, often with hidden treasures lying within, if you take the time to look.

M13, the grand globular cluster in mighty Hercules, is the perfect example. We've already tested our naked-eye prowess with M13 by trying to spot it in Chapter 2 as naked-eye Challenge 6. We now return armed with large telescopes to unlock some of its unusual structural features.

First, it doesn't take a double-digit aperture to see that the stars of M13 are distributed asymmetrically. An 8-inch, or even smaller instruments, will show how the stars appear to be arranged into curves or rows. Many have likened the cluster's appearance to that of a spider. John Herschel described the cluster as exhibiting "hairy-looking, curvilinear branches." Later, Lord Rosse saw M13 as "more distinctly separated and brighter than anticipated; singularly fringed appendages to the globular figure branching out into the surrounding space." Notes made through my 10-inch telescope at 58× recall irregular strings of stars streaming out from the cluster's dense core. Two thin threads curving away toward the west strike me as particularly conspicuous. These star-strings give the impression that M13 is hurtling through space so quickly that is it leaving a trail of stars behind in its wake.

Increasing the 10-inch's magnification to 181× reveals a surprise that goes unsuspected at lower values. The star streamers are still evident, but hidden within the core are three subtle dark lanes that seemingly join together to form the letter "Y." These unusual lanes, or propellers as many call them, were first discovered by Bindon Stoney in about 1850. Stoney was an astronomer working for Lord Rosse at Birr Castle in Parsonstown, Ireland. After Stoney's initial sighting

Figure 6.7 M13 propellers

became known, many other observers confirmed the existence of these unique dark rifts through instruments as small as 6 inches aperture. But, as photography diminished the need for accurate visual observations, the M13 propeller became lost in the glow of the intense core.

Walter Scott Houston resurrected Stoney's dark lanes in his Deep-Sky Wonders column in the June 1976 issue of Sky & Telescope magazine. Thanks to his persistence, many of today's amateurs have seen the M13 propeller. But, again, magnification is key. Too low and the lanes will remain hidden from view.

To see the lanes for yourself, wait until the cluster is high in the sky, away from any haze and light pollution, which can stifle them. Under ideal conditions, the dark lanes are evident through a 12-inch, joined at their ends to resemble the corporate logo of a famous German automobile maker – proving once again that M13 is the Mercedes-Benz of globular clusters.

While you are enjoying M13, be sure to continue on to the next challenge, which tests your ability to spot a few of the cluster's faint neighbors.

139 Galaxies around M13

	Target	Type	RA	Dec.	Constellation	Magnitude	Size	Chart
139a	NGC 6194	Galaxy	16 36.6	+36 12.0	Hercules	14.6	1.0'×0.8'	6.7
139b	IC 4614	Galaxy	16 37.8	+36 06.9	Hercules	15.4	0.8'×0.8'	6.7
139c	NGC 6196	Galaxy	16 37.9	+36 04.4	Hercules	13.9b	2.0'×1.1'	6.7
139d	IC 4616	Galaxy	16 38.0	+35 59.8	Hercules	15.4	0.6'×0.3'	6.7
139e	IC 4617	Galaxy	16 42.1	+36 41.0	Hercules	15.5	1.2'×0.4'	6.7
139f	NGC 6207	Galaxy	16 43.1	+36 50.0	Hercules	12.2b	3.3'×1.7'	6.7

Hold on a moment. We aren't done with M13 just yet. Within about a degree of the Great Hercules Globular Cluster are six faint intergalactic members of the NGC and IC listings. Can you find them all?

The brightest of the bunch is NGC 6207, a 12th-magnitude spiral set just 28' to the northeast from the heart of the globular. To find it, switch to a wide-field eyepiece that just squeezes M13's two "sentry stars," SAO 65481 to the southwest and SAO 65508 to the east, into the same field as the cluster. From the eastern sentry star, scan about half a degree north to 8th-magnitude SAO 65509. The faint glow of the galaxy lies about halfway between those two stars, and just west of a small triangle of 13th-magnitude suns. Although it measures only 3.3'×1.7' in size, NGC 6207 is bright enough to be seen in 6-inch telescopes under dark, rural sky conditions. Look for an elliptical glow oriented northeast–southwest and punctuated by a faint stellar core.

William Herschel discovered NGC 6207, but he missed IC 4617. Can you do better? You'll find it centered between NGC 6207 and M13, just touching the western side of a trapezoid of 13th-magnitude stars. This little cigar-shaped spiral is a fairly tough catch with my 18-inch from my suburban backyard, but is clearly visible through my 10-inch under better conditions. Aperture is great, but there is no substitution for dark skies.

If IC 4617 was a little too demanding, NGC 6196, the brightest of a trio of galaxies found about a degree southwest of the globular, should prove easier. Under dark skies, a 10-inch shows this 14th-magnitude target as a small, round glow highlighted by a brighter central core.

Figure 6.8 NGC 6194, IC 4614, NGC 6196, IC 4616

A second, dimmer galactic smudge is nestled 6' to the southeast. The true identity of this second object is IC 4616, although some references identify it incorrectly as NGC 6197. The confusion over its true identity dates back to German-born astronomer Albert Marth (1828–1897). Marth came to England in 1853 and eventually became British astronomer William Lassell's (1799–1880) assistant at Lassell's observatory in Malta. There, Marth went on to discover more than 600 new deep-sky objects through the observatory's 48-inch reflector between 1863 and 1865. He bumped into this galactic pair on July 9, 1864, but incorrectly recorded their positions. The errant galaxy was subsequently reidentifed as IC 4616 by French astronomer Camille Guillaume Bigourdan (1851–1932) after a search found nothing at Marth's original location. Subsequent historical studies now conclude they are one and the same.

Next, try your luck with IC 4614, set 3′ northwest of NGC 6196. Like IC 4616, IC 4614 is rated at 15th magnitude and takes a special night to be seen through even the largest backyard scopes.

The furthest west of the bunch is NGC 6194. Lying 1.1° west-southwest of M13, this little elliptical galaxy shines at 14th magnitude and spans just 1′ across. You'll need at least 150× to see its small, amorphous glow.

140 **NGC 6578**

Target	Type	RA	Dec.	Constellation	Magnitude	Size	Chart
NGC 6578	Planetary nebula	18 16.3	−20 27.1	Sagittarius	13.1p	9″	6.8

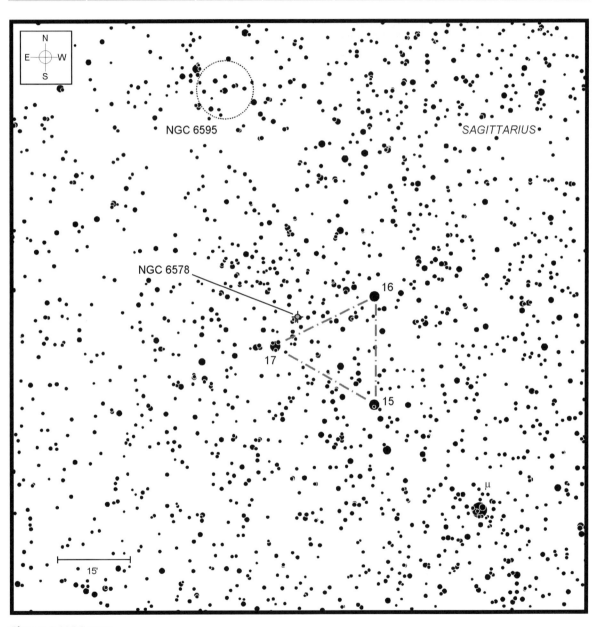

Chart 6.8 NGC 6578

Figure 6.9 NGC 6578

The next challenge is NGC 6578, a planetary nebula planted in central Sagittarius, one of the richest regions of the night sky. NGC 6578 is north-northwest of Kaus Borealis [Lambda (λ) Sagittarii] at the top of the teapot asterism. Look about 6° to the star's north for 4th-magnitude Mu (μ) Sagittarii and an equilateral triangle fashioned by 6th-magnitude 15, 16, and 17 Sagittarii, about 45′ further northeast. NGC 6578 lies 7′ northwest of 17 Sagittarii, the triangle's eastern tip.

While NGC 6578 is easy enough to locate, it is not so easy to see, thanks to an 11th-magnitude star just 20″

to the west and a small clump of 9th- to 11th-magnitude stars to the east. Try your highest usable magnification to remove the stars from the field of view, and then take a look. An oxygen-III filter will also help dim that 11th-magnitude star while also accenting the planetary. Blinking the nebula with the filter should help to pull it out from the crowd.

With averted vision, you should see that NGC 6578 has a round, mottled disk. Can you spot the uneven texture that shows up so clearly in images taken through the Hubble Space Telescope? Those show a brighter central core measuring about 6″ across surrounded by a fainter outer halo about 11″ in diameter. Hubble images also reveal a pair of what University of Washington astronomers Stacy Palen and Bruce Balick describe as "blowout bulbs"[4] passing near the nebula's 14th-magnitude central star. The authors note that "The inner core has a bright rim and a blotchy inner appearance, reminiscent of a heap of soapsuds, with faint regions surrounded by bright rims." Any hint of these fine details or the central star has eluded my eye, but you may have better luck.

Incidentally, if you are hunting for NGC 6578 by starhopping, be aware that it is plotted incorrectly on the *Sky Atlas 2000.0* star atlas. Instead, use the chart here in conjunction with another atlas, such as the *Uranometria 2000.0*, to zero in on it.

[4] S. Palen, B. Balick, A. Hajian, Y. Terzian, H. Bond, and N. Panagia, "Hubble Space Telescope Expansion Parallaxes of the Planetary Nebulae NGC 6578, NGC 6884, NGC 6891, and IC 2448," *Astronomical Journal*, Vol. 123 (2002), pp. 2666–75.

141   IC 4732

Target	Type	RA	Dec.	Constellation	Magnitude	Size	Chart
IC 4732	Planetary nebula	18 33.9	−22 38.7	Sagittarius	13.3p	10″	6.9

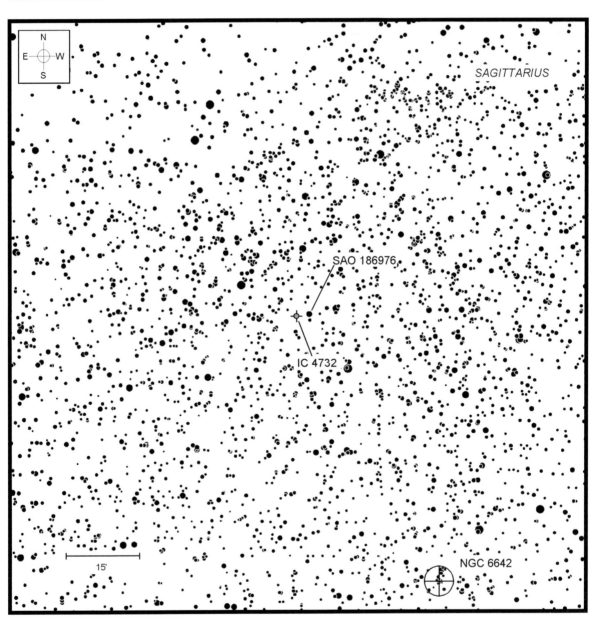

Chart 6.9 IC 4732

Figure 6.10 IC 4732

Whenever I am asked to name my favorite globular clusters, M22 in Sagittarius is always high on my list. I find it more impressive than M13 in Hercules. A 4-inch aperture is all it takes to begin to crack its stellar vault into myriad faint points around a densely packed core. In a 10- to 14-inch, it's a wondrous sight.

M22 is right in the thick of it, not far from the galactic center of the Milky Way. As such, it has lots of company. One particular planetary nebula proves a worthy adversary through 10- to 14-inch scopes: IC 4732. IC 4732 lies just 1.4° north-northwest of M22. Cataloged at photographic magnitude 13.3, its tiny disk is difficult to pick out from the mob of field stars – difficult, but not impossible.

To find IC 4732, center on M22 with a wide-field eyepiece in place. From the center of the globular, slide half a degree due north to 9th-magnitude SAO 187033, and then another half degree further northward to SAO 187032, also 9th magnitude. Finally, shift another half degree northwest to 8th-magnitude SAO 187000. By offsetting that last star toward the northwest of the eyepiece field, IC 4732 will lie close to the center, just 2′ to the east of 10th-magnitude SAO 186976.

Under a steady suburban sky several years ago, my old 13.1-inch reflector at 214× picked it out from the crowd using the same "flicker" nebula-filter method mentioned previously. I had the greatest success using an oxygen-III filter.

Regardless of the instrument and magnification used, IC 4732 will not show up as much more than a point of faintly bluish light. Its "stellar" rating of Class 1 on the Vorontsov-Velyaminov scale of planetary nebulae morphologies, detailed in Table 5.3 of Chapter 5, indicates that even the professionals cannot resolve its disk beyond a small sphere. And, while few amateurs have spotted IC 4732, even fewer can claim to have glimpsed its central star, glowing at a dismal magnitude 16.6.

Another planetary nebula just 14′ to the east-southeast of IC 4732 proves to be even more challenging. PK 10-6.2 shines at photographic magnitude 15.1 and measures only 8″ across. Rated as Class 2, PK 10-6.2 also looks like a star even when viewed at magnifications exceeding 400×. Once again, try the in-and-out filter method to identify it.

There is also a planetary nebula hidden among the stars of M22, but I doubt that PK 9-7.1 is viewable through 10- to 14-inch scopes. We'll leave that one for the next chapter.

142 Palomar 11

Target	Type	RA	Dec.	Constellation	Magnitude	Size	Chart
Palomar 11	Globular cluster	19 45.2	−08 00.4	Aquila	9.8	10.0′	6.10

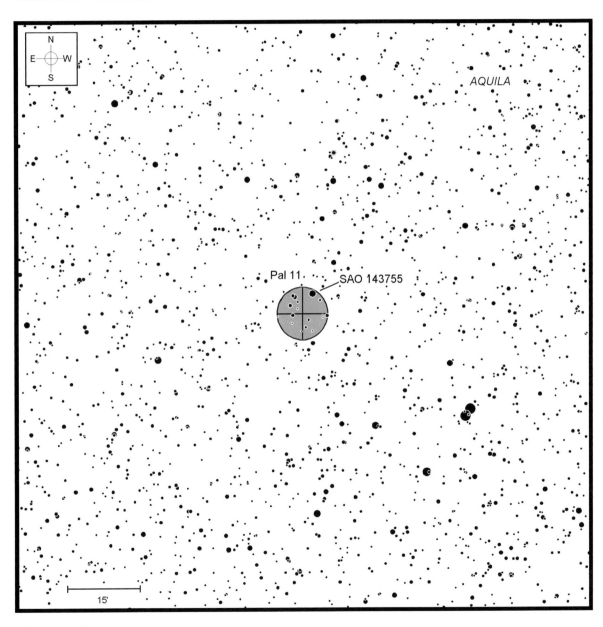

Chart 6.10 Palomar 11

Table 6.4 *Palomar globular clusters*

Palomar number	Challenge number (chapter)	Constellation	Right ascension	Declination	Magnitude	Size (')
01	181 (7)	Cepheus	03 33 23.0	+79 34 50	13.6	2.8
02	183 (7)	Auriga	04 46 05.8	+31 22 55	13.0	2.2
03	–	Sextans	10 05 31.4	+00 04 17	13.9	1.6
04	161 (7)	Ursa Major	11 29 16.8	+28 58 25	14.2	1.3
05	–	Serpens	15 16 05.3	−00 06 41	11.8	8.0
06	–	Ophiuchus	17 43 42.2	−26 13 21	11.6	1.2
07	–	Serpens	18 10 44.2	−07 12 27	10.3	8.0
08	–	Sagittarius	18 41 29.9	−19 49 33	10.9	5.2
09	–	Sagittarius	18 55 06.0	−22 42 06	8.4	5.4
10	–	Sagitta	19 18 02.1	+18 34 18	13.2	4.0
11	142 (6)	Aquila	19 45 14.4	−08 00 26	9.8	10.0
12	–	Capricornus	21 46 38.8	−21 15 03	11.7	2.9
13	145 (6)	Pegasus	23 06 44.4	+12 46 19	13.8	0.7
14	–	Hercules	16 11 04.9	+14 57 29	14.7	2.5
15	–	Ophiuchus	16 59 51	−00 32 31	14.2	3.0

http://www.astronomy-mall.com/Adventures.In.Deep.Space/obscure2.htm

When it was released half a century ago, the Palomar Observatory Sky Survey showed the universe at a level of detail never before achieved. This major photographic survey covers the sky from the north celestial pole to −30° declination and recorded stars down to an average of magnitude 22.

Astronomers at the time immediately turned to the Survey's 900-plus photographic plates to examine known objects in exquisite detail as well as to discover other sights that earlier studies had missed. We have already met a number of planetary nebulae discovered on the plates by George Abell, but now we turn to a previously unknown globular cluster that was also discovered on those plates. During a survey of the Survey, Abell, along with Halton Arp, Walter Baade, Edwin Hubble, Fritz Zwicky, A. G. Wilson, and others, cataloged 15 "unknown" Milky Way globulars.[5] Each represents a suitably daunting test for today's most diehard deep-sky observers. Table 6.4 lists the vital statistics of these clusters. Five are described in detail in this book.

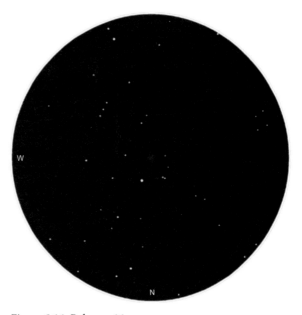

Figure 6.11 Palomar 11

Some of the magnitudes listed may lull you into the false sense that a few of the Palomar 15 must be easy. After all, the brightest of the bunch, Palomar 9 (NGC 6717), shines at magnitude 8.4; that's bright enough to be visible through 50-mm binoculars, right?

---

[5] Truth be told, two Palomar globulars – Palomar 7 and Palomar 9 – were previously known. The former is listed as IC 1276, while the latter is NGC 6717.

Not really. In the case of Palomar 9, the globular is conspicuous, but its appearance is severely muffled by 5th-magnitude Nu^2 (ν^2) Sagittarii, which lies just 1.8′ northeast of the cluster's core. That's a problem!

Eliminating Palomar 9, since it was already known before the Survey, the brightest "new" cluster in the list is Palomar 11 in Aquila. Like some other Palomar globulars, Palomar 11 is relatively nearby, about 42,000 light years away. But it's a victim of circumstance. In between us and it are rich clouds of interstellar dust that muffle the cluster's true luster. Were it not for these nearly opaque clouds, Palomar 11 would undoubtedly be far more spectacular. But, as it is, Palomar 11 is only a vague glow, a mere whisper of its true self.

Palomar 11 dwells 17° due south of brilliant Altair [Alpha (α) Aquilae]. The brightest star in the cluster's immediate area is 5th-magnitude Kappa (κ) Aquilae; aim your finderscope so that Kappa is near the field's western edge. That should bring a wide pair of 5.6-magnitude stars – 56 and 57 Aquilae – into the eastern edge of the same field. Our quest, Palomar 11, lies almost exactly between Kappa and 56/57, and just 4′ southeast of the 8.6-magnitude star SAO 143755. A close-set pair of 12th-magnitude stars lie an equal distance the northeast of the cluster's center.

I can vaguely make out Palomar 11 as a circular patch adjacent to and surrounding several very faint stars through my 10-inch reflector at 106× under dark skies. A couple of the faintest points were undoubtedly members of the cluster, although even the brightest barely break 15th magnitude. Increasing magnification to 127× helped to bring out the cluster, but experimenting with still higher power caused it to fade into the background. Therefore, use magnification judiciously here.

143 Abell 70

★
★★★
★

Target	Type	RA	Dec.	Constellation	Magnitude	Size	Chart
Abell 70	Planetary nebula	20 31.6	−07 05.3	Aquila	14.3p	42″	6.11

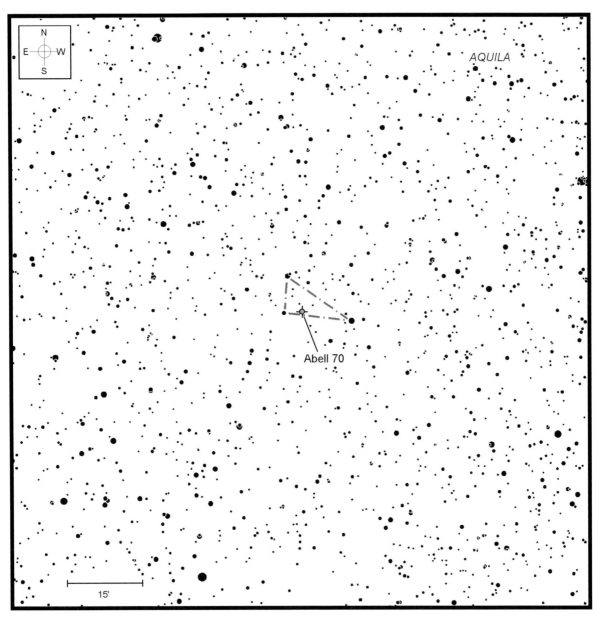

Chart 6.11 Abell 70

Of the more than 80 planetaries listed by George Abell in his 1966 paper "Properties of Some Old Planetary Nebulae," Abell 70 is one of the most unique. Actually, the planetary itself is a stereotypical example of a ring nebula, like M57, with a round shell of gas expanding away from the dim progenitor star. But look carefully and there is clearly more here than just that. Photographs show that the ring has a brightening along its northern edge. The allusion to an engagement ring is umistakable, but that's no diamond. Instead, that odd brighter segment proves to be a distant lens-shaped galaxy that just happens to lie along the same line of sight. Can you spot both through your telescope?

If you will be starhopping, zeroing in on this unlikely odd couple will take some effort, since they lie in the dark southeastern corner of Aquila. Here's one approach. Begin at the wide double star Alpha (α) Capricorni in neighboring Capricornus and scan about 3° to the northeast. Use Chart 6.11 to spot Abell 70 lying to the southeast of the halfway point between two 10th-magnitude stars and just 3' west of an 11th-magnitude sun.

Even though it shines at only magnitude 14.3, Abell 70 is large enough that high magnification is not absolutely necessary to identify its disk from among the neighboring stars. To resolve its distinctive annular shape, however, will take a slow, careful examination using averted vision and probably no less than 350×. An oxygen-III filter will also help to accentuate its subtle annularity.

Remove the filter to try to catch a glimpse of its anonymous galactic companion behind the ring's

Figure 6.12 Abell 70

northern edge, but don't be surprised if it evades detection. It did through my 10-inch, which is why it is shown only as a dotted oval in Figure 6.12. Spotting the interloper's dim presence, rated at 16th magnitude, taxes even my 18-inch reflector under 6th-magnitude skies. Undoubtedly, darker conditions would have made my quest a little easier. But even though the galaxy may come clean, even the darkest, most transparent skies probably will not help to show the planetary's central sun that started it all. It only shines at 19th magnitude.

144 Stephan's Quintet

★
★★★
★

Target	Type	RA	Dec.	Constellation	Magnitude	Size	Chart
Stephan's Quintet	Galaxy group	22 36.0	+33 57.0	Pegasus	–	~3′	6.12

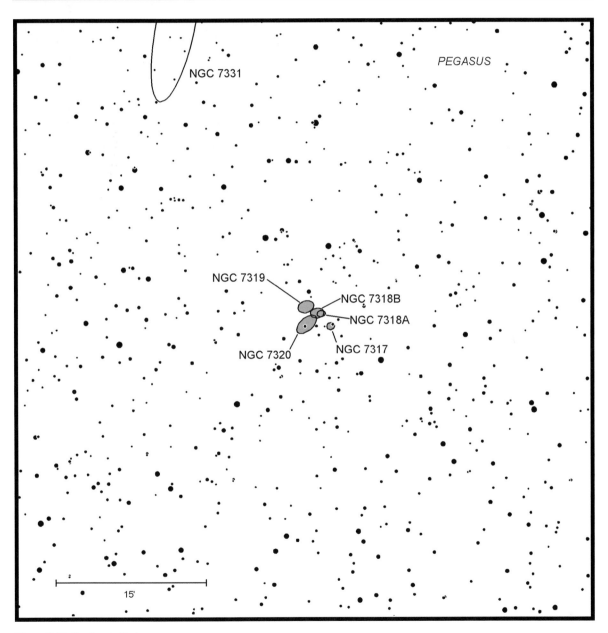

Chart 6.12 Stephan's Quintet

Table 6.5 *Stephan's Quintet*

Target	RA	Dec.	Magnitude	Size
NGC 7317	22 35.9	+33 56.7	13.6	0.8′×0.7′
NGC 7318A	22 35.9	+33 57.9	14.3b	0.8′×0.6′
NGC 7318B	22 36.0	+33 58.0	13.9b	1.4′×0.9′
NGC 7319	22 36.1	+33 58.6	13.1	1.5′×1.1′
NGC 7320	22 36.1	+33 56.9	13.2	2.3′×1.1′

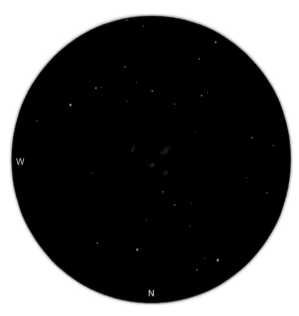

Figure 6.13 *Stephan's Quintet*

The canvas on which our picture of the universe is painted relies on the unwavering acceptance of Hubble's Law. Hubble's Law states that a relationship exists between the distance to a galaxy and the speed at which it is receding from us. The farther away a galaxy is, the greater the speed of its recession and farther its spectral lines are shifted toward the red end of the spectrum.

For Hubble's Law and the Red Shift Principle to be valid, it must work not just for a few galaxies, but for all. And indeed it does – well almost. In the observable universe, there are a few notorious exceptions to these rules. One of the best-known paradoxes is found high in the autumn sky, less than half a degree south of the bright galaxy NGC 7331 in Pegasus. Stephan's Quintet was discovered in 1877 by the director of Marseille Observatory, Édouard Stephan (1837–1923). This group has been the subject of many detailed studies and heated debates ever since.

As the name implies, five galaxies comprise Stephan's Quintet. The first, NGC 7317, has been classified as an E2 elliptical because of its slightly oval disk. Next, NGC 7318 was thought to be a single object when Stephan first spotted it, but it is now known to be two separate, overlapping systems. NGC 7318A is labeled as an E2 elliptical like NGC 7317, whereas NGC 7318B is an SBb barred spiral. NGC 7320 has also been found to be an SBb barred spiral, while NGC 7319 is a wide-armed Sd spiral. All are crammed within a tight 20′ area. Table 6.5 lists them all.

The controversy surrounding these five galaxies stems from measured differences in their spectral red shifts, indicating that they lie at radically different distances away. Four of the galaxies (NGC 7317, 7318A, 7318B, and 7319) appear to be moving away from us at an average of 6,000 km/s, placing them about 270

million light years away. The fifth, NGC 7320, has a measured red shift of only 800 km/s, indicating it to be about 35 million miles distant. What's going on here?

Further examination of detailed photographs of the group shows partial resolution of NGC 7320, with a level of detail similar to relatively nearby galaxies. The other four galaxies in the Quintet show only blurred features that seem to say they lie at much greater distances. From this evidence, along with the difference in red shifts, many astronomers feel that NGC 7320 is simply a chance foreground object superimposed in front of a more distant galaxy quartet. In fact, it turns out that its red shift matches that of NGC 7331, which means that the two may be gravitationally associated. Further studies by Mariano Moles from the Instituto de Matematicas y Fisica Fundamental in Madrid also suggest that NGC 7318B is just passing by and not bound to the group either.

Just as it challenges cosmological theories, Stephan's Quintet also challenges the observing skills of amateur astronomers. Can you spot them? The double galaxy NGC 7318A/B strikes me as the brightest in the group. Through my 10-inch reflector, NGC 7318A/B appears as a small 13th-magnitude glow measuring about $1' \times \frac{1}{2}'$ of arc across. Its twin nuclei are only visible with averted vision, and then just barely at magnifications greater than 250×. Controversial NGC 7320 seems slightly fainter than NGC 7318A/B, but over twice as large. Visually, its disk spans about $2' \times 1'$, with a faint central nucleus seen fleetingly.

Of the final two galaxies, NGC 7317 measures less than 0.5 arc-minute across and looks like a slightly fuzzy "star" even at high power. The existence of its tiny 14th-magnitude disk is further masked by the "glare" of a 12th-magnitude star found only a few seconds of arc away.

Lastly, we come to NGC 7319. Though largest of the lot, this galaxy impresses me as the hardest to see. Even though it shines at 13th magnitude, its surface brightness is very low, which makes detection difficult. A starlike central hub might be seen, but only after an extended examination with averted vision. I find it is best not to strain when trying to see faint, diffuse objects like this. Any stress will generate "noise" between the observer's eye and brain, causing enough distraction to miss a subtle target entirely.

145 Palomar 13

Target	Type	RA	Dec.	Constellation	Magnitude	Size	Chart
Palomar 13	Globular cluster	23 06.7	+12 46.3	Pegasus	13.8	0.7′	6.13

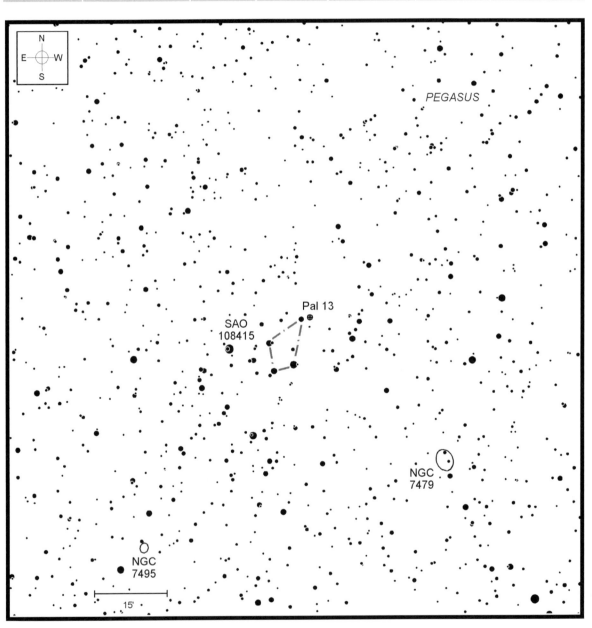

Chart 6.13 Palomar 13

Pegasus is best known to deep-sky observers as a breeding ground for faint galaxies, with more than 100 faint NGC galaxies littering this winged steed. Floating seemingly out of place among those systems is the globular cluster M15, one of the season's finest targets.

Did you know that there is a second globular within Pegasus lying just within the grasp of your 10-inch telescope? That little-known target is Palomar 13, one of those nasty globulars discovered by Abell and company while surveying the Palomar Observatory Sky Survey half a century ago.

Credit for the discovery of Palomar 13 actually goes to A. G. Wilson in 1955. Though an accomplished astronomer, Wilson was not a student of bright deep-sky objects. He nicknamed his find the "Pegasus Globular Cluster," apparently forgetting all about M15. When Abell later published the list of Palomar clusters, he credited Wilson with the discovery, but omitted his ill-considered nickname.

Although Palomar 13 only shines at 14th magnitude, zeroing in on it is simple, thanks to its position $2\frac{1}{2}°$ south-southeast of Markab [Alpha (α) Pegasi], the southwestern star in the Great Square. Follow a winding line of 7th- and 8th-magnitude stars southward to 7th-magnitude SAO 108415, and then look just to its west to find a diamond pattern of four 8th- to 10th-magnitude stars. One more hop, 9′ northwest of the diamond and you'll find an 11th-magnitude star along with Palomar 13 just a bit to its west.

As one of the smallest, faintest globular clusters known, Palomar 13 takes extra effort to be seen. With averted vision through my 10-inch reflector at 181×, I can just eek it out as a small, hazy spot. I can find no evidence of a central concentration, but instead see only a dim, featureless blur.

While most globular clusters maintain a fairly constant distance away from the Milky Way's core as they slowly orbit the galaxy, Palomar 13 swaggers and weaves drunkenly. Over the course of its 1.1 billion-year orbit, the cluster can lie far from the galactic core, as it does now, some 284,000 light years away. At other times, its highly eccentric path plunges

Figure 6.14 Palomar 13

the cluster perilously close to the heart of our galaxy, as it did as recently as 70 million years ago. With each pass, gravitational tidal forces rip and tug at the cluster's stars, pulling some away in the process. The title of a study published in 2001 by M. H. Siegel and colleagues tells the tale: "A Cluster's Last Stand: The Death of Palomar 13." In their paper, the authors report "The small size of this cluster, combined with the shape of its light profile, suggests that Palomar 13 is in the final throes of destruction."[6] Palomar 13, which has already been disrupted severely during previous passages, may be completely disemboweled the next time it passes by the center of the Milky Way.

While in the area, be sure to pay a call on two bonus galaxies. NGC 7479 shines at 11th magnitude, while NGC 7495 is about a magnitude and a half fainter. Both are shown on Chart 6.13.

---

[6] M. H. Siegel, S. R. Majewski, K. M. Cudworth, and M. Takamiya, "A Cluster's Last Stand: The Death of Palomar 13," *Astronomical Journal*, Vol. 121 (2001), pp. 935–50.

146   NGC 1 and NGC 2

	Target	Type	RA	Dec.	Constellation	Magnitude	Size	Chart
146a	NGC 1	Galaxy	00 07.3	+27 42.5	Pegasus	12.8	1.8′×1.1′	6.14
146b	NGC 2	Galaxy	00 07.3	+27 40.7	Pegasus	14.1	1.2′×0.7′	6.14

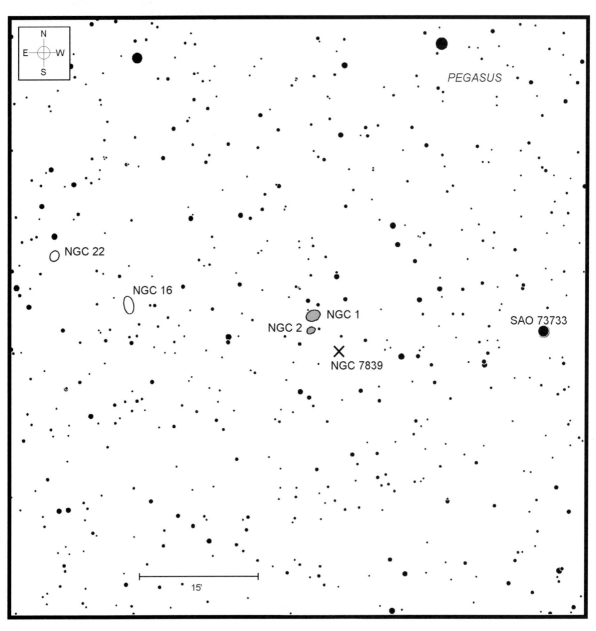

Chart 6.14 NGC 1 and NGC 2

I am always interested in seeing the first of anything, whether it's the opening day of baseball season, the first robin of spring, the first snowflake of winter, or the first object in a particular deep-sky catalog. In the case of the last of these, NGC 1, along with NGC 2, create our next challenge. When he assembled and published the *New General Catalogue* in 1888, John L. E. Dreyer decided to organize the more than 7,800 entries in order of increasing right ascension, beginning at 00 hours. In epoch 1860 coordinates, upon which the original NGC was based, this tiny pair of galaxies came in first, with NGC 1's position listed as 00h 00m 04s. But, in the ensuing years, Earth's precession – that slow, circular, 26,000-year wobbling of our rotational axis – has shifted the celestial coordinate system underneath the stars. Today, in epoch 2000 coordinates, no fewer than 30 NGC objects have "lower" right ascension values than NGC 1.

We are not going to let that little fact spoil our fun, are we? Never! NGC 1 and NGC 2 still present formidable challenges for our telescopes. Together, these spiral galaxies are located 1.4° south of Alpheratz, the star at the northeastern corner of the Great Square (although technically, Alpheratz belongs to neighboring Andromeda; hence its dual identity of Alpha [α] Andromedae). Follow a crooked line of four 6.5-magnitude stars that extends from Alpheratz to the southwest for about $1\frac{1}{2}°$. The fourth star in that line, the yellow giant SAO 73733, is $\frac{1}{2}°$ due west of our galactic pair.

NGC 1 is the brighter of the galaxies, and may actually be visible in telescopes as small as 6 inches in aperture under dark skies. My 10-inch at 58× uncovers a dim, oval disk just 2′ south of an 11th-magnitude star. By increasing magnification to 106× and using averted vision, I can just spot a stellar core in the center. The core becomes easier to see by increasing magnification three- or four-fold, but only under steady seeing.

I also find that 106× is just right for spotting the small, dim disk of NGC 2 through the 10-inch. Like

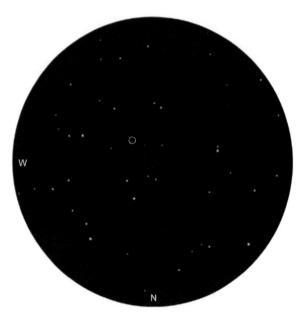

Figure 6.15 NGC 1 and NGC 2

NGC 1, NGC 2 is slightly elongated and oriented approximately northwest–southeast. NGC 2 is three times fainter than its neighbor, so only shows a faint, uniform glow. Get set to use averted vision just to spot it, regardless of magnification.

Despite their close proximity to each other, NGC 1 and NGC 2 do not constitute a true physical pair. Astronomers can tell that NGC 2 is farther away than NGC 1 by studying the red shifts in their spectra, as well as by comparing the level of structural detail visible in photographs. Today's best estimates place NGC 1 at 190 million light years away, while NGC 2's calculated distance is 329 million light years.

Incidentally, some charts also plot another target, NGC 7839, in the immediate area. Although this object can appear "nebulous" through telescopes, it turns out that NGC 7839 is nothing more than a pair of very faint Milky Way stars some 4′ southwest of NGC 2.

147 Dissecting M33

Target	Type	RA	Dec.	Constellation	Magnitude	Size	Chart
Dissecting M33	Emission nebulae and stellar associations	01 33.8	+30 39.6	Triangulum	–	–	6.15

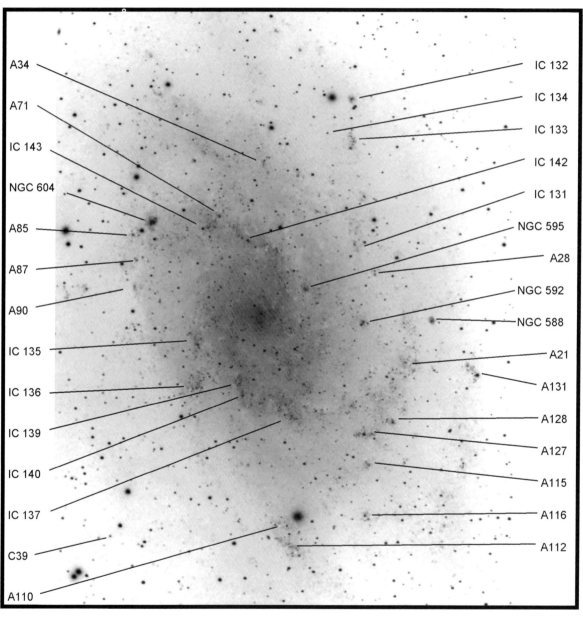

Chart 6.15 M33

Table 6.6 *Nebulae and associations in M33*

Target	Type	RA	Dec.	Magnitude	Size
NGC 588	Emission nebula	01 32.8	+30 38.9	11.5	40″
NGC 592	Emission nebula/stellar association	01 33.2	+30 38.7	13.0	21″
IC 131	Emission nebula	01 33.3	+30 45.1	12.5b	6″
IC 132	Emission nebula	01 33.3	+30 56.7	–	42″
IC 133	Emission nebula	01 33.3	+30 53.3	12.2	24″
IC 136	Stellar association	01 33.4	+30 21.0	11.0b	36″
IC 137	Stellar association	01 33.5	+30 31.3	–	24″
NGC 595	Emission nebula	01 33.6	+30 41.5	13.1	30″
IC 140	Emission nebula	01 33.7	+30 29.1	–	–
IC 139	Stellar association	01 33.8	+30 28.7	13.9b	–
IC 142	Emission nebula	01 33.9	+30 45.3	14.2	30″
IC 143	Emission nebula	01 34.1	+30 47.4	11.4b	18″
IC 135	Emission nebula	01 33.4	+30 25.5	–	36″
C39	Globular cluster	01 34.8	+30 21.9	15.9	2″

In Chapter 2, naked-eye Challenge 12 called you to spot M33, the Triangulum Spiral Galaxy, sans optical aid. We now return to that same scene, this time armed with our telescopes, to do battle again with this fascinating object. The test this time is not in seeing M33; rather, now we are going for deep-sky objects lying within.

Actually, if you succeeded in meeting small-scope Challenge 67 by spotting NGC 604, then your study of M33's internals is already underway. We're back again to dig a little deeper. NGC 604 is just one of a dozen or more targets within that distant galaxy that are potentially visible through 10- to 14-inch telescopes. All are listed in Table 6.6.

NGC 604 is found near the tip of M33's northern spiral arm, about 13′ northeast of the galaxy's core. Several other lesser splotches between the two are just detectable with careful scrutiny through medium apertures. To begin, look for an extremely faint smudge of light just 5′ west of NGC 604, on the other side of a 12th-magnitude Milky Way field star. That's IC 143, a large emission nebula that is just on the edge of visibility with dark skies.

If you missed IC 143, try to spot the somewhat brighter IC 142, slightly further to the west-southwest. IC 142 appears long and lean, extending away from a very faint star.

Figure 6.16 M33

If both prove a little too challenging, give NGC 595 a go. NGC 595 is seen as a small blotch just 5′ west of M33's galactic core. Don't be surprised if M33's nucleus, which is very bright and diffuse, overpowers this emission nebula with anything less than 200×, perhaps even 250×. Try a nebula filter to suppress the core and enhance the nebula.

The stellar association NGC 592 should be a little easier to find, since it is farther from the core. Look for a detached grayish stain 10′ due west of the galactic center. Continuing another 6′ due west puts another target, NGC 588, into view. A narrowband nebula filter may help to isolate this emission nebula from the surroundings, but, at the same time, it may suppress NGC 592 to the point of invisibility.

Two other targets lie far to the northwest of M33's core, near the 9.2-magnitude field star HD 9444. IC 132, a tiny emission nebula with a stellar core that is on the brink of visibility, is just 2′ to the star's west. IC 133, a fainter mist, is 3′ south of IC 132, on the other side of a close-set pair of 14th-magnitude field stars.

M33's southern spiral arm splits into two branches, with several *Index Catalogue* objects speckled throughout the region. Working outward, we first meet IC 135, an elongated haze about 5′ southeast of the core. Again, high magnification and averted vision will likely be needed to pull out this tough object from the bright surroundings. Some 3′ due south of IC 135 is IC 136, another demanding blur of circular light.

Hooking southwestward along the southern arm, IC 139 and IC 140 are next in line. Separated by only 1 arc-minute, both blend into a single knot at 100×, and only begin to hint at their true, dual nature at double that magnification. Finally, we reach IC 137 near the tip of the southern arm's eastern branch. Look for a very subtle glow some 9′ south-southwest of M33's core, although its large, diffuse nature often masks its existence.

148 Globular clusters in the Fornax Dwarf Galaxy

★
★
★
★

	Target	Type	RA	Dec.	Constellation	Magnitude	Size	Chart
148a	Fornax 1	Globular cluster	02 37.0	−34 11.0	Fornax	15.6	0.9′	6.16
148b	Fornax 2	Globular cluster	02 38.7	−34 48.6	Fornax	13.5	0.8′	6.16
148c	NGC 1049	Globular cluster	02 39.8	−34 15.4	Fornax	12.6	0.8′	6.16
148d	Fornax 6	Globular cluster	02 40.1	−34 25.2	Fornax	–	0.6′	6.16
148e	Fornax 4	Globular cluster	02 40.1	−34 32.2	Fornax	13.6	0.8′	6.16
148f	Fornax 5	Globular cluster	02 42.4	−34 06.2	Fornax	13.4	1.7′	6.16

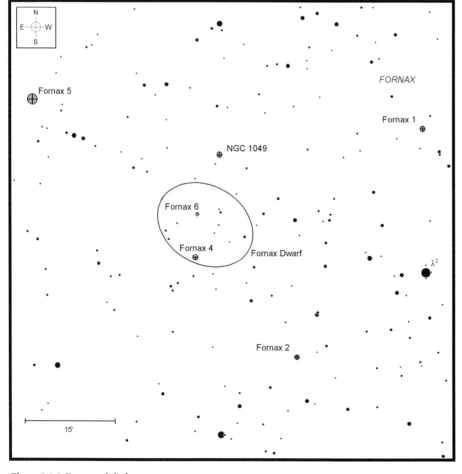

Chart 6.16 Fornax globulars

Let's begin this challenge with a riddle. What's big and round, close at hand, and yet nearly impossible to see? If you answered "the Fornax Dwarf Galaxy," then you are correct! The Fornax Dwarf, a dwarf spheroidal system, covers a $17' \times 13'$ area of our late autumn sky and lies about 530,000 light years from the Milky Way. That's well within the confines of our Local Group of galaxies. And with a magnitude rating of 9.3, it sounds like it should be bright and easy to see. But, when we look its way, it's not there. Even the best photos manage to record only an incredibly dim, elliptical haze peppered by some 19th-magnitude stars!

The Fornax Dwarf Galaxy is a paradox. Even though the galaxy itself is beyond the range of our telescopes even from the darkest observing locations, four of its six known globular clusters are within the grasp of 10-inch, or maybe 12-inch, telescopes.

Of those distant globulars, NGC 1049 is the brightest, so we will begin there. Interestingly, the Fornax Dwarf was only discovered in 1938 by Harlow Shapley, but NGC 1049 was found a century earlier by John Herschel as he cataloged the southern sky from the Cape of Good Hope. Of course, Herschel never realized the true location or distance of what he had found.

Part of the challenge posed by NGC 1049 is in locating it. Fornax is not an easy constellation to see. Your best bet is to start at the pentagon representing the tail of Cetus the Whale and drop about 35° southward along the Cetus–Eridanus border to 3rd-magnitude Beta (β) Fornacis. Binoculars will certainly help with the trip. Once at Beta, look for a small isosceles triangle just to its south formed by Eta-1 (η1), Eta-2 (η2), and Eta-3 (η3) Fornacis. Follow the triangle's "point" (Eta-1) toward the northwest to Lambda-2 (λ2) Fornacis. NGC 1049 is about $\frac{3}{4}$° northeast of Lambda-2.

Although some observers claim to have seen NGC 1049 in telescopes as small as 6 inches in aperture, it's usually considered a difficult catch in 10-inchers when viewed through suburban skies. My old 13.1-inch *f*/4.5 Newtonian showed NGC 1049 as a round glow measuring only about 1 arc-minute across and shining at about 13th magnitude. At 125×, I could just make out a vague starlike central core. Its nucleus became a little more obvious at 214×, but there was little hope of seeing any individual stars, the brightest of which shine at magnitude 18.4.

Three of the Fornax Dwarf's other globular clusters also lie within range of large backyard instruments. The brightest of these, designated Fornax 5 is found 40′ northeast of NGC 1049. In their book *Observing*

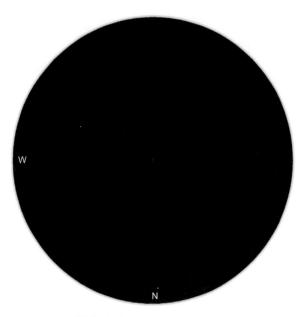

Figure 6.17 NGC 1049

Handbook and Catalogue of Deep-Sky Objects (Cambridge University Press, second edition, 2003), authors Christian Luginbuhl and Brian Skiff report that they have glimpsed both NGC 1049 and Fornax 5 as stellar points in a 6-inch telescope. Ahh, to live in Arizona! Through a 12-inch instrument, they feel that Fornax 5 may even be a little brighter than NGC 1049. Meanwhile, here on the East Coast, it struck me as a little smaller and a little fainter. What is your interpretation?

Globular cluster Fornax 4 is smaller and fainter still. Look for a tiny diffuse disk about 7′ east-southeast of an 8th-magnitude star and 18′ southeast of NGC 1049.

Although it appears the largest of the four, Fornax 2's exceedingly low surface brightness makes it difficult to confirm. Look for it about 37′ southwest of NGC 1049. Luginbuhl and Skiff tell us it is visible in their 12-inch Cassegrainian reflector at 250×, but I was never able to duplicate this feat in my 13.1-inch Newtonian here on Long Island. Maybe I should consider relocating!

This leaves us with leftover globulars that defy detection through all but the largest apertures. Fornax 1 resides 23′ due north of Lambda-2. Its 0.8′ disk rates only magnitude 15.6. Fornax 6, superimposed near the center of its parent galaxy, is even dimmer and smaller. They are plotted on Chart 6.16. Good luck spotting either of those!

149 **Arp 77**

★
★ ★
★

Target	Type	RA	Dec.	Constellation	Magnitude	Size	Chart
Arp 77	Galaxy pair	02 46.3	−30 16.4	Fornax	10.2b	12.7′ ×9.4′	6.17

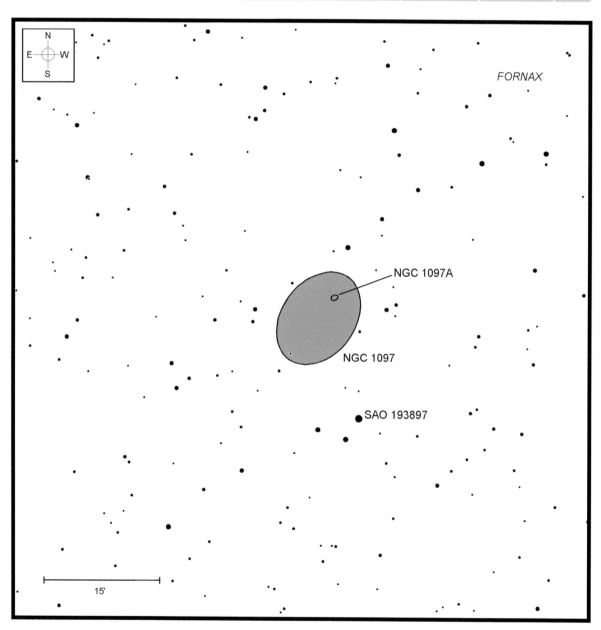

Chart 6.17 NGC 1097 and NGC 1097A

As we have already seen, Fornax the Furnace is not much of a constellation to look at without optical aid. But turn a telescope its way and this nearly empty void in the southern autumn sky comes alive with distant island universes. We visited the central portion of the Fornax Galaxy Cluster in Chapter 5 (medium-scope Challenge 115), but that is just the tip of the iceberg.

One of the most unusual objects in the region is NGC 1097, a barred spiral that has a story to tell. NGC 1097 is classified as a Seyfert galaxy. Seyfert galaxies are named after American astronomer Carl Keenan Seyfert, who was the first to study these unusually active systems. Seyfert's results, published in 1943,[7] showed that these galaxies emit more than 1,000 times the energy produced by our comparatively passive Milky Way. Jets of hydrogen, helium, nitrogen, and oxygen gas scream away from their cores at speeds of up to 9 million miles per hour (4,000 km/s). These jets are believed to originate from near the accretion disks surrounding supermassive black holes buried in their galactic cores.

In the case of NGC 1097, deep photographs record an odd X-pattern of four jets beaming away from the galaxy and extending as far as 15 arc-minutes from the core. Whether or not these are true "jets" or some sort of gravitational perturbation, perhaps from the cannibalization of a smaller galaxy sometime in the past, remains to be seen. If they are true optical jets, however, NGC 1097 is the only galaxy known to have such an extensive system.

Images revealing a wonderfully detailed star-formation ring encircling the core have been returned by the Hubble Space Telescope. The ring, which spans some 6,500 light years, is intertwined with the nucleus, while beyond are vast clouds of ionized hydrogen. The galaxy's bar extends away from this complex central structure, with two tightly wound spiral arms reaching beyond. Gas is believed to transfer from the bar to the ring to fuel the rapid star formation, which appears to be ongoing over the past 7 million years.

Whether or not the disruptive structure we see in NGC 1097 is the result of a past collision with a smaller galaxy is open to debate, but we do know that it has a close-set galactic companion today. NGC 1097A, a

---

[7] Carl K. Seyfert, "Nuclear Emission in Spiral Nebulae," *Astrophysical Journal*, Vol. 97 (1943), pp. 28–40.

Table 6.7 *Arp 77*

Target	RA	Dec.	Magnitude	Size
NGC 1097	02 46.3	−30 16.4	10.2b	12.7′×9.4′
NGC 1097A	02 46.2	−30 13.8	14.6p	0.9′×0.5′

Figure 6.18 NGC 1097 and NGC 1097A

14.6-magnitude elliptical galaxy lying just $6\frac{1}{2}'$ to its northwest, is just within the grasp of 10-inch telescopes.

To find NGC 1097, start at 3rd-magnitude Acamar [Theta (θ) Eridani] in neighboring Eridanus. Head 5° north-northwest of Acamar to the small isosceles triangle of 6th-magnitude Eta-1, Eta-2, and Eta-3 (η1, 2, 3) Fornacis, and then continue another 3° north-northwest to 5th-magnitude Beta (β) Fornacis. Keep on going in the same direction for 1° more, to 7th-magnitude SAO 193921, and then one final degree to 8th-magnitude SAO 193897. The galaxy is just 13′ north-northeast of that last star.

Once it's in view, NGC 1097 rewards patient study. At first glance with about 100×, its core appears oval, elongated northwest–southeast. A closer look reveals a stellar core surrounded by an oval glow with a spiral arm curving away from the bar's northern end. Can you spot it? Look carefully for knots of gas dotting its reach. The southern arm is there, too, but is much fainter. It does include a fairly bright patch of nebulosity about 2′ southwest of the core that might be visible. High magnification and necessarily steady seeing conditions are required to make out this level of detail, however, so wait for those precious moments when things settle down.

NGC 1097A is found just to the north of the northern spiral arm, but it's a tough catch unless you are observing under very dark skies. Even then, NGC 1097A can be difficult to confirm for most of us living in the northern hemisphere because of its far southern declination. Look for a gentle smudge barely above the background skyglow.

As you scan the full breadth of NGC 1097, keep an eye peeled for any rogue stars you may find across its face. That's because, as supernova hunters know, NGC 1097 has yielded three supernovae since 1992 – SN 1992bd, SN 1999eu, and SN 2003B. None appeared brighter than 15th magnitude, so they took a keen eye to spot.

Adding to the mystery of NGC 1097 and NGC 1097A is a study published in 1984 by Halton Arp, Raymond Wolstencroft, and X. T. He.[8] Arp had previously cataloged the pair as number 77 in his list of oddball galaxies; he and his colleagues showed that 34 quasars float within 80 arc-minutes of NGC 1097, an unexpectedly high number. Furthermore, most are situated to the north of the galaxy. None appear brighter than 18th magnitude, so we shouldn't expect to see them directly through our telescopes. Still, their presence does make me wonder what impact they may be having, if any, on what we can see.

---

[8] H. Arp, R. D. Wolstencroft, and X. T. He, "Complete Quasar Search in the NGC 1097 Field," *Astrophysical Journal*, Vol. 285 (1984), pp. 44–54.

150 Arp 41

	Target	Type	RA	Dec.	Constellation	Magnitude	Size	Chart
150	Arp 41	Galaxy pair	03 09.8	−20 34.9	Eridanus	10.5b	7.4′×6.4′	6.18

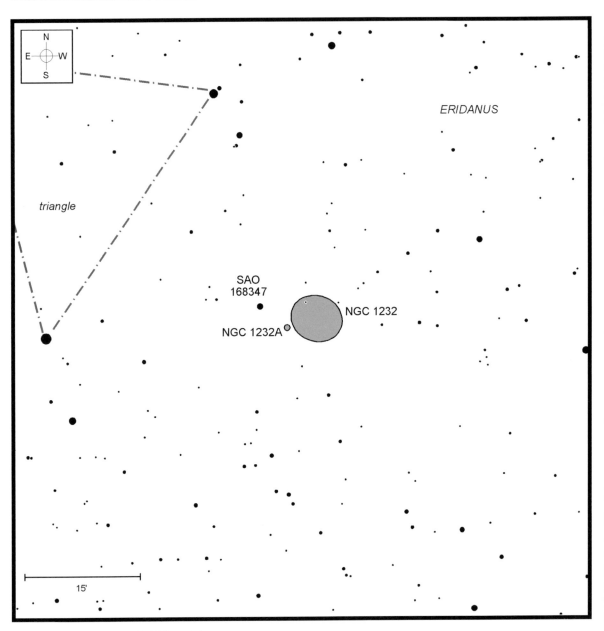

Chart 6.18 NGC 1232 and NGC 1232A

Without a doubt, the sky's most impressive galaxies through large-aperture scopes are face-on spirals. Their graceful arms sweep around the galactic centers in broad arcs, embracing the cores in a galactic pirouette. When we think of face-on spiral galaxies, two that immediately come to mind are the Whirlpool Galaxy, M51, and the Pinwheel Galaxy, M101. Both of these are in our spring sky, but there are many other fine examples of spirals scattered throughout all four seasons that reveal intricate detail to those who take the time to seek them out.

One of the most interesting, yet least observed, examples is NGC 1232 in the constellation Eridanus. Have you seen it? William Herschel himself discovered NGC 1232 on October 20, 1784, and later added it to his catalog as entry H-II-258, signifying it as the 258th example of a "faint nebula." But unless you are a devout deep-sky observer who likes to strike an independent path, the chances are that you have not seen it. After all, it is not listed in the Herschel 400 list promoted by the Astronomical League; for that matter, it is not in their follow-up Herschel II list either.

Hints of the galaxy's spiral structure are detectable through 10-inch scopes, and even in smaller apertures, from under clear, dark skies. As with M51 and, especially, M101, the trick to see it is to first move those instruments away from light pollution. Even a modicum of skyglow can soak up spiral arms as a dry sponge would a puddle of water. By freeing yourself from the extinguishing effect of society's needless waste of nighttime lighting, NGC 1232 really blossoms. It even has a hidden surprise.

Finding NGC 1232 is itself something of a challenge. There are a few different ways to approach the hunt, but I usually follow a line drawn from Betelgeuse through Rigel in Orion and then onward to the southeast for another 24° to the wedge-shaped asterism of Tau-6, Tau-7, Tau-8, and Tau-9 (τ6, 7, 8, 9) Eridani. I then ride the rapids westward past Tau-5 (τ5) to pause at Tau-4 (τ4). NGC 1232 is $2\frac{1}{2}$° northwest of Tau-4 (τ4). Bumping that way will put a triangle of 7th-magnitude stars into your finderscope's view. A 9th-magnitude star, SAO 168347, is just west of the triangle and in the same low-power eyepiece field as NGC 1232.

In that low-power eyepiece, NGC 1232 will appear as little more than a dim, ill-defined smudge. My 10-inch reflector at 58× shows that the galaxy's brighter

Table 6.8 *Arp 41*

Target	RA	Dec.	Magnitude	Size
NGC 1232	03 09.8	−20 34.9	10.5b	7.4′×6.4′
NGC 1232A	03 10.0	−20 36.0	15.2p	0.5′×0.3′

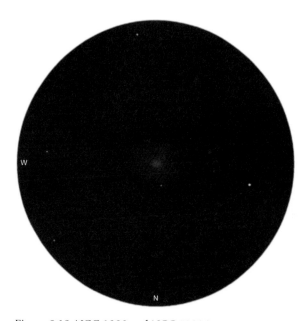

Figure 6.19 NGC 1232 and NGC 1232A

central core is surrounded by the amorphous haze of the spiral-arm halo. By swapping that eyepiece for a 7-mm (181×), using averted vision and waiting for those rare moments of steady seeing, I can see hints of the spiral arms that appear so prominent in photographs. The more obvious arm curves away from the core west to east around the northern perimeter.

Cheating a bit and calling my 18-inch into service, the same eyepiece, now magnifying the scene 294×, actually loses the spiral detail. Switching to 171× brings it back into view and even adds a blotchy texture corresponding to embedded H II regions. I have little doubt that the 10-inch could have shown this same level of detail under better conditions.

NGC 1232 is cross-listed in Halton Arp's catalog of strange and unusual galaxies as Arp 41 (Table 6.8) because of the company it keeps. As you look for hints of the spiral arms in NGC 1232, keep an eye out for the detached, circular glow a companion galaxy, NGC 1232A. Don't be put off by its listed brightness of nearly 15th magnitude; this barred spiral strikes me as perhaps two full magnitudes brighter. Its soft glow is a little to the south of the halfway point between NGC 1232 and SAO 168347.

The gravitational tug-o'-war between these two galactic contenders is playing havoc with NGC 1232's spiral arms in much the same way as NGC 5195 is disrupting M51. In many ways, the two pairs resemble each other, with a luminous bridge of interstellar matter connecting NGC 1232 with its satellite, just as with M51.

151 Abell Galaxy Cluster 426

★
★ ★
★

Target	Type	RA	Dec.	Constellation	Magnitude	Size	Chart
AGC 426	Galaxy cluster	03 18.6	+41 30.0	Perseus	–	190′	6.19

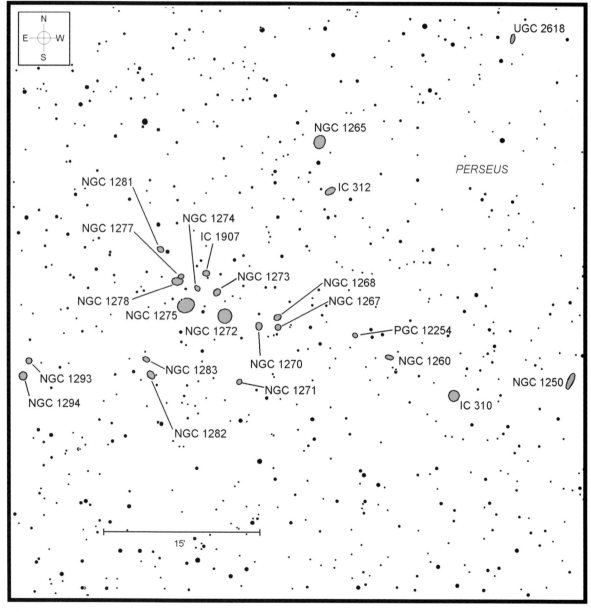

Chart 6.19 Perseus Galaxy Cluster

Ever since I got my first "good" telescope in 1971, an 8-inch reflector, I have been fascinated with the Perseus Galaxy Cluster. One reason I am so fond of this collection of more than 500 galaxies is that the cluster grows as the telescope's aperture increases. Small backyard scopes will show the two big kids on the block, NGC 1272 and NGC 1275, but even the largest amateur instruments fail to show all of the "little guys."

Also known as Abell Galaxy Cluster 426, the Perseus Galaxy Cluster lies close enough to the plane of the Milky Way that the surrounding field is strewn with nearby stardust, which creates a very pretty overall effect. There are plenty of hidden treasures scattered throughout cluster's full 190′ expanse. How many can you pick out in your telescope?

At about 230 million light years away, the Perseus Galaxy metropolis is easy to find just 2° east-northeast of the Demon Star, Algol [Beta (β) Persei]. The brightest of the bunch, NGC 1275, shines at about 12th magnitude and lies just 2′ east of an 11th-magnitude field star at the heart of the cluster. Through your telescope, as through my 10-inch, NGC 1275 will look like a small, slightly elliptical glow that is punctuated by a bright stellar core.

When we gaze upon that small blemish, we are seeing a seething system in uproar, a galaxy that is emitting tremendous amounts of X-radiation. The full story behind NGC 1275 was first unveiled in 1943, when Carl Seyfert included it in his list of galaxies with active nuclei. NGC 1275 is also included as 3C 87 in the *Third Cambridge Catalogue* of quasars and radio sources, published in 1959. And not just any radio source, mind you, but the second strongest in the entire sky – only Centaurus A (NGC 5128) is stronger.

Studies now reveal that filamentary jets of material are erupting from the core of NGC 1275 and discharging into space at greater than 5.3 million miles per hour (2,400 km/s). Hubble images reveal what the fuss is all about. We are not looking at one galaxy when we view NGC 1275; rather, we are looking at two separate galaxies that are intimately embraced by gravity. Photographs clearly show the disrupted disk of a dust-laden spiral galaxy cutting through a large elliptical galaxy at speeds approaching 7 million miles per hour (3,000 km/s). In the process, gravitational tidal forces distort each galaxy, compressing huge clouds of interstellar matter and triggering new star formation.

Table 6.9 *Abell Galaxy Cluster 426*

Object	RA	Dec.	Magnitude	Size
UGC 2598	03 14.1	+41 17.5	14.4p	1.5′×0.5′
IC 301	03 14.8	+42 13.4	14.2p	1.2′ × 1.2′
UGC 2608	03 15.0	+42 02.2	13.7p	0.9′ × 0.7′
UGC 2614	03 15.3	+42 41.8	14.3p	1.6′ × 0.7′
NGC 1250	03 15.4	+41 21.3	12.8v	2.2′ × 0.8′
UGC 2617	03 16.0	+40 53.2	13.8p	2.5′×0.8′
UGC 2618	03 16.0	+42 04.5	14.5p	1.2′ × 0.4′
IC 309	03 16.1	+40 48.3	14.5p	0.9′ × 0.9′
IC 310	03 16.7	+41 19.5	12.7v	1.4′ × 1.4′
NGC 1260	03 17.5	+41 24.3	14.3b	1.1′ × 0.6′
PGC 12254	03 17.9	+41 27.1	13.9v	0.7′ × 0.5′
IC 312	03 18.1	+41 45.3	14.4p	1.4′ × 0.7′
NGC 1265	03 18.3	+41 51.5	12.1v	1.7′ × 1.4′
NGC 1267	03 18.7	+41 28.1	14.1	0.8′ × 0.8′
NGC 1268	03 18.7	+41 29.3	14.2p	1.0′ × 0.7′
UGC 2654	03 18.7	+42 18.0	14.2p	1.4′ × 0.5′
NGC 1270	03 19.0	+41 28.2	13.1v	1.0′ × 0.8′
NGC 1271	03 19.2	+41 21.2	13.9v	0.7′ × 0.3′
NGC 1272	03 19.4	+41 29.5	11.7v	1.8′ × 1.8′
NGC 1273	03 19.4	+41 32.4	13.2v	1.0′ × 0.8′
IC 1907	03 19.6	+41 34.8	14.2v	0.9′ × 0.8′
NGC 1274	03 19.7	+41 32.9	14.0v	0.8′ × 0.4′
NGC 1275	03 19.8	+41 30.7	11.9v	2.2′ × 1.8′
NGC 1278	03 19.9	+41 33.8	12.4v	1.4′ × 1.0′
NGC 1277	03 19.9	+41 34.4	13.4v	0.8′ × 0.4′
NGC 1281	03 20.1	+41 37.8	13.3v	0.9′ × 0.4′
NGC 1282	03 20.2	+41 22.0	13.9b	1.2′ × 0.9′
NGC 1283	03 20.3	+41 23.9	13.5v	0.9′ × 0.6′
UGC 2686	03 21.0	+40 47.9	14.4	0.9′ × 0.4′
UGC 2689	03 21.5	+40 48.1	14.1	1.4′ × 0.5′
NGC 1293	03 21.6	+41 23.6	14.5b	0.8′ × 0.8′
NGC 1294	03 21.7	+41 21.6	14.3b	1.0′ × 1.0′
UGC 2698	03 22.0	+40 51.8	13.9p	1.0′ × 0.6′
UGC 2717	03 24.6	+40 41.5	14.3p	1.0′ × 0.8′
IC 320	03 26.0	+40 47.4	14.6p	1.2′ × 1.0′
UGC 2733	03 26.1	+41 15.2	14.5p	1.0′ × 0.6′

Figure 6.20 Perseus Galaxy Cluster

After NGC 1275, the next brightest member of the Perseus clan is NGC 1272. You will find it just 5′ to the west. Although NGC 1272 is listed at visual magnitude 11.7, you may feel, as I do, that its surface brightness is at least a full magnitude lower. My most pleasing view of NGC 1272 through my 10-inch is at 106×.

While its magnitude is only 13.2, NGC 1273 is actually easier to see than NGC 1272. The difference is in the apparent size. NGC 1273 is only half the diameter of its larger, but dimmer, neighbor. The resulting higher surface brightness (magnitude 11.7 versus 13.2) helps to make this little spiral an easier

catch than the larger elliptical. All three galaxies form a triangle at the center of the cluster.

A fourth very faint, distended patch of light actually turns that triangle into a parallelogram. This dimmest member of the four is actually two galaxies, NGCs 1277 and 1278, separated by less than 50″. By increasing magnification to around 175× and waiting for steady seeing, both should be distinguishable as individual objects, with the larger and brighter NGC 1278 lying southeast of NGC 1277.

In between NGC 1278 and NGC 1273, just 2.7′ northwest of NGC 1275, is NGC 1274, a tough test indeed. Notes made through my 13.1-inch at 125× recall simply a very faint, very small blur.

NGC 1270 lies further southwest of NGC 1272. That same night through my 13.1-inch, I recorded it as a "dim, poorly concentrated glow." To its west are NGC 1267 and NGC 1268, a pair of even fainter challenges.

Can you see that 10th-magnitude star 7′ north of NGC 1275? Look just 1′ to its east for dim NGC 1281. Can you spot it? Seeing its tiny disk, which measures just 0.9′×0.4′ across and shines at magnitude 13.3, is tough enough, but the scattered light from that star can really get in the way. If you have an occulting eyepiece, described in Chapter 1 under "Your eyepieces," you might give it a try here.

Once you conquer the galaxies described above, branch out on your own to find even more. Table 6.9 lists all of the galaxies in AGC 426 that are brighter than magnitude 14.5, which is a reasonable cutoff point for 10- to 14-inch telescopes, while Chart 6.19 plots the cluster's population center. As you can see, there are plenty of other, mostly smaller and fainter systems, waiting for you.

152   Abell 12

Target	Type	RA	Dec.	Constellation	Magnitude	Size	Chart
Abell 12	Planetary nebula	06 02.3	+09 39.3	Orion	13.9p	37″	6.20

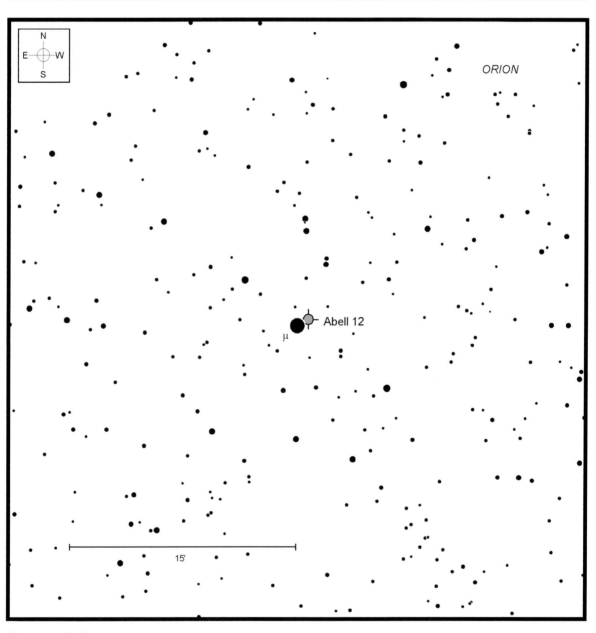

Chart 6.20 Abell 12

Deep-sky objects can be challenging for several reasons. Some are especially faint, while others are especially small, and still others are so large that they can't fit into a single eyepiece field. Or the problem might be that a particular target is so close to another, noticeably brighter, object that the light from that intruder all but obliterates the quarry. The latter problem plagues planetary nebula Abell 12. It shines at about 12th magnitude, which is not exactly bright, but is also not exceptionally dim for a telescope 10 inches or more in aperture. The problem, however, is that it is located a scant arc-minute away from 4th-magnitude Mu (μ) Orionis.

The nearness of the planetary to the star begs the question, "Are we looking at a former binary star system where one of its members is no longer with us?" According to data from the HIPPARCOS parallax-measuring satellite, we are not. Mu (μ) Orionis is 152 light years away. While the distance to Abell 12 is not as well established, most estimates place it at least 7,000 light years from Earth. When we look toward this stellar odd couple, we are just looking at a chance alignment of two objects at quite different distances.

Were it not for that interloper, Abell 12 would undoubtedly have been discovered by William Herschel and included in the NGC. As it is, however, he and his son John missed it during their intensive sky searches. Instead, George Abell was first to uncover this little bubble of expanding hydrogen gas while he searched the Palomar Observatory Sky Survey photographic plates in 1955. In the original image, Mu looked as though it was blowing a bubble while chewing gum. Only after more intensive study did Abell realize that the "Bubble Gum Nebula," as I think of it, was a separate object.

Mu (μ) Orionis is found 6° to the northeast of Betelgeuse [Alpha (α) Orionis] along the Hunter's raised arm, so Abell 12 is a snap to aim toward. To have any chance of seeing it, however, takes some effort. The light from that star, nearly 1,600 times more intense than the planetary, can completely overwhelm the tiny 37″ disk.

To have the best chance at finding objects situated like Abell 12, your telescope's optics must be clean and well collimated. If either of these criteria is not met,

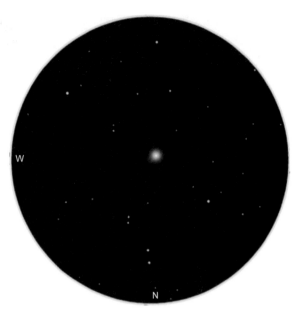

Figure 6.21 Abell 12

then glare from Mu will certainly block the planetary. Check collimation during the day so that you do not have to worry about it at night. Also remember that any sky haze will smear the starlight enough to overpower the planetary, so wait for a clear, dry winter's eve.

When all of these pre-existing conditions are met, point your telescope toward Mu with a reasonably high power eyepiece in place. Experience shows that a magnification 150× or more produces the best results. Experiment with different eyepieces as seeing conditions allow, but you will probably find that a narrowband (UHC-type) or oxygen-III filter is a must, regardless. Not only will the filter enhance the planetary, it will also help to stifle the star's light somewhat in the process.

Under optimal conditions, Abell 12 shows a perfectly round disk with sharply defined edges, especially on the side facing away from the star. My notes recall the disk as appearing uniform in texture, although some other observers report a very subtle ring-like appearance. Abell 12 has been glimpsed through telescopes as small as 6 inches in aperture, so, even if you do not have a double-digit aperture, give this one a go. You just might be surprised.

153   Sharpless 2-301

Target	Type	RA	Dec.	Constellation	Magnitude	Size	Chart
Sharpless 2-301	**Emission nebula**	07 09.8	−18 29.8	**Canis Major**	–	9.0′×8.0′	6.21

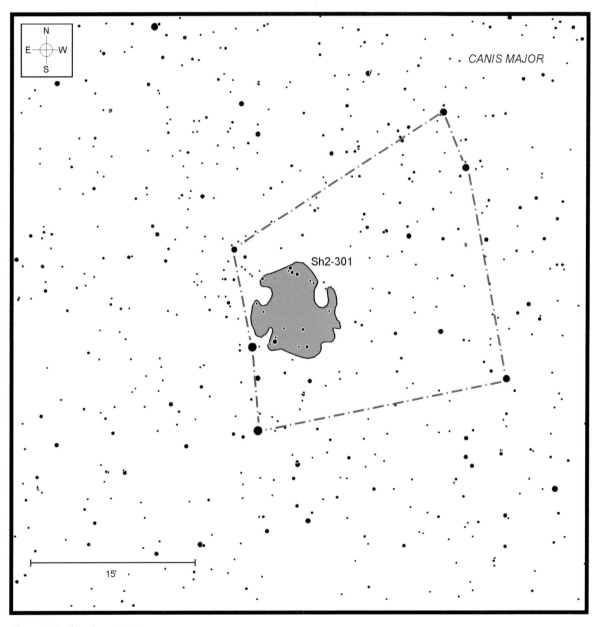

Chart 6.21 Sharpless 2-301

The 1950s was a banner decade for deep-sky catalogs. Not only did it give us such seminal works as the Abell catalogs of galaxy clusters and planetary nebulae, it also closed with the release of the second edition of Stewart Sharpless' famous catalog of emission nebulae. Sharpless had assembled his collection of objects from his research at the United States Naval Observatory's Flagstaff Station in Arizona. The "Sharpless 2" catalog,[9] a revised version of a list he first published in 1953 while at Mount Wilson Observatory,[10] lists 313 emission nebulae (Hydrogen-II regions, as Sharpless preferred to call them) that are among the most spectacular photographic sights that the Milky Way has to offer.

While a few of Sharpless's entries, such as Sh2-25 (better known as M8, the Lagoon Nebula) and Sh2-49 (M16, the Eagle Nebula), are well known to visual observers, most are among the most challenging objects to see visually.

If you have never made a concerted effort to see some of the lesser-known Sharpless objects, then this challenge, Sh2-301 in Canis Major, is a good introduction to the sport. You will find it about 6° east-southeast of Sirius [Alpha (α) Canis Majoris], within a diamond of six 6th- to 8th-magnitude stars.

Unlike many of the Sharpless objects, which can cover swaths of sky larger than the fields of many telescopes, Sh2-301 measures just $9' \times 8'$ across. That's small enough to fit easily into a single eyepiece field, yet is still large enough to be apparent if you are aimed in the right direction. My best view with my 10-inch comes at 58×, with a 22-mm eyepiece and a

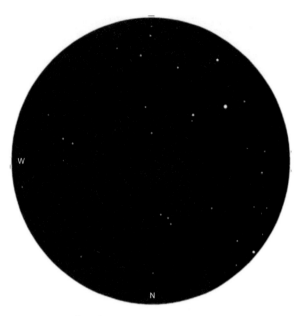

Figure 6.22 Sharpless 2-301

narrowband (UHC-style) filter in place. Without a filter, the nebula is all but impossible to see, even from a dark site. With a filter in place, however, hints of the cloud may be glimpsed from suburban settings as long as light pollution is minimal toward the south.

From dark-sky sites, Sh2-301 reveals an unusual shape that is best described as an irregular, three-lobed fog with thin lanes of dark nebulosity threaded throughout. Several stars appear superimposed on the nebula, including a small triangle of three that are almost dead center. The brightest superimposed sun is a 10th-magnitude point found toward the cloud's southeastern edge. Another tuft of brighter nebulosity appears to surround a triangle of 12th-magnitude stars at the northern border. These stars make a handy gauge for judging the full extent of the nebula.

---

[9] Stewart Sharpless, "A Catalogue of H II Regions," *Astrophysical Journal Supplement*, Vol. 4 (1959), p. 257.
[10] Stewart Sharpless, "A Catalogue of Emission Nebulae near the Galactic Plane," *Astrophysical Journal*, Vol. 118 (1953), p. 362.

154   Arp 82

	Target	Type	RA	Dec.	Constellation	Magnitude	Size	Chart
154a	NGC 2535	Galaxy pair	08 11.2	+25 12.4	Cancer	13.3	3.3′×1.8′	6.22
154b	NGC 2536	Galaxy pair	08 11.3	+25 10.8	Cancer	14.7	0.9′×0.7′	6.22

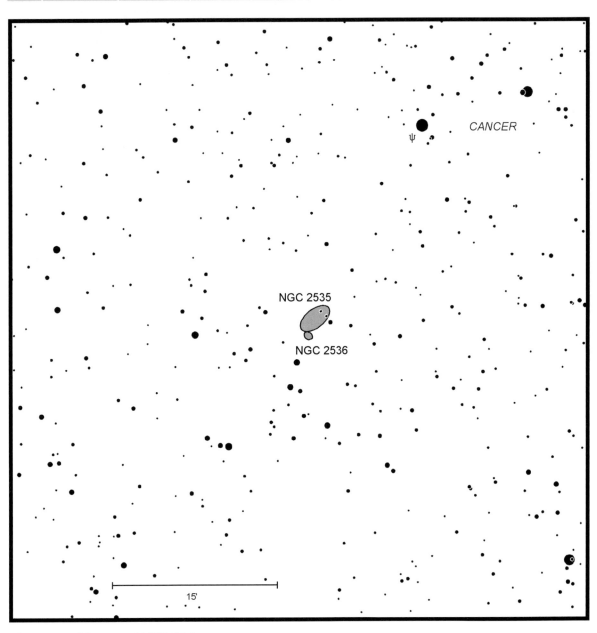

Chart 6.22 NGC 2535 and NGC 2536

The constellation Cancer the Crab may not be much to look at, but it holds some fascinating objects within its emaciated body. Case in point: Arp 82, the 82nd entry in Halton Arp's catalog of interacting galaxies. Made up of NGC 2535 and NGC 2536, Arp 82 is a strange pair that seems to be experiencing a galactic version of arrested development. As galaxies formed in the early universe, theory says that massive amounts of nebulosity came together in quick succession, triggering vast expanses of rapid star formation. Then, as each galaxy aged, the rate of star formation slowed.

Not so with NGC 2535 and NGC 2536, however. A 2007 study led by Mark Hancock of East Tennessee State University concluded that these galaxies did not create their stars early in their existence.[11] Instead, based on observations from NASA's Galaxy Evolution Explorer, the Spitzer Space Telescope, and the Southeastern Association for Research in Astronomy Observatory at Kitt Peak, Arizona, Hancock and colleagues concluded that star formation was triggered much later in their lives when waves of new stars suddenly began to pop into existence. Today, we see NGC 2535, the larger galaxy, connected by a bridge of material to NGC 2536, its smaller companion. A second long tail of interstellar material sweeps away from NGC 2535 directly opposite NGC 2536.

As we study individual stars within each galaxy, we find few that are greater than about 2 billion years old. That's a small fraction of the universe's estimated age of 13.7 billion years. Apparently, before the galaxies swung past each other about 2 billion years ago, they were both mostly nebulosity. Star formation was accelerated only after the gravity of one galaxy swirled up the material in the other. A second close passage around 2 million years ago resulted in a second burst of activity. Why the galaxies in Arp 82 didn't begin to form stars earlier like other galaxies remains the stuff of future studies.

To see this unusual pair for yourself, begin at 4th-magnitude Kappa (κ) Geminorum and scan eastward about one finder field to 6th-magnitude Psi (ψ) Cancri. Arp 82 is 21′ southeast of Kappa, next to a line of 12th- and 13th-magnitude stars.

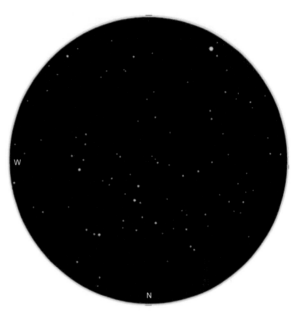

Figure 6.23 NGC 2535 and NGC 2536

That line of stars makes an excellent reference to estimate the apparent size of each galaxy as well as their separation. In photos of the area, the full span of NGC 2535, arms included, matches the length of that line of faint stars very closely, although in deep photos the spiral arm opposite NGC 2536 actually curves completely around the northwestern end of the line. Whether or not that effect can be seen visually, however, remains doubtful.

The view through my 10-inch reflector at 106× is an interesting study in surface brightness versus magnitude. Although NGC 2535 has the brighter magnitude value, its larger apparent size causes the resulting surface brightness to be lower than "fainter" NGC 2536. As a result, NGC 2536, although nearly stellar in appearance at that magnification, impresses me as a bit brighter. Larger NGC 2535 appears slightly oval and oriented northeast–southwest. Its weak concentration only hints at a centralized core, although photos show a sharp nucleus surrounded by an active ring of star formation. Despite the complex nature of its spiral arms, no hint of structure was seen with the 10-inch. Indeed, even my 18-inch offered little help beyond brightening up the galaxies to some extent. Perhaps even larger instruments can reveal the complex nature of these galaxies that images show so spectacularly.

[11] M. Hancock, B. J. Smith, C. Struck, M. L. Giroux, P. L. Appleton, V. Charmandaris, and W. T. Reach, "Large-Scale Star Formation Triggering in the Low-Mass Arp 82 System: A Nearby Example of Galaxy Downsizing Based on UV/Optical/Mid-IR Imaging," *Astronomical Journal*, Vol. 133 (2007), pp. 676–93.

155 The major satellites of Uranus

Target	Type
Uranus: Major satellites	Year-round

If you were like me, then one of the first things you saw through a telescope was Jupiter and its four Galilean moons. I was absolutely amazed with the idea that I could come back after only a few hours and see that the moons had moved with respect to the planet and each other. Imagine Galileo's astonishment when he saw this wondrous sight for the first time.

The Galilean satellites can be seen in a pair of steadily held binoculars. Even Saturn's largest moon, Titan, can be spotted with little effort through the smallest backyard scopes. What about the moons orbiting the planet Uranus? Have you ever seen any of them? Not possible, you say?

Of the 27 known satellites in the Uranian family, four stand out, just as the four Galilean satellites do among the Jovian clan. William Herschel discovered the first two Uranian moons on January 11, 1787, six years after he had discovered the planet itself. The next two remained undetected until the British astronomer William Lassell (1799–1880) spotted them on October 24, 1851. It is these four that we hope to catch through our own telescopes.

The four major moons of Uranus – Titania, Oberon, Ariel, and Umbriel – take their names from the writings of William Shakespeare and Alexander Pope. Oberon was the King of the Fairies in *A Midsummer Night's Dream*, while Titania was his queen. Ariel was the leading sylph in Alexander Pope's poem *The Rape of the Lock* and, coincidentally, also the spirit who serves Prospero in Shakespeare's *The Tempest*. Finally, Umbriel was named for the "dusky melancholy sprite" in *The Rape of the Lock*.

All four are made of rock mixed with a frozen cocktail of ammonia, methane, and water ice. Like the major moons of Jupiter, Saturn, and Neptune, Titania, Oberon, Ariel, and Umbriel all orbit Uranus almost

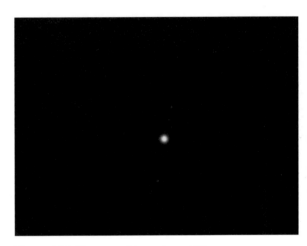

Figure 6.24 Uranus and its major satellites

exactly in the plane of the planet's equator. Owing to the odd, sideways tilt of Uranus's rotational axis, however, this can cause the moons to appear at odd angles with respect to the planet itself. As the planet orbits the Sun, the paths followed by the moons appear to change shape and orientation over time. Currently, they appear oriented northwest–southeast, tracing out narrow clockwise paths around the planet.

Uranus's two largest moons, Titania and Oberon, each measures about 900 miles (1,500 km) in diameter. Photos taken during the 1986 Voyager 2 flyby show that Titania's battered surface is crisscrossed by valleys and faults that stretch for hundreds of miles. Although many impact craters are seen, there are fewer than on Oberon. Together, the fewer craters and many fault lines tell us that Titania's surface is younger than Oberon's. Since both moons are the same physical age, this means that Titania must have once experienced a higher level of geologic activity than its neighbor.

Neither Titania nor Oberon ever exceeds 14th magnitude. Of the two, Oberon is easier to spot since it orbits farther from Uranus and, therefore, can appear farther away from the planet's glare. Even at greatest elongation, however, it is always less than an arc-minute away from the planet.

Ariel, measuring 720 miles (1,158 km) across, resembles Titania in Voyager photos, with an interweaving network of valleys and faults threaded across its frigid surface. These faults, as well as those on Titania, may be the result of the rapid cooling of the moon's hot core shortly after formation.

Finally, Umbriel (727 miles, or 1,170 km, in diameter) bears a strong resemblance to Oberon. Its dark surface is scarred by many impact craters, but

lacks the faults and valleys that characterize Titania and Ariel.

Umbriel and Ariel get about as bright as Titania and Oberon but, because they orbit even closer to the planet, present an even greater challenge to observers. Ariel, the closest, is never more than 15″ from Uranus.

For the best chance at seeing these four challenging targets, wait for Uranus to be near opposition, when its distance away from Earth is least. Then, patiently wait until each satellite is at its greatest elongation point from the planet. To help you find out when that will occur, several popular software programs, including *Guide* by Project Pluto, plot the location of each Uranian moon for any date and time. *Sky & Telescope* magazine's website also has an excellent JavaScript utility that will show each moon's position in real time.

156 Triton: Largest satellite of Neptune

Target	Type	Magnitude
Neptune: Triton	Year-round	13.5

Nearly two magnitudes fainter than Uranus, Neptune poses a fun challenge for observers with binoculars and small telescopes. But let's take that one step better. The idea of seeing not only the planet, but its largest moon, Triton, is a test that few amateurs have ever tried and even fewer have succeeded in mastering.

Triton was spotted in 1846 by William Lassell, not long after Neptune itself was discovered. The name comes from the son of the Greek god of the sea, Poseidon, which in turn became the Roman god Neptune.

Triton measures 1,600 miles (2,700 km) in diameter and orbits Neptune at an average distance of 220,000 miles (354,760 km). Its trajectory is unique among the Solar System's large satellites in that it is circling its home world backwards. That is, it orbits in the opposite direction of Neptune's rotation. More than likely, that means Triton experienced a different origin elsewhere in the Solar System than Neptune, perhaps within the Kuiper Belt. Triton was only later captured into orbit by the gravity of the eighth planet.

Despite the fact that Neptune is much farther from Earth than Uranus, Triton is actually slightly brighter than the largest moons orbiting Uranus. That unexpected result comes from Triton's unusually high reflectivity, or albedo. Triton's albedo is rated 0.75, which means that its bright surface reflects 75% of the sunlight striking it. By comparison, Oberon has an albedo of 0.31. Our own Moon's albedo is even lower, only 0.12.

The stark reflectivity of Triton also means that it absorbs very little solar energy, what there is of it at a mean distance of 2.8 billion miles (4.5 billion km) from the Sun. As a result, Triton's surface temperature is estimated to be just 35 K (−235 °C, −391 °F). At this extreme temperature, tremblingly close to absolute zero, methane, nitrogen, and carbon dioxide all freeze solid to the moon's surface.

Figure 6.25 Neptune and its satellite Triton

When that surface was studied for the first and only time during the Voyager 2 flyby of 1989, Triton revealed a large polar cap covering nearly one-third of its hemisphere. Along the edge of that cap, Voyager detected unusual ice volcanoes, fountains of liquid nitrogen or methane compounds erupting from deep beneath the surface.

Triton never strays more than 17″ from Neptune, which means that very high magnification is needed for any hope of separating the satellite from the planet's glare. Fortunately, Triton is a magnitude brighter than Ariel, while Neptune is two magnitudes fainter than Uranus, so the glare factor is not quite as extreme. Still, the same advice applies: wait for Neptune to reach opposition, and then wait for Triton to be at its greatest elongation point from the planet. Again, several popular software programs, such as Project Pluto's *Guide*, plot the location of Triton with respect to Neptune for any date and time. Print a finder chart before heading out and the king of the sea just might moon you.

7
Monster-scope challenges

15-inch and larger telescopes

This final chapter throws down the gauntlet. Today's largest backyard telescopes, defined here as 15 inches in aperture or larger, have the capacity to show us sights that only a few years ago were either unknown entirely or, at the very least, not understood. All require good viewing conditions, whether that is excellent transparency, steady seeing, freedom from light pollution, or a combination of all three.

How many of these ultimate challenges can *you* see?

157 Galaxies beyond M44

Target	Type	RA	Dec.	Constellation	Magnitude	Span	Chart
Galaxies beyond M44	Galaxy collection	08 40.4	+19 40	Cancer	–	66′	7.1

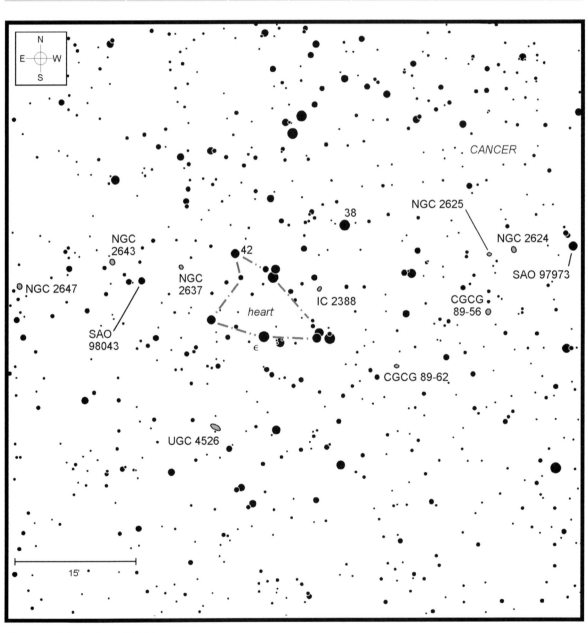

Chart 7.1 Galaxies beyond M44

Table 7.1 *Galaxies beyond M44*

Target	RA	Dec.	Magnitude	Size
NGC 2624	08 38.2	+19 43.6	14.6	0.7′×0.5′
NGC 2625	08 38.4	+19 43.0	14.5p	0.4′
CGCG 89-56	08 38.4	+19 35.8	15.2	0.7′×0.2′
CGCG 89-62	08 39.2	+19 28.9	15.6	0.4′
IC 2388	08 39.9	+19 38.7	15.7	0.5′×0.3′
UGC 4526	08 40.9	+19 21.3	14.8p	1.4′×0.2′
NGC 2637	08 41.2	+19 41.5	15.4	0.5′×0.4′
NGC 2643	08 41.9	+19 42.1	15.6	0.7′×0.4′
NGC 2647	08 42.7	+19 39.0	15.1	0.7′×0.6′

One of my favorite binocular open clusters in the entire sky is M44, the Beehive Cluster or Praesepe, in Cancer the Crab. It's a wonderful target through just about any pair of binoculars. Even the smallest, cheapest pair will show a rich vault of stars. Nine of the brightest stars near the center of the cluster form a distinctive V asterism that is sometimes called the Heart of the Crab. The Heart points toward the southwest and always attracts attention.

Hidden among the stars of M44 are no fewer than eight distant galaxies. Until 1987, most of us knew nothing of them. That was the year when the *Uranometria 2000.0* star atlas was published. It showed the sky to a depth never before captured in a convenient star atlas format, and immediately shed light on thousands of objects that no amateurs, except possibly for a few extreme deep-sky hunters, even knew existed.

The biggest problem in spotting the galaxies beyond the Beehive is not their dimness, although that is clearly a factor. That dilemma pales compared with the predicament caused by the cluster's stars. Those stars shine as brightly as 6th magnitude, so they easily out-dazzle these puny 14th- and 15th-magnitude galaxies. Although a 15-inch or larger scope will be needed to see all of the galaxies listed in Table 7.1, the brightest two or three can be captured with a 10-inch under dark skies.

German-born astronomer Albert Marth discovered the first five background galaxies in 1864 while working as William Lassell's assistant at Lassell's observatory in

Malta. Personally, I find NGC 2647 the easiest to spot. Look for it about 1′ northeast of a 13th-magnitude star along the cluster's eastern flank. At 206×, my 18-inch displays a faint, round glow no more than 20″ across, as shown in Figure 7.1. I've never seen any hint of a central core, but others have noted a stellar nucleus, so be sure to check for yourself. The largest backyard telescopes might also pick up a pair of very faint stars to either side of the galaxy.

Scan westward across the full breadth of M44 for NGC 2624, another galaxy bright enough to be seen in 10-inch scopes. Look for it 8′ to the east of 8th-magnitude SAO 97973 at the cluster's western edge. Through my 18-inch at 206×, this tiny spiral galaxy appears as a faint, circular glow measuring perhaps 25″ across and slightly brighter toward the center. A very faint field star lies nearby to the southwest, while another galaxy, NGC 2625, is about 3′ to the east-southeast.

NGC 2625, an elliptical galaxy, is smaller and fainter than its neighbor, but I can still make it out with direct vision in the 18-inch at 206×. Using averted vision, however, reveals a very faint star on the western edge of the galaxy – don't mistake it for a supernova! Interestingly, some catalogs list NGC 2625 as more than a magnitude fainter than NGC 2624. My own experience, however, points to it being no more than half a magnitude dimmer, so take those numbers under advisement.

Moving southward, CGCG 89-56, a pair of close-set galaxies listed in the *Catalogue of Galaxies and Clusters of Galaxies* by Fritz Zwicky, is visible as a very small, dim blur. The Digital Sky Survey shows two small, edge-on spirals oriented east–west and north–south but, try as I might through my 18-inch, I cannot resolve them.

Another member of Zwicky's catalog, CGCG 89-62, lies about 3′ west-northwest of a pair of 10th- and 11th-magnitude stars set southwest of the Heart. I have never been able to see it convincingly, despite being able to see a 15th-magnitude star just 1′ to its east. Perhaps you will have better success.

IC 2388 is found 1.6′ to the south of a 10th-magnitude star that marks the tip of an isosceles triangle near the Heart of the Crab. Marth missed it that night in 1864, but it is visible through my 18-inch under 5th-magnitude skies, so don't shy away from it just because it is not in the NGC.

Figure 7.1 Galaxies beyond M44

Continuing eastward, we come to NGC 2637, a fairly easy catch because of its position comparatively far from any bright cluster stars. The nearest bright stars are 42 Cancri, 7′ to its west-northwest, and SAO 98043, 5′ south-southeast. Averted vision is still required to see this little system but, with patience, it should come through pretty clearly.

Finally, try to spot NGC 2643 about 1′ northwest of an 11th-magnitude cluster star. NGC 2643 is the toughest of the NGC galaxies here. I can barely make it out through the 18-inch at 206× by using averted vision, and then only as a very dim, very small smear no more than 15″ across.

158   Leo I

★
★ ★
★

Target	Type	RA	Dec.	Constellation	Magnitude	Span	Chart
Leo I	Galaxy	10 08.5	+12 18.5	Leo	9.8	9.8′×7.4′	7.2

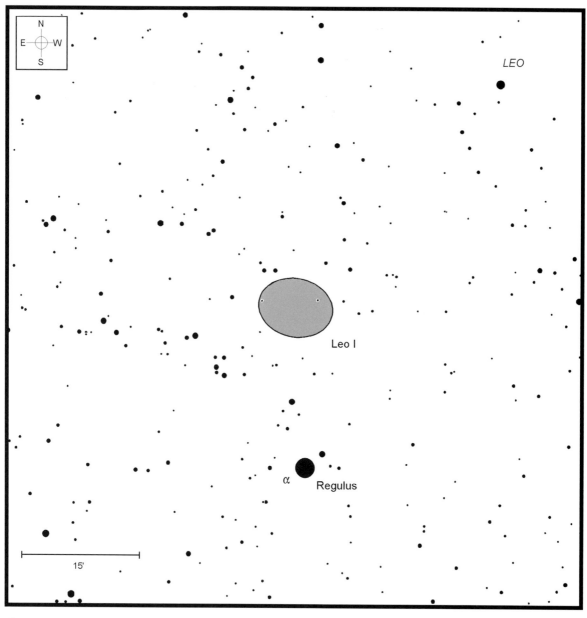

Chart 7.2 Leo I

Six decades ago, while scanning a Palomar Sky Survey plate of the area around brilliant Regulus in Leo the Lion, astronomers Robert Harrington and A. G. Wilson noticed a faint blur of light just $\frac{1}{3}°$ north of the star. They may have thought at first that the glow was just an internal lens flare caused by stray starlight, but it soon became apparent that they had discovered something very real. We know their find today as Leo I, one of many dim dwarf spheroidal galaxies orbiting the Milky Way. Leo I has a total mass equivalent to only about 20 million solar masses. That's a drop in the galactic bucket compared with the Milky Way, which is comparable to about 600 million solar masses.

Dwarf spheroidal galaxies are something of an enigma. Like elliptical galaxies, they show very little evidence of nebulosity or star formation. As with many dwarf galaxies, the stars in Leo I contain a very small proportion of heavy elements: that is, elements heavier than hydrogen and helium. That tells us that the stars are very old, since heavier elements are in abundance in young stars. Dwarf galaxies do, however, contain unusually high amounts of dark matter. Indeed, pound for pound, dwarf spheroidal galaxies have more dark matter than any other type of galaxy in the universe.

Studies conducted by Jaroslaw Klimentowski of the Nicolaus Copernicus Astronomical Center in Warsaw, Poland, and colleagues suggest that dwarf spheroidal galaxies begin life embedded in halos of dark matter.[1] Over time, the dwarf galaxies orbit around and pass near or through the larger galaxies to which they are gravitationally bound. During each close passage, the dwarfs are stripped of some of their original mass, including star-forming nebulosity. While these kamikaze maneuvers have stripped away the clouds of interstellar matter, the dark matter halo appears to remain largely unaffected.

There are two dwarf spheroidal galaxies within the constellation of Leo, appropriately dubbed Leo I and Leo II. Let's take them in order. Finding Leo I is no problem at all. Simply point toward Regulus and look 20′ due north.

What's that, you don't see it? I'm not surprised. Although Leo I shines at an integrated magnitude of 11,

Figure 7.2 Leo I

its surface brightness is closer to 15th. That depressing number coupled with the blinding glare from Regulus is enough to hide Leo I from view. In fact, it did for all the classical observers, such as Messier, Méchain, and the Herschels.

Spotting Leo I successfully takes a little planning. First, switch to a high enough magnification so that Regulus can be moved out of the field. Avoid the temptation to use too high a power, however, since the galaxy's soft glow is easily dispersed into the background. Leo I spans about 10′, so its span appears to extend about a quarter of the way back toward Regulus.

Keep in mind that, after you aim at Regulus, your "observing eye" will no longer be fully dark adapted. Therefore, aim toward Regulus with your *other* eye and shift it out of the field before switching back. Now, compare the view with Figure 7.2. Can you see the pair of 12th-magnitude stars just off the northeastern edge of the galaxy, as well as the triangle of 12th-magnitude stars off its northwestern edge? Position them toward the edge of the field, and then slowly sweep back and forth for the galaxy's vague oval glow. Remember, it will appear quite large in the field.

The best eyepiece for Leo I through my 18-inch reflector has proven to be a 10-mm Tele Vue Radian. The combination yields 206× with a real field of view of approximately 17′. Although Leo I fills a good part of

[1] Jaroslaw Klimentowski, Ewa L. Lokas, Stelios Kazantzidis, Lucio Mayer, and Gary A. Mamon, "Tidal Evolution of Disky Dwarf Galaxies in the Milky Way Potential: The Formation of Dwarf Spheroidals," *Monthly Notices of the Royal Astronomical Society*, Vol. 378 (2007), pp. 353–68.

the view, there is still enough open sky around the edge to identify the galaxy.

While you are ferreting out Leo I, keep an eye out for IC 591, a small spiral galaxy just 15′ to the west. Look for a tiny, dim smudge just west of a very faint field star.

Using the right eyepiece and knowing the field will help you add this dwarf spheroidal to your list of conquered challenges with comparative ease. But don't get too cocky. Spotting its sibling, Leo II, our next target, is an even greater challenge.

★
★
★
★

159 Leo II

Target	Type	RA	Dec.	Constellation	Magnitude	Size	Chart
Leo II	Galaxy	11 13.5	+22 09.2	Leo	11.9	10.1′ × 9.0′	7.3

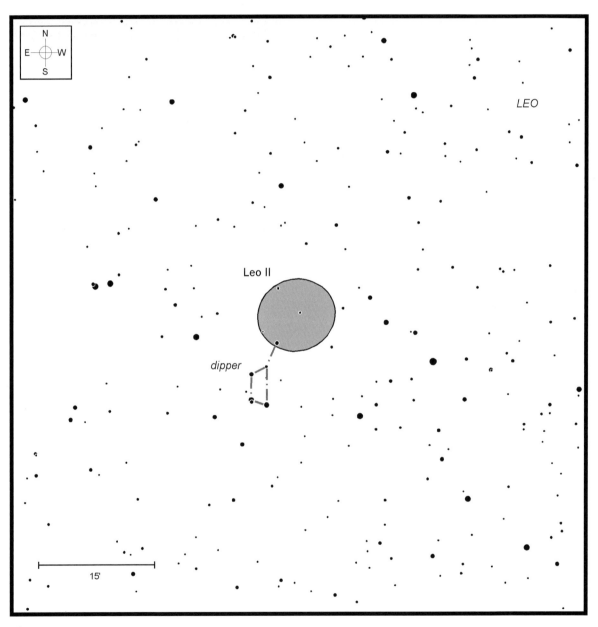

Chart 7.3 Leo II

The team of Harrington and Wilson identified a second dwarf spheroidal galaxy within Leo on the Palomar Sky Survey and announced both discoveries in 1950.[2] Later studies found that Leo II is approximately 760,000 light years away from our Milky Way. That's nearly four times farther than the Large and Small Magellanic Clouds. (Leo I is even more remote, some 900,000 light years away.)

In 2007, a team of Japanese astronomers using the 8.2-meter Subaru Telescope atop Mauna Kea, Hawaii, found that, like Leo I, Leo II is dominated by old stars and has very little interstellar gas and dust. Their results showed that the stars in the outer portions of the galaxy contain very little metal. The lower the metal content in a star, the older it is assumed to be, since metals are only formed in the cores of massive stars. When those stars explode, their metallic cores seed nearby nebulae, ultimately to end up in future generations of stars. The stars in Leo II, lacking those metals, are thus comparatively old. Interestingly, however, stars towards the inner regions of the galaxy show a greater abundance of metals, and so must be comparatively young. From this, the astronomers concluded that most stars in Leo II formed about 8 billion years ago, and that the process started from the outside and moved in toward the center. It ceased some 4 billion years ago, except, strangely, at the very center of the galaxy.

To pin down this second dwarf spheroidal, start at Zosma [Delta (δ) Leonis] at the tail end of the lion. Viewing through your finder, hop a degree north-northeast to an 8th-magnitude star and then another 45′ farther still to a pair of close-set 9th-magnitude suns. With an eyepiece producing about a half-degree real field, slide those two stars toward the

Figure 7.3 Leo II

eastern edge of the view. As you do, a reddish 8th-magnitude star should slip into the western edge. In between, look for an asterism of 11th- to 13th-magnitude stars that resemble the bright stars of the Pleiades – that is, a tiny, short-handled dipper. The bowl is aimed toward the south, while the handle extends northward. Using this tiny dipper pattern as a guide, look 4′ northwest of the handle star for a soft glow. That's Leo II.

It took several failed attempts before I finally located Leo II through my 18-inch telescope on an especially clear evening a few springs ago. At 171×, it appeared as a very faint, oval disk extending perhaps 6′ × 4′, or about half of its full size in photographs. Increasing magnification to 206× revealed a brighter core at the heart of the galaxy (Figure 7.3) that went unnoticed at the lower power.

[2] R. G. Harrington and A. G. Wilson, "Two New Stellar Systems in Leo," *Publications of the Astronomical Society of the Pacific,* Vol. 62, No. 365 (1950), p. 118.

160   Hickson Compact Galaxy Group 50

Target	Type	RA	Dec.	Constellation	Magnitude	Span	Chart
Hickson Compact Galaxy Group 50	Galaxy group	11 17.1	+54 55.3	Ursa Major	–	<1'	7.4

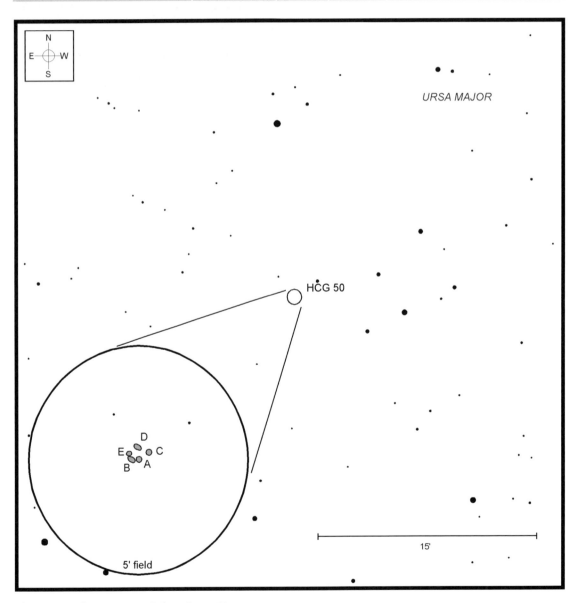

Chart 7.4 Hickson Compact Galaxy Group 50

Table 7.2 *Members of HCG 50*

Target	RA	Dec.	Magnitude	Size
HCG 50A	11 17.1	+54 55.0	18.7b	0.2′×0.2′
HCG 50B	11 17.1	+54 55.0	18.9b	0.3′×0.1′
HCG 50C	11 17.1	+54 55.3	19.6b	0.2′
HCG 50D	11 17.1	+54 55.4	19.5b	0.3′×0.1′
HCG 50E	11 17.1	+54 55.2	20.0b	0.2′×0.1′

Figure 7.4 *Hickson Compact Galaxy Group 50*

Observing compact galaxy groups from Paul Hickson's 1982 study is an interesting challenge for owners of the largest backyard telescopes. Most are at the edge of detection, even from dark sites, and so offer great tests of our observing skills as well as the quality of our instruments' optics.

I especially enjoy the hunt for number 50 in his list of 100. Hickson Compact Galaxy Group 50 (HCG 50) is comprised of five dismally faint galaxies crammed into an incredibly tight 42″. Table 7.2 lists each.

Why is this my favorite Hickson challenge? Largely for the company it keeps. HCG 50 is just $\frac{1}{3}^{\circ}$ east-southeast of M97, the famous Owl Nebula, which is fascinating in its own right through large apertures. I rarely head out on a spring night when I don't at least stop by to pay the Owl a quick visit. If the sky is clear enough, I'll often continue on to HCG 50. If I can make out even the faintest hint of it, then I know the night is, in fact, quite special.

To spot HCG 50, begin by placing M97 in the center of a medium-power eyepiece; I tend to favor a 12-mm eyepiece (171×) in my 18-inch. Moving the Owl off to the northwestern edge of the field brings a distinctive asterism of four stars in the shape of the Hercules keystone into view. HCG 50 is just 9′ east of the keystone's center. For scale, each side of the keystone measures between 3′ and 4′ in length.

From my suburban backyard observatory, the best I can report of this challenge is seeing a slightly elongated smudge in the right spot that is just barely above the background at 171× and 206× (Figure 7.4). I have never been able to resolve the individual galaxies, even at 300× or more. The smudge I saw was the combination of HCG 50A and 50B, the two brightest in the group. HCG 50C and 50D are both fainter and evaded my quest, as did the dimmest, HCG 50E.

The fact that I can see any evidence of the group's existence through a 5th-magnitude sky attests to the fact that we should not be put off by faint magnitude values. Notice how the magnitude for each galaxy here is its "b" or "blue magnitude." That is often how galaxies are listed, but it can be deceiving. In general, so-called blue magnitudes are biased toward lower values than visual magnitudes. A target with a blue magnitude of, say, 18 may appear closer to magnitude 16 visually.

Studies reveal that HCG 50 is an incredible 1.7 billion light years away. No wonder it's so tough – but at the same time, what a great challenge!

161 Palomar 4

Target	Type	RA	Dec.	Constellation	Magnitude	Size	Chart
Palomar 4	Globular cluster	11 29.3	+28 58.4	Ursa Major	14.2	1′	7.5

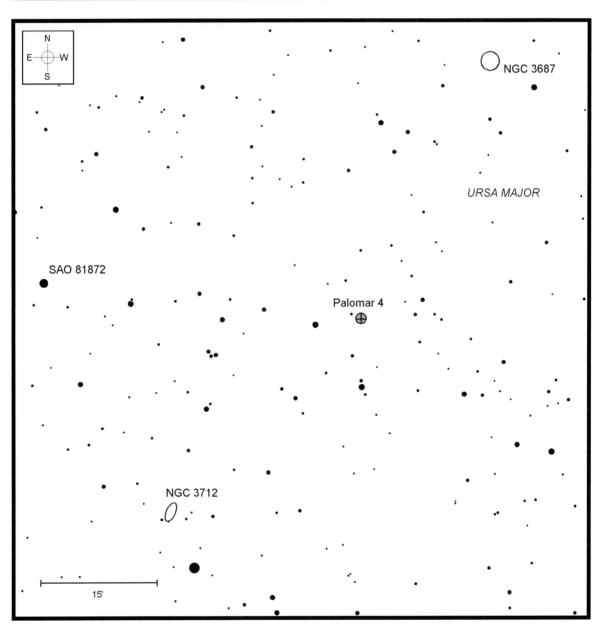

Chart 7.5 Palomar 4

When you think of deep-sky objects in Ursa Major, you probably don't think of globular clusters. Galaxies, sure! Planetary nebulae, well, there's the Owl Nebula. But globular clusters? Probably not.

Well, guess what? There is actually a renegade globular cluster within the Great Bear. Today, we know it as Palomar 4, the fourth entry in the list of 15 challenging globular clusters discovered on the plates of the Palomar Sky Survey of 60 years ago. This particular globular was discovered by Edwin Hubble in 1949 and confirmed a year later by A. G. Wilson. Based on early observations, Palomar 4 was initially misclassified as a dwarf spherical galaxy, in part because of its remote distance. Palomar 4 is estimated to be 356,000 light years away from the Sun. That's farther away than the two Magellanic Cloud satellite galaxies. More recent studies, however, prove that Palomar 4 is, indeed, a member of the Milky Way's family of globular clusters, albeit the most secluded, save for Arp-Madone 1 in the southern constellation Horologium.

Figure 7.5 Palomar 4

Palomar 4 lies near the star Alula Australis [Xi (ξ) Ursae Majoris], the Bear's big toe on its hind leg. Alula Australis, together with its twin star just to the north, Alula Borealis [Nu (ν) Ursae Majoris], is located 10° south of Phecda [Gamma (γ) Ursae Majoris]. Be sure to pause at both on your way to Palomar 4, as each is an attractive binary system. Nu's golden primary is accompanied by a 10th-magnitude companion 7″ to its south-southeast. Xi is tougher thanks to its 4.3- and 4.8-magnitude suns being separated by only about $1\frac{1}{2}''$ at present. Incidentally, Xi was the first binary star whose components were proven to be physically related. William Herschel drew that conclusion in 1802 after he had found each star changed orientation relative to the other after a span of 22 years. A quarter century later, the French astronomer Felix Savary determined that the stars orbited each other in just under 60 years.

Palomar 4 is parked about $3\frac{1}{2}°$ southeast of Alula Australis. To find it, move $2\frac{1}{2}°$ southeastward from the star to a westward-aimed isosceles triangle of 7th-magnitude stars. Extend the triangle's eastern side southward to 7.8-magnitude SAO 81872, and then turn toward the west, passing a 9.3-magnitude star 12′ later and coming to a 9.8-magnitude star in another 24′. Palomar 4 is just 6′ further west of that last star. Look for its faint glow just to the west-southwest of a 13th-magnitude field star.

The cluster's far-flung distance coupled with its inherently weak stellar concentration conspires to dilute it to nothing more than a very faint glow measuring just an arc-minute or so across. From my suburban backyard observatory, my 18-inch reflector revealed only the faintest hint of the cluster at 206×, and then only fleetingly with averted vision (Figure 7.5). From darker sites in North Carolina and California, respectively, experienced deep-sky observers Kent Blackwell and Stephen Waldee have independently seen Palomar 4 through 10-inch telescopes. Congratulations to both; that's quite an accomplishment.

If searching for Palomar globulars is your thing, check large-scope Challenge 142 for a complete listing of the 15 objects in that elite list.

162   Copeland's Septet

Target	Type	RA	Dec.	Constellation	Magnitude	Span	Chart
Copeland's Septet	Galaxy group	11 37.9	+21 58.9	Leo	–	5′	7.6

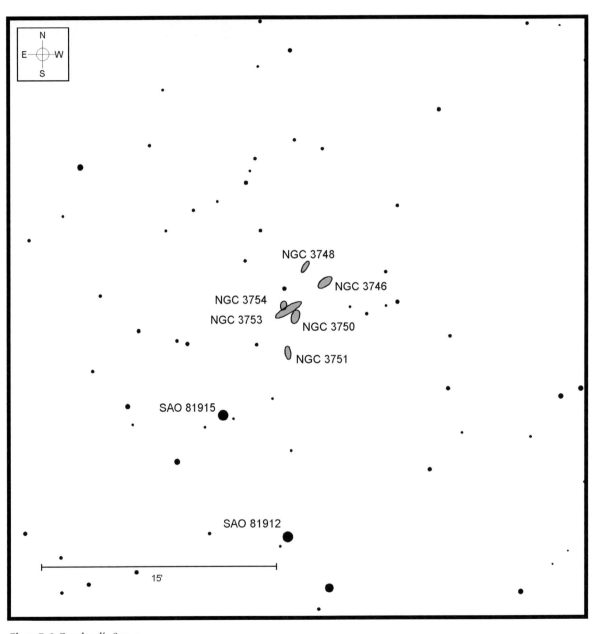

Chart 7.6 Copeland's Septet

Table 7.3 *Copeland's Septet*

Galaxy	RA	Dec	Magnitude	Size
NGC 3745	11 37.7	+22 01.3	15.2	0.4′×0.2′
NGC 3746	11 37.7	+22 01.0	14.2	1.1′×0.5′
NGC 3748	11 37.8	+22 01.6	14.8	0.8′×0.3′
NGC 3750	11 37.9	+21 58.4	13.9	0.9′×0.5′
NGC 3751	11 37.9	+21 56.2	13.9	0.9′×0.4′
NGC 3753	11 37.9	+21 58.9	13.6	1.9′×0.5′
NGC 3754	11 37.9	+21 59.1	14.3	0.6′×0.4′

Figure 7.6 *Copeland's Septet*

Copeland's Septet is formed by NGCs 3745, 3746, 3748, 3750, 3751, 3753, and 3754, all packed into a tight, 5 arc-minute circle. The group was discovered by Ralph Copeland, an assistant to William Parsons, Earl of Rosse, Ireland, on February 9, 1874, through the Earl's famous 72-inch "Leviathan of Parsonstown" reflector at Birr Castle. At the time, it was the largest telescope in the world.

No doubt Copeland was thrilled at his find of seven uncharted galaxies. Anxious to record his discovery, he noted the position of each galaxy carefully . . . and incorrectly. In his excitement, Copeland messed up. From what historians tell us, Copeland confused nearby reference stars, misplacing the location of his seven galaxies. Not knowing this, John Dreyer used Copeland's incorrect data when he compiled the *New General Catalogue* 14 years later. As a result, when other observers subsequently tried to find Copeland's Septet, as the group was later nicknamed, they found nothing.

The mystery of the Septet gone astray has subsequently been solved, but some observing handbooks from before the 1980s still note it as "missing." It was found in time for Paul Hickson to include it as 57 in his study of 100 compact galaxy groups. Table 7.3 lists the seven members.

To zero in on the group's correct location, begin at 5th-magnitude 92 Leonis, found about $5\frac{1}{2}°$ north from Denebola [Beta (β) Leonis]. From there, shift 45 arc-minutes farther northwest to a pair of 7th-magnitude suns. Copeland's Septet lies another 9′ to their northwest, and should be in the same field of view. Use about 100× to find the field and then increase the magnification as needed to identify individual galaxies. I've had the greatest success around 200×.

My first encounter came in 1988 through my old 13.1-inch reflector. NGC 3753 impressed me as the brightest of the group, although at 13th magnitude that's a relative term. My notes recall an elongated object punctuated by a brighter core. A smaller, fainter galaxy just to its northeast, NGC 3754, went undetected in my 13.1-inch. I finally nabbed it recently with my 18-inch reflector. NGC 3750, just 7′ southwest of NGC 3753, was also visible faintly as a tiny patch of light less than 1 arc-minute across.

If you thought NGC 3750 and 3753 were demanding, then you're going to have an even tougher time with NGCs 3745, 3746, and 3748. This galactic trio lies about 3′ further northwest and is set in an elongated triangle. The brightest of the three, NGC 3746, shines weakly at magnitude 14.2, while NGC 3748 comes in at magnitude 14.8 and NGC 3745 at magnitude 15.2. Even my 18-inch has trouble spotting those last two, but supernova hunters take note. NGC 3746 played host to two faint supernovae in 2002 and 2005 (SN 2002ar and SN2005ba, respectively). Will it spawn another?

The seventh galaxy in the septet, NGC 3751 lies alone, 3′ south of NGC 3753. It impresses me as about as bright as NGC 3748, but a little smaller.

★★★★★ 163 Abell Galaxy Cluster 1367

Target	Type	RA	Dec.	Constellation	Magnitude	Span	Chart
AGC 1367	Galaxy cluster	11 44.5	+19 50	Leo	–	100.8′	7.7

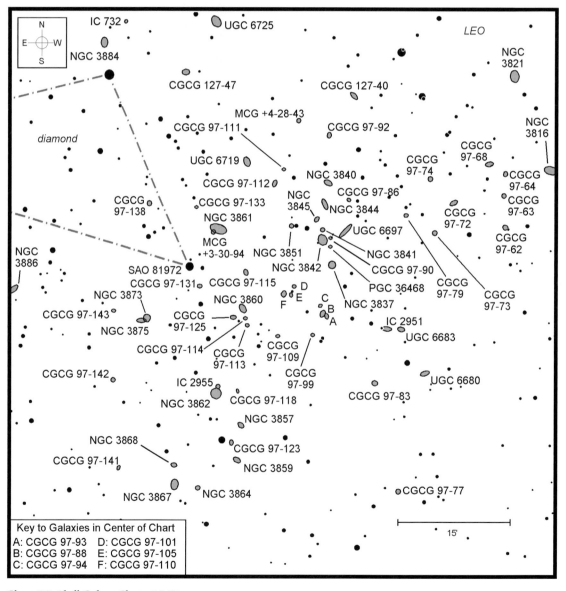

Chart 7.7 Abell Galaxy Cluster 1367

AGC 1367 was first cataloged as a single system by
George Abell more than half a century ago and known
informally as the Leo Cluster. Its gravitational grasp
reaches across a huge region some 100 million light
years wide and holds 117 individual galaxies lying an
average of 300 million light years away. That places it at
about the same distance away as AGC 1656, the Coma
Cluster, which we will wrestle with next as
monster-scope Challenge 164. Even at this great
distance, AGC 1367 still covers 101 arc-minutes of
our sky, an area equal to two Full Moons stacked
side by side.

   Together, AGC 1367 and AGC 1656 form the Coma
Supercluster, the closest massive cluster of galaxies to
the Virgo Supercluster, of which our Milky Way is an
outlying member. The Coma Supercluster appears to lie
at the center of a galactic filament known as the Great
Wall. The Great Wall, discovered in 1989 by Margaret
Geller and John Huchra,[3] is one of the largest known
structures in the universe and one of the first to indicate
the web-like nature of the universe on the grandest

Figure 7.7 *Abell Galaxy Cluster 1367*

scale. We now understand that galaxies appear to lie in
groups and clusters, which in turn form superclusters
and, beyond that, filaments. Filaments are long,
thread-like structures that wrap around and form the
boundaries of huge empty voids of space. As to why
that is, well, that will have to wait for further study,
although it probably has to do with the intervention of
dark matter.

   Of the galaxies in AGC 1367, 21 found their way
into the *New General Catalogue*. The remaining 80-plus
members were discovered afterward and are
inventoried in a variety of catalogs, including the *Index
Catalogue* (IC), the *Morphological Catalogue of Galaxies*
(MCG), the *Principal Galaxy Catalog* (PGC), the *Uppsala
General Catalogue of Galaxies* (UGC), the Zwicky
Catalogue of Galaxies and of Clusters of Galaxies (CGCG),
and others. As might be expected, the NGC galaxies,
being the first discovered, are typically easier to spot
than the rest. But even some of the NGCs can give you a
run for your money.

   To locate AGC 1367, slip 6° north of Denebola
[Beta (β) Leonis] to 5th-magnitude 93 Leonis. As chart
7.7 shows, 93 lies just on the eastern edge of the galaxy
cluster. Through your telescope, look for three 7th- and
8th-magnitude stars to 93's west that, together with 93,

form a crooked diamond asterism. The western star in
the diamond, SAO 81972, is just 8′ away from the
cluster's center.

   Use that star to orient yourself to the field. Begin
with the brightest galaxy of the bunch, NGC 3842, a
12th-magnitude elliptical found 19′ to the star's
west-northwest. There, you should see a small, round
disk highlighted by a brighter, nearly stellar core. As
Figure 7.7 shows, a 15th-magnitude field star just
touches NGC 3842's eastern edge; don't mistake it for a
new extragalactic supernova.

   Without moving the field, scan the area
immediately around NGC 3842 for several fainter
galactic systems. Only 1.3′ to its north is NGC 3841, a
smaller, fainter spiral galaxy. Another 1.7′ north, we
come to another spiral, NGC 3845, a bit larger, but also
a bit fainter than NGC 3841. Shifting 3.6′ east-southeast
of NGC 3845, look for the dim glow of NGC 3851 just
beyond a 12th-magnitude star. NGC 3851, a tiny
elliptical galaxy, might well look like another faint
star at first, since it only measures about 30″
across.

   Wending another 2.4′ north-northwest of NGC
3845 brings us to NGC 3844, an S0 lenticular galaxy.
Visually, NGC 3844 impresses me as a small, round
disk, but photographs record fainter feathery extensions
that protrude to the northeast and southwest of the
galactic nucleus.

[3] M. J. Geller and J. P. Huchra, "Mapping the Universe," *Science*,
Vol. 246 (1989), p. 897.

Table 7.4 *Members of AGC 1367 brighter than magnitude 16 (bold entries are discussed above)*

Galaxy	RA	Dec	Magnitude	Size (')
NGC 3805	11 40.7	+20 20.7	13.6p	1.4×1.1
CGCG 97-55	11 41.4	+19 57.8	15.1p	0.6×0.5
NGC 3816	11 41.8	+20 06.2	13.5b	1.9×1.1
CGCG 97-64	11 42.2	+20 05.9	15.9p	0.7×0.4
NGC 3821	11 42.2	+20 19.0	13.7b	1.6×1.3
CGCG 97-62	11 42.2	+19 58.6	15.4p	0.7×0.3
CGCG 97-63	11 42.3	+20 02.9	15.6p	0.6×0.3
CGCG 97-68	11 42.4	+20 07.2	14.6b	1.1×0.6
CGCG 97-72	11 42.8	+20 01.9	14.9b	1.0×0.4
CGCG 97-73	11 42.9	+19 58.0	15.5p	0.7×0.7
CGCG 97-74	11 43.0	+20 05.2	15.3p	0.6×0.3
UGC 6680	11 43.0	+19 39.0	15.0p	1.3×0.5
CGCG 97-79	11 43.2	+20 00.3	15.7	0.5×0.3
UGC 6683	11 43.3	+19 44.9	15.4p	1.0×0.3
CGCG 97-77	11 43.3	+19 23.1	15.1p	0.6×0.4
IC 2951	11 43.4	+19 45.0	14.5b	1.4×0.6
CGCG 97-83	11 43.5	+19 37.7	15.1	0.8×0.7
CGCG 127-40	11 43.7	+20 16.4	15.3	1.2×0.3
CGCG 97-86	11 43.8	+20 02.4	15.7	0.5×0.2
UGC 6697	**11 43.8**	**+19 58.2**	**14.1b**	**2.5×0.4**
NGC 3837	**11 43.9**	**+19 53.7**	**14.3b**	**1.0×1.0**
PGC 36468	**11 44.0**	**+19 56.2**	**15.3**	**0.4×0.2**
CGCG 97-90	**11 44.0**	**+19 57.2**	**15.3**	**0.3**
CGCG 97-92	11 44.0	+20 11.1	15.5	0.7×0.4
NGC 3840	**11 44.0**	**+20 04.6**	**14.5b**	**1.2×0.7**
CGCG 97-93	11 44.0	+19 46.7	15.5p	0.8×0.4
NGC 3844	**11 44.0**	**+20 01.8**	**14.9b**	**1.5×0.4**
CGCG 97-88	11 44.0	+19 47.1	15.2p	1.0×0.4
NGC 3842	**11 44.0**	**+19 57.0**	**12.8b**	**1.4×1.2**
NGC 3841	**11 44.0**	**+19 58.3**	**14.6b**	**0.6×0.6**
CGCG 97-94	11 44.1	+19 48.1	15.7	0.3×0.2
NGC 3845	**11 44.1**	**+19 59.8**	**15.0b**	**0.8×0.5**
CGCG 97-99	11 44.1	+19 44.3	15.7	0.5×0.5
MCG +4-28-43	11 44.3	+20 13.0	15.5	0.6×0.4
CGCG 97-101	11 44.3	+19 50.7	15.4	0.5×0.4
NGC 3851	**11 44.3**	**+19 58.8**	**15.4**	**0.6×0.4**
CGCG 97-105	11 44.3	+19 49.6	15.5	0.5×0.3

Table 7.4 *(cont.)*

Galaxy	RA	Dec	Magnitude	Size (')
CGCG 97-111	11 44.4	+20 06.5	15.5	0.5×0.5
CGCG 97-110	11 44.4	+19 49.7	15.5	0.8×0.3
CGCG 97-109	11 44.5	+19 44.1	15.1p	0.4×0.3
CGCG 97-112	11 44.5	+20 04.6	14.9	1.0×0.4
CGCG 97-113	11 44.8	+19 45.5	15.6	0.4×0.3
UGC 6719	11 44.8	+20 07.5	14.4b	1.4×0.9
CGCG 97-115	11 44.8	+19 52.6	15.4p	0.8×0.4
CGCG 97-114	11 44.8	+19 46.4	15.3p	0.4
NGC 3860	11 44.8	+19 47.7	14.2b	1.4×0.7
NGC 3857	11 44.8	+19 32.0	15.1b	1.0×0.5
CGCG 97-118	11 44.9	+19 36.6	15.7	0.5×0.2
NGC 3859	11 44.9	+19 27.3	14.8b	1.2×0.3
CGCG 97-125	11 44.9	+19 46.6	15.7p	0.8×0.5
CGCG 97-123	11 44.9	+19 29.6	15.9	0.7×0.3
IC 2955	11 45.1	+19 37.2	15.1b	0.6×0.4
NGC 3861	**11 45.1**	**+19 58.4**	**13.5b**	**2.4×1.2**
NGC 3862	11 45.1	+19 36.4	13.7b	1.4×1.4
UGC 6725	11 45.1	+20 26.3	13.9p	1.6×1.2
MCG +3-30-94	**11 45.1**	**+19 58.0**	**15**	**0.6×0.3**
CGCG 97-131	11 45.2	+19 50.7	14.8b	0.7×0.5
NGC 3864	11 45.3	+19 23.5	15.1p	0.7×0.5
CGCG 97-133	11 45.3	+20 01.4	15.6p	0.3
CGCG 127-47	11 45.4	+20 19.5	15.4p	1.0×0.7
NGC 3867	11 45.5	+19 24.0	14.2b	1.4×1.0
NGC 3868	11 45.5	+19 26.7	15.3p	0.8×0.4
CGCG 97-138	11 45.7	+20 01.9	14.4p	0.7×0.7
NGC 3873	11 45.8	+19 46.4	13.9b	1.0×1.0
NGC 3875	11 45.8	+19 46.1	14.9p	1.4×0.4
IC 732	11 46.0	+20 26.3	15.1	0.5×0.3
CGCG 97-141	11 46.1	+19 26.3	15.6	0.4×0.4
CGCG 97-143	11 46.1	+19 47.4	15.3	0.5×0.3
CGCG 97-142	11 46.1	+19 38.1	15.7	0.6×0.2
NGC 3884	11 46.2	+20 23.5	13.5b	1.3×0.9
NGC 3886	11 47.1	+19 50.2	14.1p	1.5×0.6
CGCG 97-152	11 47.7	+19 56.4	15.5b	0.9×0.3
CGCG 97-155	11 48.1	+20 00.4	14.8p	0.7×0.5

More northward still, we come to NGC 3840 at the end of this short line of galaxies at the core of AGC 1367. Despite its relatively large size, NGC 3840 impresses me as tougher to see than some of the others here because of the resulting low surface brightness.

Moving back to home base NGC 3842, look for irregular galaxy UGC 6697, a very faint, very long, and very thin dagger of light 3' to its west-northwest. At 14th magnitude, nabbing this unique object may take a good dose of patience.

A pair of faint field stars is set between NGC 3842 and UGC 6697. Try to spot a faint glow next to the star closest to NGC 3842. See anything? If so, you have conquered CGCG 97-90, which, at magnitude 15.3 and only 0.3' across, is not an easy task. Neither is seeing PGC 36468, 1.3' southwest of NGC 3842.

If some of these systems have been too much for you, the elliptical galaxy NGC 3837 should return your confidence. At magnitude 14.3, NGC 3837 is the second brightest member of the cluster. Still, it may require some hunting before its small, round disk becomes evident. It only measures 1' × 1' across, so will probably require 300× or more magnification to tell it apart from a random foreground star.

From NGC 3842, head east for about 15' to NGC 3861, which is another of the cluster's brightest members. It lies just $6\frac{1}{2}'$ northwest of SAO 81972, our reference star. Photographs show that the arms of this spiral galaxy are wrapped around its core like a hula hoop. The largest telescopes may be able to pick out that very faint outer halo, although most of us will have to settle for seeing only the small, elongated core surrounding a pinpoint nucleus. A very faint companion, 15th-magnitude MCG +3-30-94, lies just off its southeastern edge.

All of the galaxies detailed here, as well as others in the NGC and IC listings, are plotted on Chart 7.7. How many other members in AGC 1367 can you spot? Table 7.4 lists those that shine brighter than 16th magnitude. To find them, however, you will first need a series of highly detailed charts scaled large enough to prevent overcrowding. Computer programs such as *MegaStar* (Willmann-Bell) or *Guide* (Project Pluto) prove invaluable for tasks like this.

164 Abell Galaxy Cluster 1656

★★★
★★
★

Target	Type	RA	Dec.	Constellation	Magnitude	Span	Chart
AGC 1656	Galaxy cluster	12 59.8	+27 58	Coma Berenices	–	224′	7.8

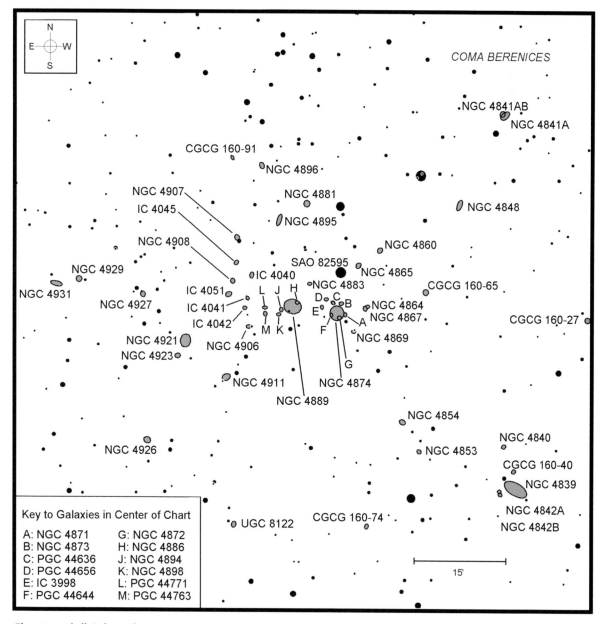

Key to Galaxies in Center of Chart

A: NGC 4871   G: NGC 4872
B: NGC 4873   H: NGC 4886
C: PGC 44636   J: NGC 4894
D: PGC 44656   K: NGC 4898
E: IC 3998   L: PGC 44771
F: PGC 44644   M: PGC 44763

Chart 7.8 Abell Galaxy Cluster 1656

As challenging as AGC 1367 is, it pales against its big brother, the Coma Galaxy Cluster, AGC 1656. Spanning an area more than twice the size of the Leo Cluster and holding more than 800 galaxies brighter than photographic magnitude 16.5, AGC 1656 is a real galactic forest that will take great patience to make our way through. There is no rushing this one. If you do not have enough time to devote to the task, it is best to push on to the next challenge and come back to this one when you do. In fact, you will never get through this huge collection of galaxies in one sitting. Or even two, three, or four sessions, for that matter. AGC 1656 could well take *years* before every galaxy in view is recorded and identified.

Where, oh where, to begin? Making our way through the cluster may take time, but finding it does not. The geometric center of AGC 1656 lies $2\frac{3}{4}°$ west of Beta (β) Comae, the star at the right angle of the constellation's triangular outline. The cluster covers the $3\frac{3}{4}°$ gap between 31 Comae at its western edge and 41 Comae just inside its eastern border, and is centered very near 7th-magnitude SAO 82595.

Table 7.5 lists 74 of the cluster's galaxies that are photographic magnitude 15 or brighter, as well as a select few fainter members included because of their proximity to other, more prominent constituents. That's your assignment. There simply is not enough room to describe them all here, but here are a few highlights to kick off your project.

We will begin with NGC 4874 and NGC 4889 at the heart of the cluster (Figure 7.8). Both are bright enough to be within the range of 8-inch instruments, although glare from SAO 82595 will hamper their visibility. NGC 4889, the more obvious of the two, is a giant elliptical galaxy in the spirit of the monster galaxy M87 in the Virgo cluster. But, while M87 is some 60 million light years way, NGC 4889 is projected to be 300 million light years distant. Taking that into account, NGC 4889 is actually the more luminous of the two. As with M87, radio astronomers have detected strong emissions emanating from the core, the signature of a massive black hole. Through the largest backyard telescopes, NGC 4889 displays a fairly bright, oval disk elongated east-to-west and surrounding a brighter core.

Although it is a little smaller and fainter than its neighbor, elliptical galaxy NGC 4874 is no slouch. You'll find it just 7′ west of NGC 4889 and 6′ due south of SAO 82595. Again, we find a strong radio source buried deep within the core of the galaxy. Visually,

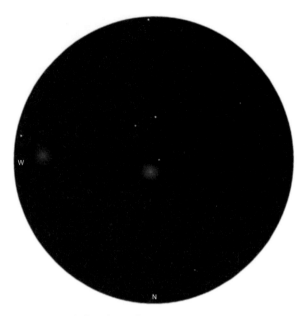

Figure 7.8 Abell Galaxy Cluster 1656

NGC 4874 strikes me as more than 0.1 magnitude fainter than NGC 4889, despite the listing. Look for a round glow that draws to a brighter central core.

The best view of both galaxies comes at 200× or more, since the resulting narrow field helps to move that distracting star out of view. Once the glare is gone, all of a sudden a swarm of smaller, fainter companions begins to dot the field. With averted vision, I can count no fewer than nine island universes. NGC 4886 is superimposed over the northwestern edge of NGC 4889; NGC 4883 about 4′ to the northwest of its core; NGC 4894 just 2′ east-southeast of the core, and NGC 4898 a little farther to the southeast still. On dark nights, PGC 44763 and PGC 44771 pose even greater challenges. Look for the PGC pair 5′ east of NGC 4889. Beyond them, lie IC 4042, IC 4041, and IC 4040, along with NGC 4906.

That covers the eastern half of that single eyepiece field; NGC 4874 and its brood still lie in the western half. Two galaxies drift nearby. Can you spot NGC 4872, which appears superimposed on, or perhaps immersed in, the larger galaxy's southwestern edge? NGC 4871 is also close at hand, just 1′ west the galaxy's core. Finally, watching these two smaller companions is NGC 4873. Look for it just 1.5′ to the northwest. All three shine between 13th and 15th magnitude and are smaller than 1 arc-minute across.

Table 7.5 *Members of AGC 1656 brighter than magnitude 15 (bold entries are discussed above)*

Galaxy	RA	Dec	Magnitude	Size
NGC 4715	12 49.9	+27 49.3	14.1p	1.4′×1.2′
NGC 4728	12 50.5	+27 26.2	14.5p	0.8′×0.7′
NGC 4738	12 51.1	+28 47.3	14.2b	2.3′×0.4′
UGC 8017	12 52.9	+28 22.3	14.6b	1.0′×0.4′
CGCG 159-104	12 53.3	+27 05.7	14.8p	0.8′×0.6′
UGC 8025	12 54.0	+29 36.1	14.8p	2.1′×0.4′
NGC 4789A	12 54.1	+27 08.9	13.9b	3.0′×2.1′
NGC 4789	12 54.3	+27 04.1	13.1b	1.9′×1.5′
NGC 4793	12 54.7	+28 56.3	12.3b	2.9′×1.4′
NGC 4798	12 54.9	+27 24.7	14.2b	1.4′×1.0′
NGC 4807	12 55.5	+27 31.3	14.5b	1.0′×0.8′
IC 3900	12 55.7	+27 15.0	15.0p	0.8′×0.5′
NGC 4816	12 56.2	+27 44.7	13.8b	1.2′×1.0′
CGCG 160-27	12 56.5	+27 56.4	14.5	0.8′×0.8′
NGC 4819	12 56.5	+26 59.2	14.1b	1.8′×1.4′
NGC 4827	12 56.7	+27 10.7	13.9b	1.4′×1.2′
NGC 4839	**12 57.4**	**+27 29.9**	**13.0b**	**4.0′×1.9′**
CGCG 160–40	12 57.4	+27 32.7	14.9p	0.5′×0.3′
NGC 4841A	12 57.5	+28 28.6	13.8b	1.6′×1.0′
NGC 4840	**12 57.5**	**+27 36.6**	**14.7b**	**0.7′×0.6′**
NGC 4841B	12 57.5	+28 28.9	13.6b	1.0′×0.7′
NGC 4842A	**12 57.6**	**+27 29.5**	**14.2**	**0.5′**
NGC 4842B	**12 57.6**	**+27 29.1**	**15.4**	**0.4′×0.2′**
UGC 8076	12 57.8	+29 39.3	15.0b	1.0′×0.6′
UGC 8080	12 58.0	+26 51.6	15.0p	1.0′×0.6′
NGC 4848	12 58.1	+28 14.5	14.4b	1.9′×0.6′
NGC 4849	12 58.2	+26 23.8	13.9p	1.8′×1.3′
CGCG 160-62	12 58.3	+29 07.7	15.0	0.8′×0.6′
CGCG 160-65	12 58.5	+28 00.8	14.1	1.0′
NGC 4853	**12 58.6**	**+27 35.8**	**14.4b**	**0.7′×0.6′**
NGC 4854	**12 58.8**	**+27 40.5**	**14.9b**	**1.0′×0.6′**
NGC 4859	12 59.0	+26 48.9	14.6b	1.6′×0.7′
NGC 4860	12 59.1	+28 07.4	14.6b	1.0′×0.7′
NGC 4864	**12 59.2**	**+27 58.6**	**14.6b**	**0.7′×0.7′**
CGCG 160-74	12 59.2	+27 24.1	15.0p	0.6′×0.4′
NGC 4867	**12 59.3**	**+27 58.2**	**15.4b**	**0.6′×0.5′**
NGC 4865	12 59.3	+28 05.1	14.6b	0.9′×0.5′
				(cont.)

Table 7.5 *(cont.)*

Galaxy	RA	Dec	Magnitude	Size
NGC 4869	12 59.4	+27 54.7	14.8b	0.7′×0.7′
NGC 4871	**12 59.5**	**+27 57.3**	**15.2b**	**0.7′×0.5′**
NGC 4873	**12 59.5**	**+27 59.0**	**15.1b**	**0.7′×0.5′**
NGC 4872	**12 59.6**	**+27 56.8**	**15.4b**	**0.6′×0.4′**
NGC 4874	**12 59.6**	**+27 57.6**	**12.6b**	**2.3′×2.3′**
PGC 44636	**12 59.6**	**+27 59.2**	**14.8**	**0.4′×0.3′**
PGC 44644	12 59.7	+27 57.2	14.4	0.4′×0.2′
PGC 44656	**12 59.7**	**+27 59.7**	**15.2**	**0.4′**
IC 3998	**12 59.8**	**+27 58.4**	**15.6b**	**0.6′×0.4**
NGC 4883	**12 59.9**	**+28 02.1**	**15.4b**	**0.6′×0.4′**
NGC 4881	13 00.0	+28 14.8	14.6b	1.0′×1.0′
NGC 4892	13 00.5	+26 53.8	14.9p	1.7′×0.5′
NGC 4886	**13 00.5**	**+27 59.3**	**13.8**	**0.6′×0.6′**
NGC 4889	**13 00.1**	**+27 58.6**	**12.5b**	**2.8′×2.2′**
NGC 4894	**13 00.3**	**+27 58.0**	**15.1**	**0.6′×0.3′**
NGC 4895	13 00.3	+28 12.1	14.2b	2.0′×0.7′
NGC 4898	**13 00.3**	**+27 57.3**	**14.5b**	**0.5′×0.5′**
PGC 44763	**13 00.5**	**+27 57.3**	**15.4**	**0.5′×0.3′**
PGC 44771	**13 00.5**	**+27 58.3**	**15.2**	**0.6′×0.3′**
NGC 4896	13 00.5	+28 20.8	14.9b	1.0′×0.6′
IC 4040	**13 00.6**	**+28 03.6**	**15.4b**	**1.0′×0.3′**
IC 842	13 00.7	+29 01.2	14.7b	1.4′×0.6′
NGC 4906	**13 00.7**	**+27 55.6**	**15.1b**	**0.5′**
IC 4041	**13 00.7**	**+27 59.8**	**15.3b**	**0.5′×0.3′**
IC 4042	**13 00.7**	**+27 58.3**	**15.3b**	**0.7′×0.5′**
IC 4045	13 00.8	+28 05.4	15.0b	0.7′×0.4′
NGC 4907	13 00.8	+28 09.5	14.4b	0.9′×0.7′
UGC 8122	13 00.8	+27 24.3	15.0p	1.0′×0.5′
NGC 4908	13 00.9	+28 02.6	14.7b	0.8′×0.6′
CGCG 160-91	13 00.9	+28 21.9	14.8p	0.7′×0.5′
IC 4051	13 00.9	+28 00.5	14.2b	1.0′×0.8′
NGC 4911	13 00.9	+27 47.4	13.6b	1.4′×1.0′
NGC 4922	13 01.4	+29 18.5	13.9	1.5′×1.2′
NGC 4921	13 01.4	+27 53.2	13.0b	2.0′×1.7′
NGC 4923	13 01.5	+27 50.8	14.7b	0.9′×0.6′
IC 843	13 01.6	+29 07.8	14.6p	1.3′×0.5′
IC 4088	13 01.7	+29 02.6	14.7b	1.8′×0.5′

Table 7.5 *(cont.)*

Galaxy	RA	Dec	Magnitude	Size
NGC 4926	13 01.9	+27 37.5	14.0b	$1.2' \times 1.0'$
NGC 4927	13 01.9	+28 00.3	14.7p	$0.9' \times 0.7'$
NGC 4929	13 02.7	+28 02.7	14.4p	$1.0' \times 1.0'$
NGC 4931	13 03.0	+28 01.9	14.5b	$2.0' \times 0.6'$
NGC 4944	13 03.8	+28 11.1	13.8b	$1.8' \times 0.6'$
NGC 4952	13 05.0	+29 07.4	13.4p	$1.8' \times 1.2'$
NGC 4957	13 05.2	+27 34.2	14.0b	$1.4' \times 1.2'$
NGC 4961	13 05.8	+27 44.0	14.0b	$1.7' \times 1.1'$
CGCG 160-136	13 06.0	+29 16.7	14.9p	$0.6' \times 0.6'$
NGC 4966	13 06.3	+29 03.7	14.0b	$1.2' \times 0.8'$
UGC 8195	13 06.4	+29 39.5	14.9	$1.6' \times 0.2'$
MCG +5-31-132	13 06.6	+27 52.4	15.0	$0.8' \times 0.4'$
NGC 4971	13 06.9	+28 32.8	14.4p	$0.8' \times 0.8'$
NGC 4983	13 08.5	+28 19.2	14.9b	$1.1' \times 0.7'$
UGC 8229	13 08.9	+28 11.1	14.3b	$1.4' \times 0.9'$

You might be able to spot a lone island universe floating a third of the way from NGC 4874 back toward NGC 4889. That's IC 3998. This extremely faint, round system forms a close pair with PGC 44652 just $1'$ to its south-southeast. A number of other members of the PGC listing also hover near NGC 4874, including PGC 44636 and PGC 44656 to its north. None of these marginal objects has a surface brightness greater than perhaps half a magnitude above background, so you may need to boost magnification even further to maximize image contrast.

Expanding the view a little, look for the interacting pair of NGC 4864 and NGC 4867 about $5'$ west of NGC 4874. At $200\times$, I can only make out a very dim, distended blur, but increasing to $300\times$ helps resolve the individual cores, which are separated by only $35''$.

From NGC 4889, scan slowly toward the southwest to find NGCs 4853 and 4854, a pair of tiny 14th-magnitude spirals separated from one another by $5'$ and lying just to the north of an 8.5-magnitude field star. Slide $17'$ due west of that star to find the elliptical galaxy NGC 4839 and another horde of fainter companions. NGC 4839 shines at about 13th magnitude and is the brightest galaxy in this quadrant of the cluster. Through my 18-inch at $206\times$ with averted vision, it appears as a soft, oval glow tilted southwest–northeast around a brighter faint core. Photographs expose a small companion superimposed just to the southwest of the core, but I have never been able to resolve it individually. Averted vision may, however, reveal several smaller, fainter galactic smudges in the same field, including the double system NGC 4842A/4842B just $2\frac{1}{2}'$ to the southeast. NGC 4840 is also in the picture, $7'$ north-northeast of NGC 4839.

This brief summary is not even the tip of the galactic iceberg that awaits the largest backyard telescopes here. Hundreds of faint fuzzies can be found within the $4°$ span of AGC 1656. But, to hunt them down, you will need to use several detailed charts that show the area at a sufficiently large enough scale and then plot field stars faint enough to wade your way through this intergalactic maze. Use the charts here to identify as many as you can but, as mentioned previously, to go really deep into a galactic ocean like AGC 1656, you need customized charts for your specific telescope and eyepieces using a computer program like *MegaStar* or *Guide*.

165 Abell Galaxy Cluster 2065

Target	Type	RA	Dec.	Constellation	Magnitude	Span	Chart
AGC 2065	Galaxy cluster	15 22.7	+27 43	Corona Borealis	–	22.4'	7.9

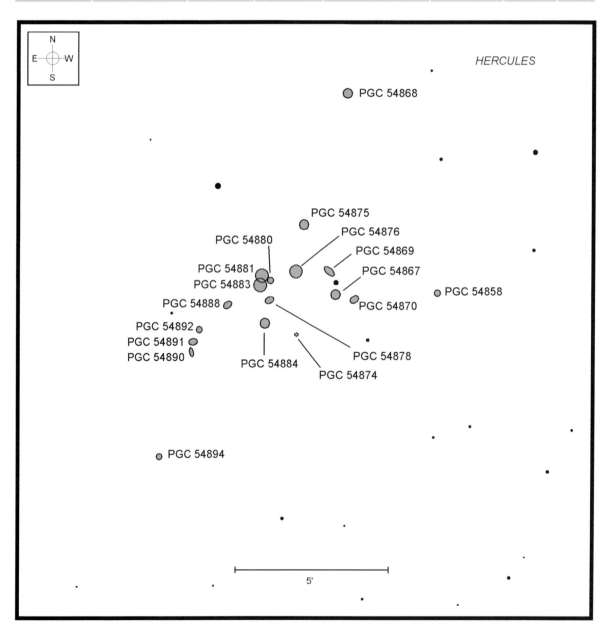

Chart 7.9 Abell Galaxy Cluster 2065

The last two challenges were just down the block, a mere 300 to 400 million light years away, compared with the next one. This time out, our telescopes transport us across an incredible 1.2 billion light years to Abell Galaxy Cluster 2065. More than 400 galaxies are huddled within AGC 2065 but, because of that incomprehensible distance, the light from these massive collections of stars has dwindled to nothing more than the faintest whisper. Seeing even the slightest hint of AGC 2065 takes more than just aperture; it also takes supremely dark skies, a trained eye, and an accurate chart of the region.

Like the other AGC galaxy clusters that we have met, AGC 2065 was discovered on the Palomar Sky Survey photographic plates taken in the 1950s and is one of more than 2,700 such clusters published by George Abell in 1958.[4] This particular cluster presents one of the most daunting challenges in Abell's list. None of the galaxies shines brighter than 15th magnitude; most are only 17th, or even 18th, magnitude.

Very little has been written about AGC 2065 in amateur observing guides. One of the first to mention it was *Burnham's Celestial Handbook*, where it was described as "one of the most remarkable of all such aggregations" in the sky. The cluster was also included in Volume 5 of the classic *Webb Society Deep-Sky Observer's Handbook* published in 1982, but the description did not encourage observation. "Because of its staggering distance it is beyond the reach of most amateur astronomers, being undetectable in 16-inch reflectors even under good sky conditions." Despite that dire statement, the location of AGC 2065 is plotted just west of Alpha (α) and Beta (β) Coronae Borealis on *Sky Atlas 2000.0*, an atlas that usually restricts its deep-sky targets to those visible through 10-inch and smaller apertures.

Its position so near the Northern Crown's two brightest stars makes zeroing in on it a piece of cake. Finding any evidence of its existence is something else, however. To try your luck, drop $1\frac{1}{2}°$ southwestward from Beta to 7th-magnitude SAO 83797. AGC 2065 is centered just 20′ south of that yellow giant sun, wedged in between two 10th- and 11th-magnitude Milky Way stars.

_____
[4] G. O. Abell, "The Distribution of Rich Clusters of Galaxies," *Astrophysical Journal Supplement*, Vol. 3 (1958), pp. 211–88.

Figure 7.9 Abell Galaxy Cluster 2065

Choose an eyepiece that will produce about 200× through your telescope. Sit down at the telescope, breathe, and relax! Standing and straining to see these faint targets will only lead to disappointment. Focus your attention on the 11th-magnitude field star shown on the chart, and then look just 1.4′ to its east. Can you spot PGC 54876 (Figure 7.9), which most agree is the easiest galaxy in the cluster to see? Using averted vision, try to catch a fleeting glimpse of its round, evenly illuminated disk. Its small size may disguise it as a star at lower magnification, but at 200× or more its slight fuzziness will give away its true nature.

Glance another 1′ to the east for an even fainter glow. Although it will probably look like a single object, this second target is a galactic pair, PGC 54881 and PGC 54883. By using averted vision through my 18-inch at 294×, I can tell that this object is about double the size of PGC 54876. Try as I might, I have never been able to resolve them as two separate systems. Others, undoubtedly observing under better sky conditions, report success.

PGC 54891 is also doable with patience. Its extremely faint, round glow appears just 1.2′ southwest of a 14.3-magnitude star. If you manage to spot it, you

Table 7.6 *Members of AGC 2065 (bold entries are discussed above)*

Galaxy	RA	Dec	Magnitude	Size
PGC 54858	15 22.1	+27 42.2	–	0.2′
PGC 54870	15 22.3	+27 42.0	18.8	0.3′×0.2′
PGC 54867	15 22.3	+27 42.1	–	0.3′
PGC 54869	15 22.3	+27 42.8	–	0.4′×0.2′
PGC 54868	15 22.3	+27 48.4	17.5	0.3′
PGC 54874	15 22.4	+27 40.9	–	0.1′
PGC 54876	**15 22.4**	**+27 42.8**	**15.1**	**0.4′**
PGC 54875	**15 22.4**	**+27 44.3**	**16.9**	**0.3′**
PGC 54884	15 22.5	+27 41.2	–	0.3′
PGC 54878	15 22.5	+27 41.9	–	0.3′×0.2′
PGC 54883	**15 22.5**	**+27 42.4**	**–**	**0.4′**
PGC 54880	15 22.5	+27 42.6	–	0.2′
PGC 54881	**15 22.5**	**+27 42.7**	**–**	**0.4′**
PGC 54891	**15 22.6**	**+27 40.7**	**16.3**	**0.3′×0.2′**
PGC 54892	15 22.6	+27 41.1	–	0.2′
PGC 54888	15 22.6	+27 41.8	–	0.3′×0.2′
PGC 54894	15 22.7	+27 37.1	–	0.2′
PGC 54890	**15 22.7**	**+27 40.3**	**17.6**	**0.3′×0.1′**

might also notice a second faint, slightly elongated speck just to its south. That's PGC 54890. PGC 54875 may also be visible, although it appears fainter than those mentioned above. Look for it 2′ northeast of the 11th-magnitude star and about $1\frac{1}{2}°$ north of PGC 54876.

How many other galaxies in AGC 2065 are visible through the largest amateur telescopes? Some online accounts report seeing dozens. To find out for yourself, Table 7.6 lists 18 possibilities. Most do not have reliable magnitude estimates, however, so approach the listing with caution.

Regardless of how many of these distant galaxies you can find, be it only 1 or 18, keep in mind what you are seeing. Bob King, a devout deep-sky addict from Duluth, Minnesota, captures the experience best by offering this thought: "When the light I saw left the galaxy cluster, complex cells had evolved on Earth but not a single multi-cellular organism swam about. In the time it took to get here, the tree of life sent out branches in many directions including the twig of humanity. What a privilege to be here now, over a billion years later, and catch a few photons from four ghosts in Corona Borealis."

166 Seyfert's Sextet

★
★
★
★

Target	Type	RA	Dec.	Constellation	Magnitude	Span	Chart
Seyfert's Sextet	Galaxy group	15 59.2	+20 45	Serpens	–	1′	7.10

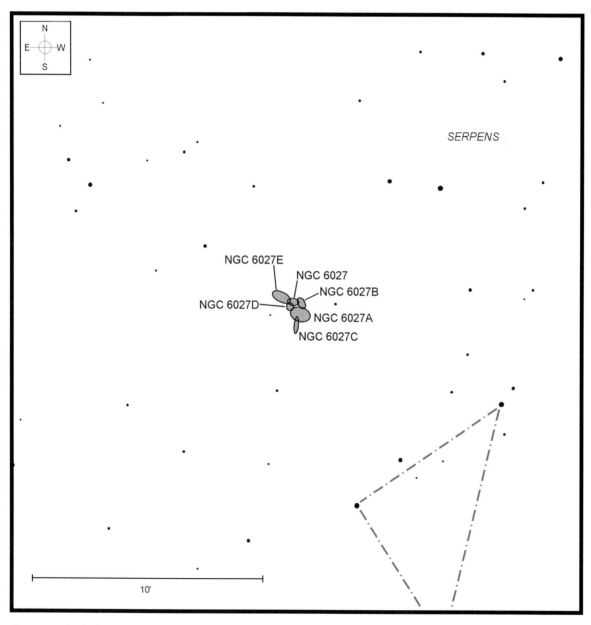

Chart 7.10 Seyfert's Sextet

Seyfert's Sextet, known to many as Hickson Compact Galaxy Group 79, is a tight gathering of galaxies in the northern corner of Serpens Caput. Serpens Caput is the western segment of this bisected constellation, marking the triangular head of the serpent that Ophiuchus is handling. Observing Seyfert's Sextet has been one of my pet projects for years. It's a fun little galactic rat pack for late spring and early summer outings before plunging headlong into the summer Milky Way.

The nickname "Seyfert's Sextet," however, is inaccurate for a couple of reasons. First, technically speaking, Karl Seyfert did not discover Seyfert's Sextet. Instead, the group's primary galaxy, NGC 6027, was found by Édouard Stephan, director of the Marseilles Observatory, while observing through the observatory's 31.5-inch (80-cm) reflector in June 1882. That was the same telescope that Stephan was using when he found another, more famous galactic bunch, his namesake Stephan's Quintet in Pegasus (large-scope Challenge 144).

Stephan's notes record only a single object, however, although he also noted two very faint stars as "involved." In historical retrospect, those two stars were actually two of the other galaxies that probably went unrecognized because of their small size. Seyfert discovered their true nature, and also spotted several additional members while scrutinizing a photographic plate taken of the region at Harvard College Observatory in 1951. Seyfert later reported his findings in a short paper entitled "A Dense Group of Galaxies in Serpens,"[5] where he also noted that the group was 26.6 million light years away.

Although five auxiliary NGC numbers – NGC 6027A through 6027E – were assigned to the galaxies recorded on the Harvard plate, Seyfert questioned whether or not he was looking at a group of six galaxies, or perhaps only four or five. He acknowledged that his colleague Walter Baade, a staff member of Mount Wilson Observatory, believed that the objects labeled as NGC 6027C and 6027D were actually tidal anomalies created by the interactions among the others.

It turns out that Baade was correct. Seyfert's Sextet does not contain six galaxies. It contains only five. He was wrong, however, about which were tidally created mirages and which were actual galaxies. NGC 6027C

---

5 C. Seyfert, "A Dense Group of Galaxies in Serpens," *Publications of the Astronomical Society of the Pacific*, Vol. 63, No. 371 (1951), p. 72.

Table 7.7 *Seyfert's Sextet*

Galaxy	RA	Dec	Magnitude	Size
NGC 6027	15 59.2	+20 45.8	15.3b	0.5′×0.3′
NGC 6027A	15 59.2	+20 45.3	14.8b	0.9′×0.6′
NGC 6027B	15 59.2	+20 45.8	15.3	0.5′×0.3′
NGC 6027C	15 59.2	+20 44.8	16.5	0.7′×0.2′
NGC 6027D	15 59.2	+20 45.6	16.5b	0.3′×0.3′
NGC 6027E	15 59.2	+20 46.0	14.4b	0.9′×0.4′

Figure 7.10 *Seyfert's Sextet*

and NGC 6027D are bona fide galaxies. Studies conducted with the Hubble Space Telescope show quite clearly that NGC 6027E is not a separate galaxy, but instead a gravitationally created plume extending away from NGC 6027. Such distortions are common features of galaxy groups as they swirl around each other and draw closer as time goes on. Ultimately, after hundreds of millions of years doing a galactic pirouette, these galaxies will ultimately merge to form a single, giant elliptical galaxy. That is, all will except NGC 6027D. This spiral galaxy is in the background and just happens to lie along the same line of sight.

Numbers and names aside, Seyfert's Sextet can be found 2° east-southeast of 5th-magnitude Rho (ρ) Serpentis, itself 3° north of the Serpent's triangular head. As a reference, look for a right triangle of 9th- and

11th-magnitude stars 10′ southwest of the group, and a closer pair of 11th-magnitude stars 7′ to its northwest.

NGC 6027 should be a fairly easy catch in your scope, but the rest will take some effort to see. Under suburban skies, my 18-inch reflector at 171× shows it as a dim glow, slightly elongated east-to-west and accented by a very dim central core (Figure 7.10), but that's about it. There is no sign of the dimmer group members until the magnification is increased to 294×, when the very faint glimmer from NGC 6027A can be seen just 36″ to the south-southwest. NGC 6027B can also be suspected with averted vision just 20″ west of NGC 6027. Darker skies are needed to confirm it with direct vision, as well as to suspect even the slightest hint of the other members of the bunch.

167   Abell Galaxy Cluster 2151

★
★
★
★

Target	Type	RA	Dec.	Constellation	Magnitude	Span	Chart
AGC 2151	Galaxy cluster	16 05.2	+17 44.0	Hercules	–	56′	7.11

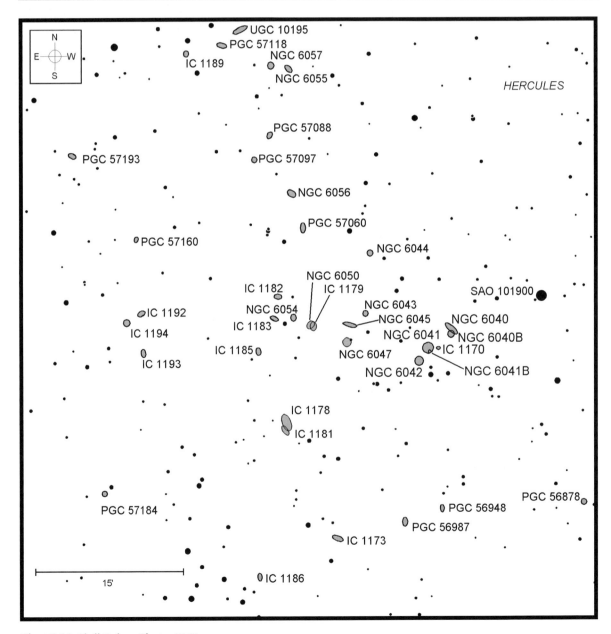

Chart 7.11 Abell Galaxy Cluster 2151

If you enjoy looking at huge collections of galaxies, then the entire area spanning northern Serpens Caput and western Hercules is for you. This next challenge focuses in on the central members of the Hercules Galaxy Cluster. Also known as Abell Galaxy Cluster 2151, the Hercules Galaxy Cluster packs more than 100 individual systems into an area just 1° across. Several are bright enough to be seen in 6- and 8-inchers, but it takes the largest telescopes to get a feel for the true immensity portrayed here.

The Hercules Galaxy Cluster presents an intriguing exception to a universal rule. Most dense galaxy clusters are dominated by one or more giant elliptical galaxies, with only a smattering of spirals and irregulars scattered throughout. The Hercules Galaxy Cluster, however, is ruled by spirals with nary a giant elliptical in the bunch. A wide variety of smaller elliptical and lenticular galaxies are also found within, but none have obtained the mass and dimensions of, say, M87, which stands center stage in the Virgo cluster.

AGC 2151 is the richest member of the Hercules Supercluster. The Hercules Supercluster covers more than 5° of the summer sky and forms part of the "Great Wall," as do the Leo (AGC 1367) and Coma Berenices (AGC 1656) galaxy clusters, which we met earlier in this chapter as monster-scope Challenges 163 and 164, respectively.

Locating AGC 2151 is easy thanks to a few nearby helpers. As with the last challenge, begin your quest at the triangular head of the Serpent, Serpens Caput. Look about 4° due east for a very slender triangle, formed by 5 Herculis at the apex and the pair of Kappa (κ) and 8 Herculis at the base. A 7th-magnitude star, SAO 101900, is just 37′ east of 5 Herculis and 20′ due west of the galaxy cluster; the star is just off the right (western) border of Chart 7.11.

Making your way through AGC 2151 takes a slow, concerted effort, so be prepared to take your time. Scanning slowly eastward from SAO 101900, we first meet the interacting pair of spirals NGC 6040 and NGC 6040B. NGC 6040 looks like a faint, elongated blur extending southwest/northeast, while NGC 6040B, just 26″ to the south, is more difficult to detect.

Can you also spot a trio of faint galaxies in the same eyepiece field (Figure 7.11), just 3′ to the southeast and near a tiny triangle of faint stars? NGC 6041 shows off an oval disk that will probably appear evenly illuminated at first. Take a closer look, however, and you might be able to resolve two very faint stellar cores.

Figure 7.11 Abell Galaxy Cluster 2151

If so, you have also detected NGC 6041B, an interacting galaxy that overlaps the southwestern edge of NGC 6041.

Two other close-at-hand targets include NGC 6042 set $1\frac{1}{2}'$ southeast and IC 1170 just 1′ to its west. NGC 6042[6] is about as bright as NGC 6041, but is smaller and more circular, while IC 1170 is fainter and nearly stellar in appearance. You'll definitely need good conditions for that last one!

A pair of 13th-magnitude field stars leads us from NGC 6041 to NGC 6045, an edge-on spiral that shows a cigar-shaped form in my 18-inch at 206×. Photographs reveal that a small companion galaxy lies on the eastern tip of NGC 6045, but I have never been able to spot it. Two nearby galaxies are clear, however. NGC 6043 is found 2′ to the north, while NGC 6047 is 2′ to the south. Both are very small and very faint, and circular or perhaps slightly oval in shape. Photographs of NGC 6043 disclose that a small, extremely faint companion galaxy also overlaps its southeastern edge, but I know of no visual observation confirming it. NGC 6047 is also overlapped, but by a 13.5-magnitude Milky Way star, not another galaxy. Together, the two remind me of a very faint comet, with a stellar nucleus accompanied by

---

[6] NGC 6042 is thought to be one and the same as NGC 6039, the duplicity caused by an error in its recorded position by Lewis Swift upon his discovering the galaxy in 1886.

Table 7.8 *Members of AGC 2151 brighter than magnitude 16 (bold entries are discussed above)*

Galaxy	RA	Dec.	Magnitude	Size
PGC 56878	16 03.5	+17 28.8	15.0	0.5′
PGC 56915	16 04.0	+17 17.0	15.6	0.6′ × 0.4′
NGC 6040	**16 04.4**	**+17 45.0**	**14.1**	**1.4′ × 0.5′**
NGC 6040B	**16 04.4**	**+17 44.5**	**14.9b**	**0.6′**
PGC 56948	16 04.5	+17 28.2	16.0b	0.7′ × 0.4′
IC 1170	**16 04.5**	**+17 43.3**	**15.5**	**0.4′ × 0.2′**
NGC 6041	**16 04.6**	**+17 43.3**	**14.3b**	**1.0′ × 1.0′**
NGC 6041B	**16 04.6**	**+17 43.1**	**16.6P**	**0.2′**
NGC 6042	**16 04.7**	**+17 42.1**	**14.9b**	**0.8′ × 0.8′**
PGC 56987	16 04.8	+17 26.9	14.8	0.8′ × 0.5′
NGC 6043	**16 05.0**	**+17 46.5**	**14.2**	**0.5′ × 0.5′**
NGC 6044	**16 05.0**	**+17 52.2**	**15.3b**	**0.5′ × 0.5′**
NGC 6045	**16 05.1**	**+17 45.5**	**14.9b**	**1.3′ × 0.3′**
NGC 6047	**16 05.1**	**+17 43.8**	**13.5**	**0.8′ × 0.8′**
IC 1173	16 05.2	+17 25.4	14.6	1.0′ × 0.5′
IC 1179	**16 05.4**	**+17 45.3**	**15.4**	**0.8′ × 0.6′**
NGC 6050	**16 05.4**	**+17 45.5**	**14.7**	**0.8′ × 0.6′**
PGC 57060	16 05.4	+17 54.6	16.0	0.9′ × 0.3′
NGC 6054	**16 05.5**	**+17 46.2**	**15.1**	**0.7′ × 0.5′**
NGC 6056	**16 05.5**	**+17 57.8**	**13.8**	**0.9′ × 0.5′**
NGC 6055	16 05.5	+18 09.6	14.8b	0.9′ × 0.5′
IC 1178	**16 05.6**	**+17 36.3**	**14.1**	**1.6′ × 0.8′**
IC 1181	**16 05.6**	**+17 35.6**	**15.2**	**1.0′ × 0.5′**
IC 1182	**16 05.6**	**+17 48.1**	**14.2**	**0.7′ × 0.5′**
IC 1183	**16 05.6**	**+17 46.1**	**14.2**	**0.8′ × 0.4′**
NGC 6057	16 05.7	+18 09.9	15.7b	0.6′ × 0.6′
PGC 57088	16 05.7	+18 03.3	14.8	0.7′ × 0.4′
IC 1186	16 05.7	+17 21.8	15.5b	0.7′ × 0.4′
IC 1185	**16 05.7**	**+17 43.0**	**13.8**	**0.7′ × 0.5′**
PGC 57097	16 05.8	+18 01.0	15.2	0.5′
UGC 10195	16 05.9	+18 13.2	15.6b	1.5′ × 0.4′
PGC 57118	16 06.0	+18 11.8	14.6	1.0′ × 0.4′
PGC 57136	16 06.2	+17 18.1	15.3	0.6′ × 0.3′
IC 1189	16 06.2	+18 11.0	14.5	0.6′ × 0.4′
IC 1193	16 06.5	+17 42.8	14.2	0.7′ × 0.5′
IC 1192	16 06.6	+17 46.6	14.9	0.8′ × 0.3′
PGC 57160	16 06.6	+17 53.5	15.3	0.5′ × 0.3′

Table 7.8 *(cont.)*

Galaxy	RA	Dec.	Magnitude	Size
IC 1194	16 06.7	+17 45.7	14.2	0.6′ × 0.6′
PGC 57184	16 06.8	+17 29.6	15.7b	0.5′
PGC 57193	16 07.0	+18 01.4	16.0b	0.8′ × 0.3′
PGC 57214	16 07.4	+17 39.8	15.1	0.9′ × 0.6′

a faint tail (the galaxy) extending to the southeast. NGC 6044 is found 6′ north of NGC 6043 and just 1.4′ to the northeast of a 13th-magnitude star. My notes recall a very faint, very small, round object.

If that galaxy left you wanting, try NGC 6056, another 9′ farther northeast. At magnitude 13.8, NGC 6056 is one of the brightest members of AGC 2151. Look for a broad oval disk that draws to a slightly brighter core.

Back to NGC 6045, where we find a classic pair of interacting spiral galaxies, NGC 6050 and IC 1179, just 4′ to its east. The two appear to just touch each other at their northern rims in photographs and trail away toward the southeast and due south, respectively. The galaxies are so close to each other, however, that I have never been able to resolve them individually even at 411×. Instead, they remain a collective mass that is very faint, round in shape, and quite diffuse. That is also how their discoverer, Lewis Swift, first described them in 1886. He never recognized their duality either.

Continuing another 2′ to the east, we find NGC 6054, a very faint, very small galaxy that was also discovered by Swift in 1883. You will find it 1′ northwest of a 12th-magnitude star. If you also see a second blur 1′ to the northeast of that star, then you have spotted IC 1183. Interestingly, Dr. Harold G. Corwin, Jr., of the California Institute of Technology, believes Swift actually found IC 1183, and not NGC 6054. Corwin, an expert on the history and ambiguity of the *New General Catalogue*, points out that Swift's own words describe his new object as being southwest of the neighboring star, not southeast. I will stick with tradition here, but with the caveat that IC 1183 may actually be NGC 6054, and vice versa.

IC 1185 lies 3′ south-southeast of IC 1183, while IC 1182 is 2′ to its north-northwest. I have been able to make out the former as a small, round blur through my 18-inch at 206× under suburban skies, but the latter has always escaped unfound.

Finally, another pair of commingling galaxies, IC 1178 and IC 1181, lies 9′ due south of that 12th-magnitude field star near NGC 6054. IC 1178 is the brighter of the two here, showing an oval disk with a brighter central core. IC 1181 is just 40″ south of that core, but I am afraid that I have never been able to identify it individually. Perhaps you can under darker, steadier skies.

Table 7.8 lists these galaxies as well as all within AGC 2151 rated at 16th magnitude or greater. See how many you can spot on a clear summer's eve.

168   UKS 1

Target	Type	RA	Dec.	Constellation	Magnitude	Size	Chart
UKS 1	Globular cluster	17 54.5	−24 08.7	Sagittarius	17.3	2′	7.12

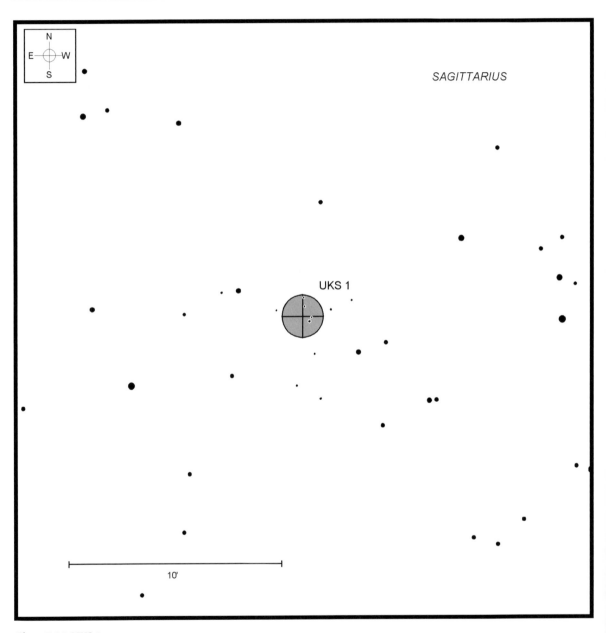

Chart 7.12 UKS 1

Thanks to their intense cores, most globular clusters punch through light pollution better than some other types of deep-sky object, such as emission nebulae and galaxies. There are exceptions to every rule, however, and here we find a beauty!

Can we see the invisible? UKS 1, more formally known as UKS 1751-241, certainly deserves that foreboding description. After all, how could it escape detection on the National Geographic Society–Palomar Observatory Sky Survey charts as well as on images obtained as part of the Sloan Digital Sky Survey (SDSS)? Their limiting magnitude approaches 22nd magnitude on average.

Well, somehow, it did. UKS 1 first came to the astronomical world's attention in 1980 when Matthew Malkan, D. E. Kleinmann, and Jerome Apt published a research paper in the *Astrophysical Journal* entitled "Infrared Studies of Globular Clusters near the Galactic Center."[7] In their opening paragraphs, the authors note that UKS 1 "was discovered by [Andrew J.] Longmore and [Timothy G.] Hawarden (private communication) on IV-N infrared plates taken with the 48-inch (1.2-meter) UK Schmidt telescope; on these plates it appears as a small round, hazy patch, no brighter than . . . ~16 mag." *Infrared* plates? That explains it. Infrared radiation can better penetrate obscuring dust clouds that might otherwise hide more distant objects at visible wavelengths. Since the Palomar Survey photos recorded images in the blue and red ends of the visible spectrum only, UKS 1 remained undetected. It has subsequently been well imaged as part of the Two Micron All-Sky Survey (2MASS).

As is the case with so many challenging objects, UKS 1 is a victim of circumstance. Plot its coordinates on any star chart and it quickly becomes apparent why. UKS 1 is in Sagittarius a scant 5° from the center of the Milky Way, where there is more than enough cosmic dust to block our view. It is estimated that all of the intervening dust diminishes the brightness of UKS 1 by 11 visual magnitudes. In other words, if we had a clearer view of UKS 1, it would shine as brightly as some of the far more noteworthy Messier globulars. But, as it is, this "bulge globular" confounds our best attempts.

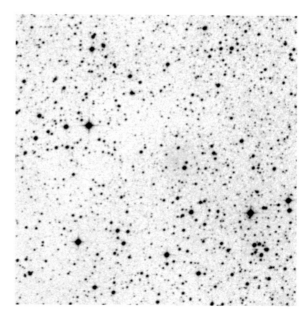

Figure 7.12 UKS 1

If UKS 1 is invisible on the Sky Survey images, how could we hope to see it visually? That was the question asked by deep-sky superobserver Barbara Wilson from Houston, Texas. They do everything in a big way in Texas, and Wilson is no exception. Like many of us, she became intrigued by the presence of this little globular when she first noticed it on chart 339 of the *Uranometria 2000.0* star atlas. She writes, "Naively, I originally had thought any object plotted on Uranometria would be visible in a 20-inch!" She quickly found out how wrong that assumption was.

After four years of failed attempts, and only after referencing a CCD image of the field taken with the 18-inch telescope at the Houston Museum of Natural Science's George Observatory, Wilson spotted her quarry. It took the full aperture of fellow amateur Larry Mitchell's 36-inch reflector operating at 663× to make out the cluster's feeble photons set between two 11th-magnitude field stars. What a triumph for a visual observer, to see something that was missed by the original Palomar Sky Survey as well as the Digital Sky Survey! She summarizes the experience and offers some sage advice for anyone trying to duplicate this achievement:

[7] M. Malkan, D. Kleinmann, and J. Apt, "Infrared Studies of Globular Clusters near the Galactic Center," *Astrophysical Journal*, Vol. 237 (1980), pp. 432–7.

UKS 1 is just east of a pair of unequal brightness stars about 14–15th mag, the south star being the brighter. If you use the distance that the two stars are separated by as a gauge, UKS 1 is slightly farther east than that separation, but equidistant between them. It was important to keep my averted vision exactly concentrated on the right place, else the object was invisible. I don't think it will ever be visible in a 20-inch unless we can develop infrared eyes.

To read her full account of the observation, visit the "Adventures in Deep Space" website, astronomy-mall.com/Adventures.In.Deep.Space/obscure.htm.

169 PK 9-7.1

Target	Type	RA	Dec.	Constellation	Magnitude	Size	Chart
PK 9-7.1	Planetary nebula	18 36.4	−23 55.3	Sagittarius	15.0p	9″	7.13

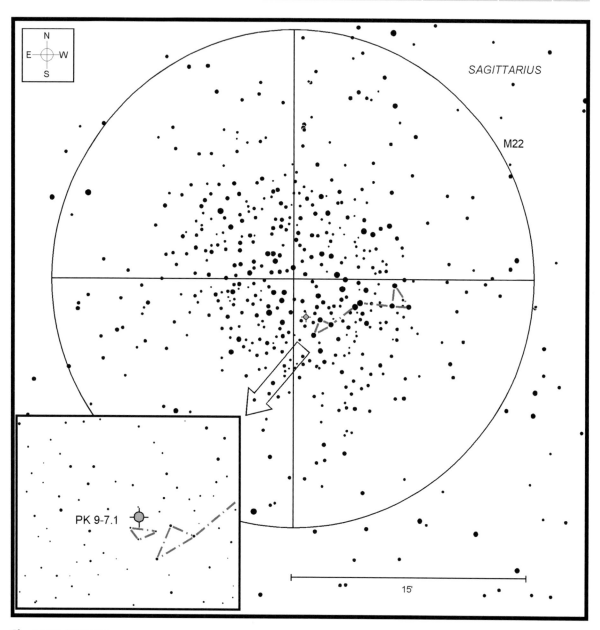

Chart 7.13 PK 9-7.1

What is your favorite globular cluster? Apart from those two southern-hemisphere beauties, Omega (ω) Centauri and 47 Tucanae, my answer has to be M22 in Sagittarius. Admittedly, the star chains and "propeller" formation within M13 in Hercules (large-scope Challenge 138) are visually intriguing. But there is just something about the remarkable richness of M22 and its surrounding star field that calls to me.

A buried treasure remained hidden between all of M22's stars until 1985, when IRAS, the InfraRed Astronomy Satellite, uncovered a mysterious infrared source. It was subsequently included as IRAS 18333-2357 in the *IRAS Point Source Catalog*, but it would be another four years before research by F. C. Gillett, G. H. Jacoby, R. R. Joyce, and J. G. Cohen[8] would reveal the unexpected origin of the radiation. It was a planetary nebula. What's the big deal? Planetary nebulae are commonplace. That's true, they are – except in globular clusters. Even today, only four are known to exist.

Our unusual friend within M22 now goes by any of three designations: IRAS 18333-2357; GJJC-1, after the initials of the four researchers; and PK 9-7.1 in the second (2000) edition of the Perek–Kohoutek *Catalogue of Galactic Planetary Nebulae*. Here, if for no other reason than for consistency with some other challenges, the M22 planetary will be referred to by its PK alias. Other sources, especially online ones, prefer GJJC-1. Either way, we are talking about the same object.

And that object is one tough challenge to find. Of the handful of amateurs who have reported spotting PK 9-7.1, their accounts seem to have four things in common. First, all but one were using 20-inch-plus apertures. Next, each observer had at hand a series of finder charts showing the area around the target in increasing detail. All were also using what I would call "crazy high magnifications," that is, in excess of 600×. In some cases, magnification exceeded 900×. That means the final common ingredient had to be steady seeing. Even minor air turbulence at these apertures and magnifications will quickly turn stars into mush.

Requirement one, aperture, is up to you, but I can give you a hand with the charts. Chart 7.13 shows the

Figure 7.13 PK 9-7.1

central core of M22 with several key asterisms that will be used to zero in on the planetary's location. Begin by locating the right triangle of 11th-magnitude stars shown along the western edge of the cluster. Follow the base of the triangle 1.1′ eastward to a close-set pair of stars at the end of a diagonal line. Extend a line southeastward from those stars to an equilateral triangle. If you can find that triangle, then you are getting very near the planetary. Just beyond the eastern tip of the triangle is another, smaller threesome of 13th-magnitude stars. PK 9-7.1 is 8″ north of that triangle's easternmost star. Be careful not to mistake a dim field star just beyond the planetary for the planetary itself. Unless you can see both clearly, then the chances are good that you are only seeing the star and not PK 9-7.1, which is fainter.

To confirm the find, try using a narrowband filter. Owing to the faintness of the planetary, reports I have seen suggest threading the filter onto your eyepiece rather than flipping it in and out, as we have done previously. The filter should shrink the bloated stars, while showing that the planetary is very slightly oval.

Studies reveal that PK 9-7.1 is a true member of M22, not just a line-of-sight superimposition. It's an odd little beast for more reasons than its location. Spectroscopic analyses also disclose that, unlike most planetaries, PK 9-7.1 is depleted of hydrogen and helium. Rather, the only prominent lines in its spectrum are those of doubly ionized oxygen (oxygen III) and neon (neon III). Its shape is also asymmetrical,

[8] F. C. Gillett, G. H. Jacoby, R. R. Joyce, J. G. Cohen, G. Neugebauer, B. T. Soifer, T. Nakajima, and K. Matthews, "The Optical/infrared Counterpart(s) of IRAS 18333-2357," *Astrophysical Journal*, Vol. 338 (1989), pp. 862–74.

shaped like a half moon probably due to gravitational compression from interaction with its packed surroundings. This would also explain the lack of lighter elements, as they would have been stripped away by the process.

For amateurs who try to spot this difficult object, planetary nebula observer par-excellence Doug Snyder offers an excellent discussion, as well as first-hand reports and finder charts for the task on his website, blackskies.org.

170	**IC 1296**

Target	Type	RA	Dec.	Constellation	Magnitude	Size	Chart
IC 1296	Galaxy	18 53.3	+33 04.0	Lyra	14.8p	0.9′×0.5′	7.14

171	**M57 central star**

Target	Type	RA	Dec.	Constellation	Magnitude	Size	Chart
M57 central star	Central star	18 53.6	+33 01.7	Lyra	15.2	n/a	7.14

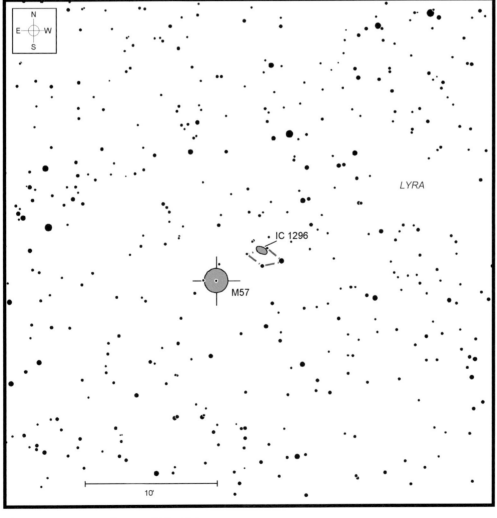

Chart 7.14 M57 central star and IC 1296

Spotting the Ring Nebula, M57, in Lyra already challenged us in Chapter 3 as binocular Challenge 30. We now return to spot the progenitor sun that started it all some 6,000 to 8,000 years ago. Seeing the Ring's central star is one of those tests that every visual observer needs to take at one point or another.

Looking through observing handbooks, reading online deep-sky logs, speaking with friends and colleagues whom I consider to be seasoned veterans, and my own personal experience all seem to show that spotting the central star takes nothing short of "the perfect storm." Unless everything comes together just right, the star will remain hidden from view.

You might be wondering what all the fuss is about. After all, the star is listed as 15th magnitude, which is dim, but within the grasp of 15-inch telescopes, perhaps even less under dark, transparent skies. So then, why is the central star so difficult through even the largest backyard scopes?

Here's an interesting observation that I have noticed time and again when trying to see the central star. Seeing the star requires transparent skies, but not necessarily dark skies. Many amateurs equate one with the other, that dark skies are transparent skies, and vice versa. If you are among those who believe dark skies are also transparent skies, I'd invite you to revisit that discussion in Chapter 1, where the two terms are defined.

But, on the topic here, I have seen the central star on several occasions through my 18-inch reflector from my naked-eye limiting magnitude 5 suburban backyard, but have missed it entirely on many other occasions using the same equipment from markedly darker sites. Why? Those other sites were darker (i.e., less light pollution), but the sky was not as transparent. The increased level of haze lowered the contrast between the star and the surrounding nebulosity just enough to mask the star.

That brings us to the second ingredient to seeing the star: seeing. Without steady seeing conditions, atmospheric turbulence will blur the star just enough to blend its already low-contrast glow into the Ring's donut hole.

Without both conditions – exceptional seeing and transparency – the central star will evade even the most careful search. But it still takes more than these. Your telescope's optics must be clean, as well. Any contamination, notably skin oils on the eyepiece's eye lens, will be enough to lose the star.

Figure 7.14 M57 with central star and IC 1296

A faint, far-off barred spiral galaxy floats in the same field as you try to spot the Ring's central star. Can you also spot IC 1296? It's a tougher task than its 14th-magnitude rating would imply. That's because, as we have seen so often before, the galaxy's surface brightness skews the integrated magnitude. In 15-inch-plus telescopes, 14th-magnitude galaxies are fairly routine. That's assuming their light is concentrated evenly across their disk. In the case of IC 1296, the central hub of the galaxy is nearly stellar in appearance, while its broad spiral arms are unusually faint.

IC 1296 is just 4′ northwest of M57, near a diamond of four 11th- to 14th-magnitude stars (Figure 7.14). More specifically, it is positioned 20″ southeast of the star at the diamond's northern facet. Proper magnification, in addition to dark skies and properly collimated optics, are key to spotting its dim glow.

I can probably count on one hand the number of times in the past half-dozen years when I have seen both of these challenges through my 18-inch scope from my suburban backyard observatory. Summer haze, air turbulence, and light pollution quickly extinguish both. But, on those rare evenings when the humidity is low, the seeing is calm, and the Ring is high in the sky, the elusive central star and its tiny galactic companion shine through. Indeed, under superior skies, telescopes with apertures as small as 11 inches have shown this little guy, so be sure to give it a go.

172   Terzan 7

Target	Type	RA	Dec.	Constellation	Magnitude	Size	Chart
Terzan 7	Globular cluster	19 17.7	−34 39.5	Sagittarius	12.0	1.2′	7.15

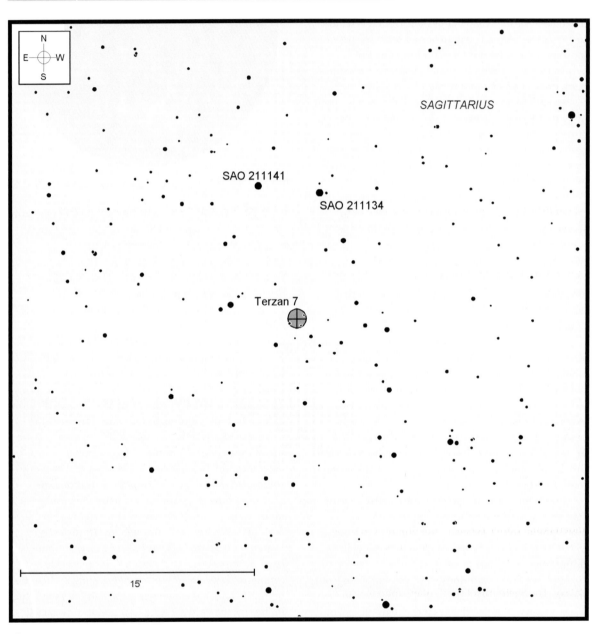

Chart 7.15 Terzan 7

Table 7.9 *Terzan globular clusters*

Name	RA	Dec.	Apparent magnitude	Size	Distance (kpc)	Brightest member star
Terzan 1	17 35.8	−30 28.9	15.9	2.4	5.6	
Terzan 2	17 27.6	−30 48.1	14.3	1.5	8.7	
Terzan 3	16 28.7	−35 21.2	12.0	3.3	7.5	15.0
Terzan 4	17 30.7	−31 35.7	16.0	0.7	9.1	
Terzan 5	17 48.1	−24 46.8	13.9	2.1	10.3	20.5
Terzan 6	17 50.8	−31 16.5	13.9	1.2	9.5	20.5
Terzan 7	19 17.7	−34 39.5	12.0	1.3	23.2	15.0
Terzan 8	19 41.8	−34 00.0	12.4	3.3	26.0	15.0
Terzan 9	18 01.6	−26 50.4	16.0	1.5	6.5	
Terzan 10	18 02.9	−26 04.0	14.9	0.3	5.7	19.7
Terzan 11	18 12.3	−22 44.5	15.6	1.5	4.8	18.5

As if the Palomar globular clusters, which were introduced in Chapter 6, weren't tough enough, the Terzan list of 11 globulars present an even more daunting test for the largest amateur telescopes. These clusters were discovered during an infrared investigation of the plane of the Milky Way by the French astronomer Agop Terzan in 1968. All of Terzan's globulars lie near the plane of our galaxy, and as such may be seriously impacted by obscuring dust clouds. Infrared energy pierces those clouds more effectively than visual wavelengths, which allowed Terzan to detect their feeble light.

The original list published by Terzan[9] included 12 entries, but there was a problem that went uncorrected until after publication. It turned out that the same globular was listed twice, as both Terzan 5 and Terzan 11. To correct the mistake, most subsequent references have simply edited out the error and renumbered the cluster that was originally Terzan 12 as Terzan 11. Table 7.9 reflects the corrected list of Terzan globulars.

Seeing all 11 Terzan globulars is a difficult project even for deep-sky diehards. To kick off that undertaking, I offer Terzan 7, the brightest of these demanding targets. Studies show that Terzan 7 is actually not one of the Milky Way's brood of globulars, but instead belongs to the Sagittarius Dwarf Spheroidal Galaxy. The Sagittarius Dwarf, often abbreviated

SagDEG in literature, is 65,000 light years away, making it the second closest satellite of the Milky Way.

Although its location on the opposite side of the galactic core from Earth kept SagDEG hidden from view until 1994, bits and pieces of it were detected long before then. The first piece of the SagDEG puzzle was found way back on July 24, 1778, by none other than Charles Messier himself. That was the night he discovered M54. M54 is now believed to be one of four members of the dwarf's family of known globular clusters, and not associated with the Milky Way after all. Terzan 7 is the second brightest of the quartet.

Besides its unique location, Terzan 7 stands out as one of the youngest globular clusters ever discovered. Nearly all of the Milky Way's globulars formed at least 12 billion years ago. A study published in 2007,[10] however, concludes that the stars in Terzan 7 are only 7.5 billion years old. This finding was based on the greater quantity of metals detected in the spectra of Terzan 7's stars versus those within Milky Way globulars. Heavier elements, such as metals, are found in greater abundance in younger stars than in older stars. That's because those elements can only be created in the cores of massive stars and in supernovae

[9] Agop Terzan, "Six nouveaux amas stellaires (Terzan 3−8) dans la region DU centre de la Voie lactee et les constellations DU Scorpion et DU Sagittaire,", *Comptes Rendus de l'Académie des sciences*, Ser. B, Vol. 267 (1968), pp. 1245−8.

[10] Doug Geisler, George Wallerstein, Verne V. Smith, and Dana I. Casetti-Dinescu, "Chemical Abundances and Kinematics in Globular Clusters and Local Group Dwarf Galaxies and Their Implications for Formation Theories of the Galactic Halo," *Publications of the Astronomical Society of the Pacific*, Vol. 119, No. 859 (2007), pp. 939−61.

Figure 7.15 Terzan 7

explosions. Stars in most globulars are first generation, and so were formed before metals even existed.

Locating Terzan 7 takes a good view to the south, since it lies nearly 35° south of the celestial equator.

Begin your quest from the semicircle of stars that form Corona Australis, found directly south of Sagittarius. Aim toward Alpha (α) and Gamma (γ) Coronae Australis through your finder. Notice that they form the base of a long, slender triangle with 6th-magnitude SAO 211175 lying 3° to their northeast. Terzan 7 is less than a degree further north-northeast. Look for a close-set pair of 9th-magnitude stars (SAO 211134 and 211141) as a reference. The globular is a degree to their south and 2′ away from a 12th-magnitude sun.

Viewing through my 18-inch at 206×, I can see Terzan 7 as a very faint glow measuring only about 1′ across (Figure 7.15). Increasing magnification to 294× reveals some very faint stars around the edges of the cluster, hinting at the object's real personality. The brightest stars in Terzan 7 are 15th magnitude, bringing them just within the grasp of our telescopes.

If you were successful at bagging Terzan 7, then try your luck with Terzan 8. Terzan 8 is 5° to the east and is also believed to belong to SagDEG. It is rated just a half magnitude fainter than Terzan 7, but because it appears twice as large, the resulting surface brightness is considerably lower. Good luck on that one!

173 Simeis 57

★
★
★
★

Target	Type	RA	Dec.	Constellation	Magnitude	Size	Chart
Simeis 57	Emission nebula	20 16.2	+43 41.2	Cygnus	–	23′×4′	7.16

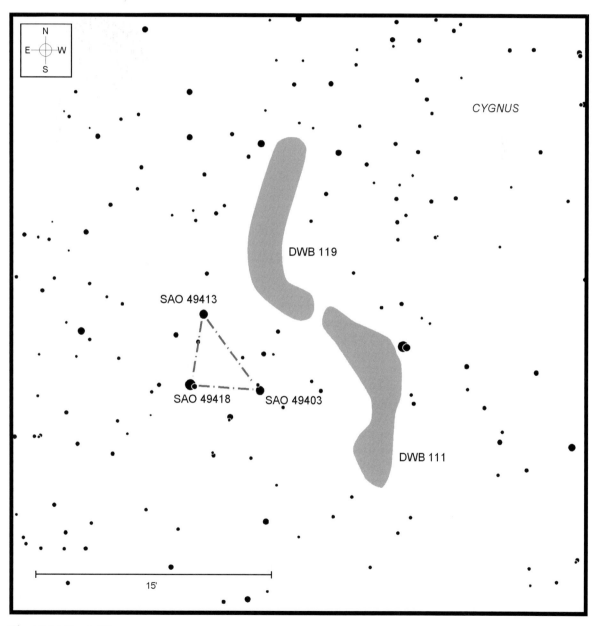

Chart 7.16 Simeis 57

Simeis 57 is one of the most intriguing emission nebulae in the late summer sky, yet it is almost unknown to visual observers. Photographers, however, know it as a pair of opposing arcs of reddish light spinning symmetrically away from a common middle, one extending to the north, the other to the south. Its unusual appearance has led to its two nicknames: the Propeller Nebula or the Garden Sprinkler Nebula.

The entire complex was assigned Simeis 57 when it was discovered in the early 1950s by G. A. Shajn and V. E. Hase at the Crimean Astrophysical Observatory at Simeis, Russia. Their results were published in the observatory's *Bulletin* (in Russian, *Izvestiya Krymskoi Astrofizicheskoi Observatorii*), although they did not become widely known outside of the Soviet Union, probably due to the Cold War raging at the time.

Later, portions of Simeis 57 were assigned separate designations in various catalogs. The propeller's southern blade is listed as DWB 111, after a 1969 article by H. R. Dickel, H. Wendker, and J. H. Bieritz that appeared in the journal *Astronomy & Astrophysics*. The same article listed the northern blade as DWB 119. Fainter sections were assigned other DWB numbers, although, for our purposes here, we will concentrate on trying to see the propeller itself. That's tough enough.

As with so many emission nebulae (or hydrogen-II regions, if you prefer), the Propeller Nebula is all but invisible to the eye alone under normal conditions. Its primary emissions lie in the red portion of the visible spectrum, where our eyes are all but blind under dim light conditions. And they don't come much dimmer than Simeis 57.

The blades of the propeller span about 20′, so, in order to squeeze both into the same view, select an eyepiece with a real-field coverage of at least half a degree. A modern, ultra-wide design with an 80°-plus apparent field is better than, say, a more traditional Plössl or Erfle, since their wide fields also produce a higher magnification for the given field. That's an important consideration, since higher magnification will generate better image contrast.

To boost contrast further, experiment with various nebula filters. I don't wish to plant any preconceived prejudices in your mind, but narrowband (UHC-type) and oxygen-III filters seem to offer little positive effect on the Propeller. On the other hand, a hydrogen-beta (Hβ) filter, which rarely seems to help objects beyond

Figure 7.16 Simeis 57

the Horsehead Nebula, usually proves to be the top choice here. But, again, try each filter in your collection and see which produces the best results.

The Propeller is 5° southwest of Deneb [Alpha (α) Cygni], and just to the west of a right triangle of the 7th-magnitude stars SAO 49403, 49413, and 49418. While that triangle is obvious in the 8×50 finderscope attached to my 18-inch, the Propeller itself takes better skies than I can hope for from my suburban observatory. Under naked-eye limiting magnitude 6.5 skies, however, the 18-inch at 94× and with an Hβ filter in place reveals a very soft glow after a concentrated search.

Of the two blades, the northern component, DWB 119, impresses me as slightly more obvious. It lies just northwest of the triangle. My notes recall the softest of glows, a gentle, concave arc opening toward the west (Figure 7.16). Two close-set 12th-magnitude stars appear centered along the length of the arc, while an 11th-magnitude star marks its northern tip.

The southern blade (DWB 111) is a tougher catch. Look for a close pair of 9th-magnitude stars just to its west; they make a handy reference marker in much the same way as 52 Cygni does for the NGC 6960 segment of the Veil Nebula. DWB 111 is a mirror image of DWB 119, with its curve opening to the east, more or less toward the right triangle of stars.

174   Pease 1

★
★★
★

Target	Type	RA	Dec.	Constellation	Magnitude	Size	Chart
Pease 1	Planetary nebula	21 30.0	+12 10.4	Pegasus	14.9p	1″	7.17

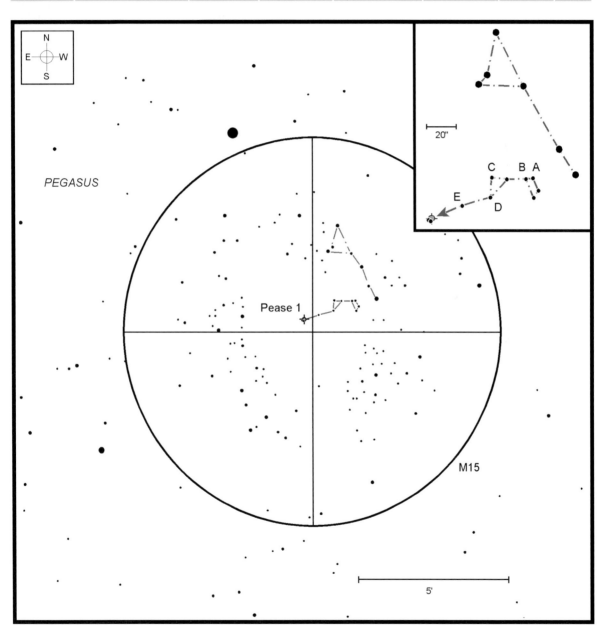

Chart 7.17 Pease 1

Monster-scope Challenge 169 tested us to find a planetary nebula buried in globular cluster M22. It was noted there that M22 is one of only four globulars known to have planetary nebulae residing within, a rare combination indeed. The best-known example of a planetary/globular pairing is Pease 1 found within M15 in Pegasus.

Pease 1 was the first planetary nebula to be detected within a globular cluster. That occurred in 1927, when Francis Gladheim Pease first spotted it on a photograph taken with the 100-inch Hooker reflector at Mount Wilson Observatory in California. In his report the following year in the *Publications of the Astronomical Society of the Pacific*,[11] Pease observed that "through the 'Pulkowa ultra-violet' color filter, the star Küstner No. 648, photographic mag. 13.78, appeared very bright as compared with the surrounding stars." Those surrounding stars appeared identical to Küstner 648 in visible-light images, which triggered Pease's curiosity. He returned to M15 a year later to conduct spectroscopic studies of this intriguing little find. The results were unmistakable. Pease had detected a planetary nebula.

The Küstner referred to in Pease's statement was German astronomer Friedrich Küstner, who had published a paper[12] seven years earlier that detailed the internal structure of M15. Küstner inventoried many of the cluster's stars, but failed to recognize the unique qualities of star number 648. That's why, although the proper designation of the planetary is Pease 1 (or if you prefer PK 65-27.1 in the Perek–Kohoutek catalog), some references note it as K648.

Is Pease 1 a true member of M15 or just a chance alignment? Such a nonexistent duo is not without precedent; recall the planetary nebula NGC 2438 against the open cluster M46 in winter's Puppis. Studies of the radial velocities of the nebula and stars within M15 show that all are traveling at nearly the same speed, which implies that they lie at the same distance. Supporting this are additional spectroscopic analyses that demonstrate the abundance of elements in Pease 4, and a notable lack of metals, matching those of the

Figure 7.17 Pease 1

cluster stars. Based on these results, it is now generally conceded that Pease 1 is gravitationally bound within M15. The question of why the gravity from nearby cluster stars has not conspired to disperse the cloud, however, is still undetermined.

Finding M15 is easy, since it is just 4° northwest of Enif [Epsilon (ε) Pegasi], the star at the tip of Pegasus's nose. Look for a 6th-magnitude star just to the cluster's west, which helps to mark its location. In fact, through finderscopes, the pair almost seems to create a double star, although M15 will look fuzzy even at low powers.

Spotting Pease 1 within that vast stellar metropolis is another matter altogether. While the nebula is bright enough to be visible under naked-eye limiting magnitude 5.0 skies, trying to pick it out from among all of the stars is the challenge. Fortunately, unlike PK 9-7.1 centered in M22, Pease 1 is slightly offset from the cluster's dense center. Chart 7.17, adapted from

[11] F. G. Pease, "A Planetary Nebula in the Globular Cluster Messier 15," *Publications of the Astronomical Society of the Pacific*, Vol. 40, No. 237 (1928), p. 342.
[12] Friedrich Küstner, *Der kugelfoermige Sternhaufen Messier 15* (*The Globular Star Cluster Messier 15*), Veröffentlichungen der Universitäts-Sternwarte zu Bonn, No. 15 (Bonn: F. Cohen, 1921).

Doug Snyder's excellent website, blackskies.org, will help us in our quest.

Pease 1 takes no less than 300× to be seen, so, unless the sky is both exceptionally transparent and steady, wait for another night. Take your time to orient the chart to match what you see through your eyepiece and to get a feel for the scale of the chart versus what your telescope is showing you.

With everything set, look about halfway out toward the northwestern edge of the cluster for the highlighted trapezoid of 14th-magnitude stars, labeled A on the chart.[13] See it? If so, then look just to its east for a right triangle of faint stars, shown as B, C, and D. Extend a line from star A in the trapezoid through star D to arrive at Star E found about 20 arc-seconds to the southeast. Continue that line past star E another 28″ to a small knot of stars.

Once there, call in an oxygen-III filter. By passing the filter in and out of view, the stars in the cluster will dim enough for the planetary to stand out from the crowd. Doing just that allowed me to see the planetary from naked-eye limiting magnitude 5.0 skies, but it remained perfectly stellar through my 18-inch even at 514×. Using the in-and-out filter method should leave no doubt in your mind as to which point is the planetary.

---

[13] The star labels on the chart here match the convention shown on Snyder's website to make cross-reference a little easier.

175   Elephant's Trunk Nebula

Target	Type	RA	Dec.	Constellation	Magnitude	Size	Chart
Elephant's Trunk Nebula	Reflection nebula	21 39.0	+57 30.0	Cepheus	–	13′×5′	7.18

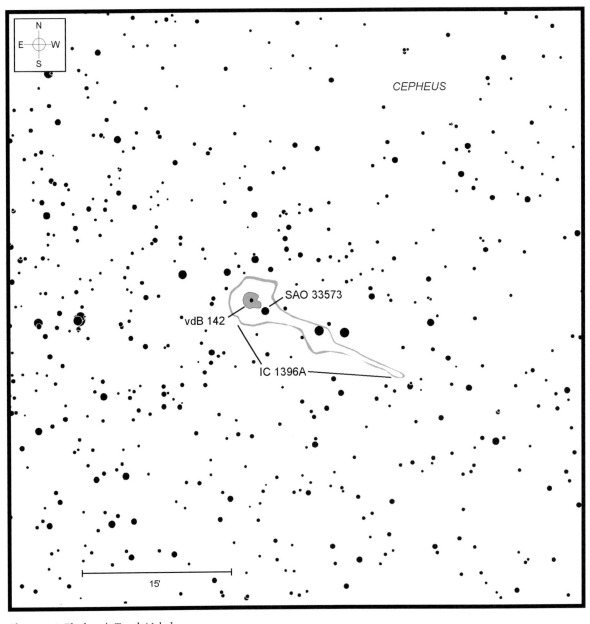

Chart 7.18 Elephant's Trunk Nebula

Take a look at just about any star atlas and you will find a huge, irregular cloud of ionized hydrogen suspended to the south of Mu (μ) Cephei, Herschel's Garnet Star. That's IC 1396, one of the largest nebulae in the night sky. Even from a distance of 3,000 light years, this complex cocktail of bright glowing gas mixed with dark dust clouds spans 3° of our sky. At that distance, 3° translates to a linear diameter of nearly 160 light years, more than three times greater than the Orion Nebula, M42.

Seeing IC 1396, however, is a difficult chore owing to its large size and very low surface brightness. Giant binoculars show myriad stars scattered across the region, but actually seeing evidence of the cloud itself is another matter. Small scopes can resolve several double and multiple stars here, notably the triple system Struve (STF) 2816 and binary Struve (STF) 2819. But again, evidence of the cloud proves hard to pin down.

Instead of trying to see the entire H II region, this challenge focuses on a particularly intriguing feature found within. Of the many pockets of dark nebulosity swirling inside IC 1396, the most captivating has to be the twisted globule known as the Elephant's Trunk. The Elephant's Trunk Nebula, set to the east of the cloud's center, is an example of a "composite nebula," that is, a mix of all three types of intergalactic clouds: emission, reflection, and dark.

The Elephant's Trunk is often misidentified as van den Bergh 142 (vdB 142), from Canadian astronomer Sidney van den Bergh's catalog of reflection nebulae. The van den Bergh list was originally released in his paper "A Study of Reflection Nebulae" published in the *Astronomical Journal* in 1966.[14] Van den Bergh's aim back then was to create an inventory of "all BD and CD stars [*Bonner Durchmusterung* and *Cordoba Durchmusterung* astrometric star catalogs, respectively] north of declination −33° which are surrounded by reflection nebulosity." Using those catalogs, van den Bergh examined the Palomar Sky Survey photographic plates for reflection nebulae surrounding embedded stars. He ultimately found and listed 158 such objects.

But the Elephant's Trunk was not one of them. Number 142 in the van den Bergh list actually refers to

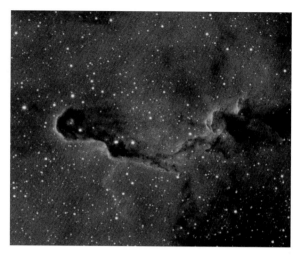

Figure 7.18 Elephant's Trunk Nebula

a small reflection nebula adjacent to the 9th-magnitude star SAO 33573 that overlaps the Elephant's Trunk, not the Trunk itself. Instead, it is correctly cited as IC 1396A.

Regardless of the catalog designation, the Elephant's Trunk certainly lives up to its nickname, with its long, prehensile silhouette stretching for about 17′ from east-northeast to west-southwest. The most obvious segment, the rounded eastern end, extends about 4′ to the east of SAO 33573. Many comment on its resemblance to one of the famous Pillars of Creation in M16. Here, as there, the nebula appears as a silhouette surrounded by the gentle glow of starlight reflecting off a multitude of cosmic dust grains.

Like the Pillars of Creation, the Elephant's Trunk contains many protostar globules. These regions of dense, cold gas and dust can be thought of as celestial maternity wards. As new stars form from the dust and hydrogen within, radiant pressure will slowly open this stellar womb to reveal a newly formed open cluster. The first signs of birth are already evident in the form of a faint double star implanted 2′ east of SAO 33573.

Despite the Trunk's length, expect to use about 100× to bring out its delicate outline (Figure 7.18). Even under the darkest skies, the nebula will still require a narrowband filter. Tapping the side of your telescope ever so slightly will also help reveal its faint outline. If you see its subtle profile, try to spot some of the subtle structural intricacies that record so beautifully in photographs.

---

[14] S. van den Bergh, "A Study of Reflection Nebulae," *Astronomical Journal*, Vol. 71 (1966), pp. 990–8.

176 Einstein's Cross

Target	Type	RA	Dec.	Constellation	Magnitude	Size	Chart
Einstein's Cross	Gravitational lens	22 40.5	+03 21.5	Pegasus	15.1p	1.1'×0.5'	7.19

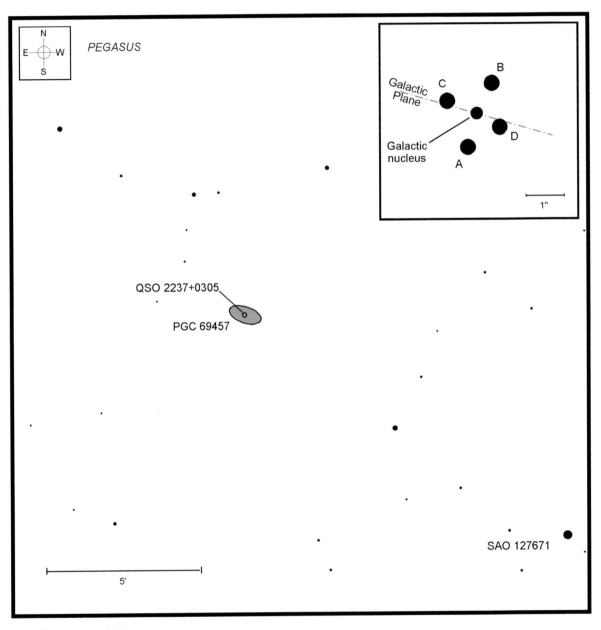

Chart 7.19 Einstein's Cross

One of the predictions of Albert Einstein's 1916 general theory of relativity was that the light from a bright, far-off source of energy would be warped, or "bent," around a massive object lying between that source and an observer. In the process, the time it takes the light to reach an observer would be altered, causing the background object to appear magnified and distorted.

That was what Einstein's theory said, but how could it be put to the test? The most massive nearby objects known at the time, such as the Sun, were also very bright. Anything that happens to lie beyond would be so faint by comparison that it would be rendered invisible.

Sir Arthur Stanley Eddington, Britain's leading astrophysicist at the time, came up with the solution: use the Sun anyway. Not just the Sun on any given day, however, but rather during the total phase of a solar eclipse, when the Moon's disk completely blocks the blinding photosphere. The upcoming eclipse of May 29, 1919, would be perfect. Not only was totality unusually long, the Sun would be located right in front of Taurus's Hyades star cluster. There would be plenty of stars in the Sun's vicinity to test Einstein's theory. Although his expedition was touch and go because of everything from clouds and rain to the clouds of World War I, Eddington's observations recorded stars beside the Sun that should actually have been positioned behind its edge at the time. Einstein was right: gravity could warp light.

This warping effect is known today as *gravitational lensing*. Photographs taken with the Hubble Space Telescope, as well as with many Earth-based instruments, show the effect well, with ghostly images of far-off quasars and galaxies floating beside foreground galaxies. Rather than creating a single image of the distant quasar, however, a gravitational lens creates multiple images. Depending on the shape of the gravitational lens (i.e., the gravitational influence on the distant light), the refracted image may be distended and bent into all sorts of odd contortions. Or, if the galaxy is positioned perfectly in line between the quasar and the Earth, then we would see a symmetrical ring of quasars.

From an aesthetic perspective, the most perfect gravitational lens is Einstein's Cross, formed by the galaxy PGC 69457 and the quasar QSO 2237+0305 in Pegasus. PGC 69457 is also known informally as Huchra's Lens after its discoverer, John Huchra,

Figure 7.19 Einstein's Cross

professor of cosmology at Harvard University. Current estimates place this small, otherwise unspectacular spiral galaxy at 400 million light years away. The quasar lurks far behind at an incredible distance of 8 billion light years. Were it not for gravitational lensing, the quasar would remain hidden by the galaxy, as the two are nearly in-line as seen from Earth. But, as it is, Huchra's lens fractures the ancient light from the quasar into four separate paths that slide around the galaxy just as water flows around a rock in a stream. The end result is not one, but four ghostly images of QSO 2237+0305 surrounding the nucleus of PGC 69457 in a practically perfect diamond pattern.

Einstein's Cross lies to the south of the Flying Horse's "head" and "neck" and due west of the Circlet of Pisces. To find it, begin at the star Biham [Theta (θ) Pegasi] and slide 5° southeast to a triangle formed by 34, 35, and 37 Pegasi. Extending a line from 35 Peg through 37 Peg five times ($2\frac{1}{2}°$) farther southeast will bring you to the orangish 8th-magnitude star SAO 127671. Centering there, look for an 11th-magnitude star 6' to the northeast. This field star is very handy for gauging distance, as Einstein's Cross is found another 6' farther to its northeast.

Although Einstein's Cross is rated at 15th magnitude, I have seen it with difficulty with my 18-inch reflector from my suburban observing site (naked-eye limiting magnitude 5.0) by using averted vision. But try as I might, even at 411× in those rare moments when seeing briefly permitted such extravagance, all I could see was the faint, nearly stellar object seen in Figure 7.19. I have never been able to separate the four quasar images from the galaxy; instead, all five remain blurred into a single object.

Some other observers have reported success seeing one or two of the lobes while viewing through larger apertures under undoubtedly superior skies. High magnification, and therefore steady seeing, are absolutely required, since the Cross has an angular size of only 1.6 arc-seconds.

177 NGC 51 group

★
★ ★
★

Target	Type	RA	Dec.	Constellation	Magnitude	Span	Chart
NGC 51 group	Galaxy group	00 14.2	+48 12.1	Andromeda	–	10′	7.20

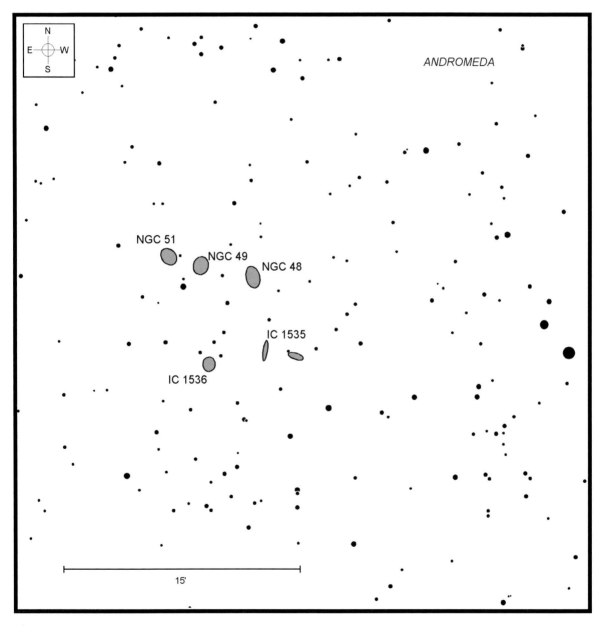

Chart 7.20 NGC 51 group

Table 7.10 *Members of NGC 51 Galaxy Group*

Galaxy	RA	Dec.	Magnitude	Size
IC 1534	00 13.8	+48 09.1	14.8p	1.0′×0.4′
IC 1535	00 13.9	+48 09.5	15.1	1.3′×0.3′
NGC 48	00 14.0	+48 14.1	14.4p	1.4′×0.9′
IC 1536	00 14.3	+48 08.6	15.4	0.9′×0.8′
NGC 49	00 14.4	+48 14.8	14.7p	1.1′×1.0′
NGC 51	00 14.6	+48 15.4	14.1p	1.2′×0.9′

Of the dozens of galaxy groups scattered around the autumn sky, the 136-million-light-year-distant NGC 51 group is one of the more difficult bunches to spot. Although they are not listed among Paul Hickson's compact galaxy groups, the six galaxies here are ideally placed near the zenith in early October and November evenings for observers at mid-northern latitudes. Its high altitude carries the group far enough above any horizon-hugging interferences that might spoil some of our other challenges. Table 7.10 lists the particulars of all six galaxies found here.

To locate this distant galactic swarm for yourself, first center your attention on 5th-magnitude 22 Andromedae in north-central Andromeda. Moving 2° to its north-northeast brings an optical double star made up of 6th- and 8th-magnitude components into view. From here, you are only minutes away (minutes of arc, that is) from the challenge, as the galaxies lie about 20′ to the stars' east.

A string of three NGC galaxies highlight the sextet. All were discovered in 1885 by American astronomer Lewis Swift through a 16-inch refractor. The easternmost of the trio is the lenticular galaxy NGC 51. Most astronomers find this to be the brightest and easiest of the bunch to spot, although it can still be tricky at photographic magnitude 14.4. Its slightly oval disk is punctuated by a brighter central core, making it possible to glimpse NGC 51 through 10-inch telescopes under superior sky conditions.

Can you spot a second, dimmer smudge 2′ to the west of NGC 51? If so, you have seen 14th-magnitude NGC 49. Look for a small, slightly oval blur with a brighter stellar nucleus that becomes evident by using

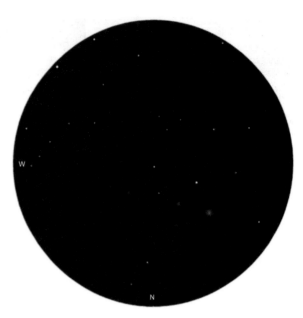

Figure 7.20 NGC 51 group

averted vision. Figure 7.20 was made on a better-than-average suburban night with my 18-inch reflector.

Finally, NGC 48 is a third faint fuzzy another $3\frac{1}{2}'$ farther west of NGC 49. Although NGC 48 is also rated at 14th magnitude, the larger apparent size of this S(B)pc peculiar barred spiral causes its surface brightness to drop, hampering detection. It displays a faint stellar core surrounded by an even dimmer galactic halo that appears slightly elongated southwest–northeast through my 18-inch at 171×.

The other three galaxies in the group all shine at about 15th magnitude. IC 1534 is seen only as an extremely faint, elliptical glow about 5′ southwest of NGC 48. Careful scrutiny with 200× or so might show its tiny disk brightening to a central core. A faint star just kisses the northeastern edge of IC 1534. IC 1535 trails close behind. This apparently coreless galaxy is also oval, but oriented perpendicular to IC 1534. It is also larger and fainter, making it a difficult test indeed. The final galaxy, IC 1536, looks almost perfectly round. Look for its tiny smudge of gray light along the southern side of a triangle of three 13th-magnitude stars.

178 Globular clusters in M31

★
★★★
★★★
★★

Target	Type	RA	Dec.	Constellation	Magnitude	Size	Chart
Globular clusters in M31	Globular clusters in galaxy	00 42.7	+41 16.1	Andromeda	n/a	n/a	7.21

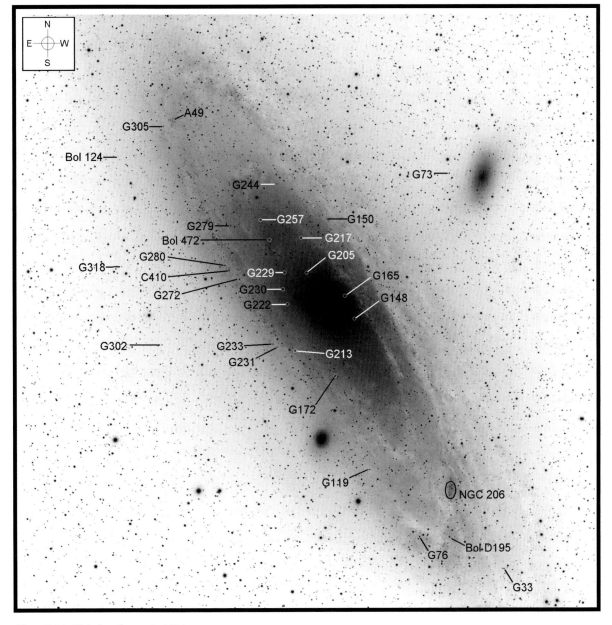

Chart 7.21 Globular clusters in M31

The Andromeda Galaxy was probably one of the first galaxies you ever saw first-hand. It was in my case. That was in 1969, and in the ensuing 40-plus years, I have grown to appreciate it as far more than just the ill-defined oval blur I drew in my logbook. Back then, the thought of seeing individual objects within M31 never crossed my mind.

That changed in 1984, when Arizona astronomer Brian Skiff penned an article in *Deep Sky* magazine. His article, simply titled "All About M31," included details for viewing many of that galaxy's open and globular star clusters. Skiff noted that a total of 355 globular clusters were identified as gravitationally bound to M31 in Paul Hodge's seminal work *Atlas of the Andromeda Galaxy*. Published three years earlier by the University of Washington Press, the atlas is long out of print. Thanks to the Internet and the charitable graces of its author, however, it remains available today via the NASA/IPAC Extragalactic Database (NED).[15]

In the ensuing years, another 150-plus members have been added to the M31 globular family. In this challenge, we will look at some of the brightest. Table 7.11, listing most that are brighter than magnitude 15.5, makes a good jumping-off point for the task at hand.

Of the globulars in the table, the brightest and largest by far is G001 (Figure 7.21). G001 lies more than $2\frac{1}{2}°$ southwest of M31's central core. That proves a real blessing, since it removes its delicate glow from the galaxy's bright spiral-arm disk.

To find G001, first find 5.3-magnitude 32 Andromedae, which is 4° south-southwest of M31 and 1.6° due east of the cluster. Center on 32 Andromedae and then follow a crooked trail of six 7th- to 9th-magnitude stars westward to SAO 53990. G001 lies 13′ farther west, just south of a slender triangle of 13th-magnitude field stars. Two fainter stars stand guard on either side of the globular, one to the northwest and the other to the southwest. At first glance, it is easy to mistake all three for a tight triple star, but the globular is easily identifiable as nonstellar at magnifications above 250×.

The remaining clusters in Table 7.11 are much smaller than G001 and are superimposed somewhere

Table 7.11 *Globular clusters in M31 brighter than magnitude 15.5 (bold entries are discussed above)*

Name*	RA	Dec.	Apparent magnitude	Size (″)
G001	**00 32.8**	**+39 34.7**	**13.8**	**36**
G033	00 39.6	+40 31.2	15.4	2.3
G064	00 40.5	+41 21.7	15.1	2.3
Bol D195	00 40.5	+40 36.3	15.2	
G073	00 40.9	+41 41.4	14.9	
G072	**00 40.9**	**+41 18.9**	**14.9**	**2.2**
G078	**00 41.0**	**+41 13.8**	**14.2**	**3.2**
G076	**00 41.0**	**+40 35.8**	**14.2**	**3.6**
G119	00 41.9	+40 47.2	15.0	2.7
G148	00 42.3	+41 14.0	15.2	2.9
G150	00 42.4	+41 32.2	15.4	2.7
G165	00 42.5	+41 18.0	15.2	2.9
G172	00 42.6	+41 03.4	15.3	2.4
G213	**00 43.2**	**+41 07.4**	**14.7**	**2.5**
G205	00 43.2	+41 21.6	14.8	2.9
G217	00 43.3	+41 27.8	15.0	2.6
G222	00 43.4	+41 15.6	15.3	3.2
G229	00 43.5	+41 21.3	15.0	3.4
G230	00 43.5	+41 18.2	15.4	2.9
G231	00 43.5	+41 07.9	16.0	2.5
G233	00 43.6	+41 08.2	15.4	2.6
Bol 472	00 43.8	+41 26.9	15.2	
G244	00 43.8	+41 37.0	15.3	2.6
G257	00 44.0	+41 30.3	15.1	3.2
G272	**00 44.2**	**+41 19.3**	**14.8**	**3.4**
G280	**00 44.5**	**+41 21.6**	**14.2**	**2.7**
G279	00 44.5	+41 28.8	15.4	4.9
G302	00 45.4	+41 06.4	15.2	2.5
G318	00 46.2	+41 19.7	15.3	
G351	00 49.7	+41 35.5	15.2	

Note: "G" numbers are from the Hodge atlas. "Bol" refers to entries in the *Revised Bologna Catalogue of M31 Globular Clusters and Candidates*.[16]

---

[15] The NASA/IPAC Extragalactic Database (NED) is operated by the Jet Propulsion Laboratory, California Institute of Technology, under contract with the National Aeronautics and Space Administration. The atlas's URL is http://nedwww.ipac.caltech .edu/level5/ANDROMEDA Atlas/frames.html.

[16] S. Galleti, M. Bellazzini, L. Federici, A. Buzzoni, and F. Fusi Pecci, "An Updated Survey of Globular Clusters in M31. II: Newly Discovered Bright and Remote Clusters (V.3.0)," *Astronomy & Astrophysics*, Vol. 471 (2007), p. 127. Catalog available at http:// www.bo.astro.it/M31/.

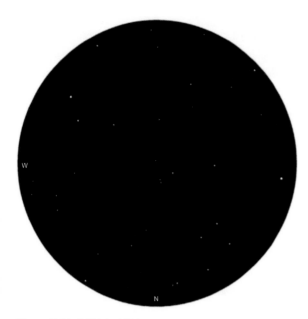

Figure 7.21 G001 in M31

on M31's disk, so they are difficult to see because of the low contrast. All need at least 300× before they appear nonstellar. Anything less and they remain as just faint anonymous "stars."

Working eastward from G001, the next stop is G076. To find it, center on 7th-magnitude SAO 36585, positioned 14' southwest of M32. Sliding southwestward 13' puts a flattened obtuse triangle of 11th- and 12th-magnitude stars in view. G076 is just

40″ southeast of the triangle's southernmost star, but take care not to mistake it for a faint Milky Way star that is settled a little farther southeast.

G078 is about half a degree north of G076 and half a degree west of M31's bright core. Look for a slightly fuzzy, 14th-magnitude point 2' east-northeast of a north–south pair of 12th-magnitude stars. If you have success with G078, then try for G072, which is another 5.4' to the north-northwest. It's about half a magnitude fainter, however, so expect it to be a tougher catch.

Some of the most challenging of M31's globulars nearly overlap the galaxy's central core. For instance, G213 is just 10' southeast of the core and is very nearly superimposed on the edge of the galaxy's halo. It lies just 1' west of an 11.5-magnitude star but, to isolate it from the bright surroundings, use the highest power you can muster.

If G213 proves a little too tough, then try these next two. Although they also suffer the effects of the bright spiral-arm halo, both are better isolated for improved contrast. G272 lies just 1.3' southeast of an 11th-magnitude star that marks the pointy apex of an isosceles triangle of 9th- and 11th-magnitude stars 20' east-northeast of the galactic core. G280 is 4' east-northeast of the apex star. Do you also notice a very soft, elongated glow just to the southwest of G280? That's another bonus object, a large open cluster of stars listed as C410 in Hodge's atlas. The overall magnitude of this difficult object is 16.1, but its brightest stars are far too faint for amateur observation.

179   IC 1613

Target	Type	RA	Dec.	Constellation	Magnitude	Size	Chart
IC 1613	Galaxy	01 04.8	+02 17.1	Cetus	9.2	16.3′ × 14.5′	7.22

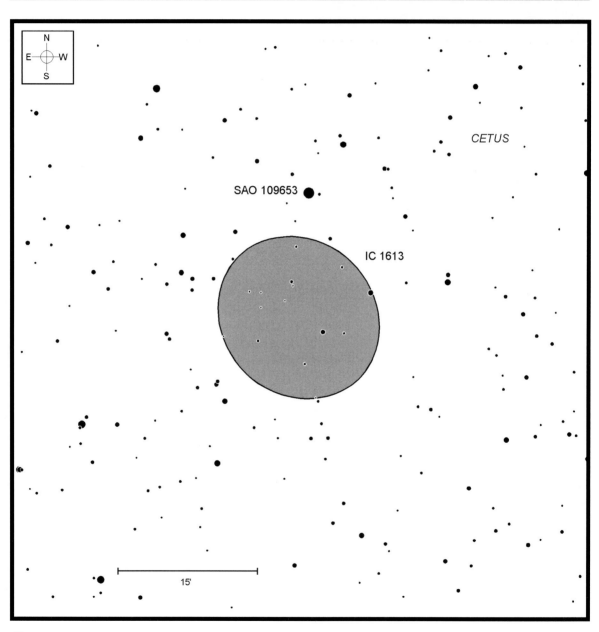

Chart 7.22 IC 1613

180   WLM and WLM-1

★
★★
★★
★

	Target	Type	RA	Dec.	Constellation	Magnitude	Size	Chart
180a	WLM	Galaxy	00 01.9	−15 27.8	Cetus	10.6	9.5′×3.0′	7.23
180b	WLM-1	Globular cluster	00 01.8	−15 27.6	Cetus	16.1	–	7.23

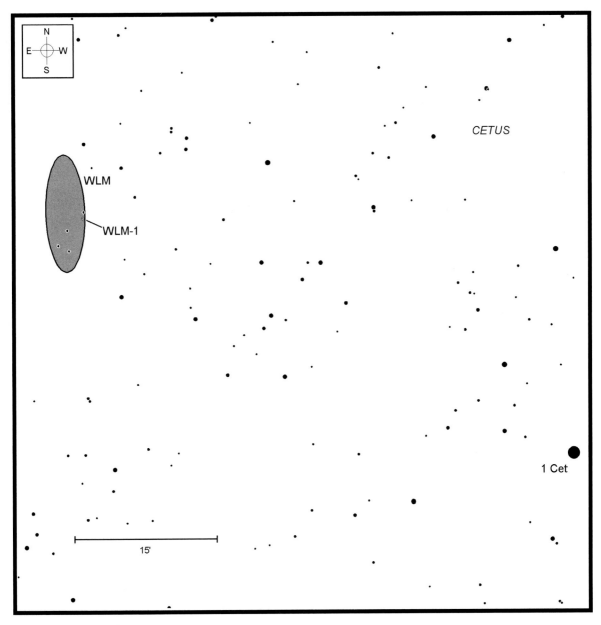

Chart 7.23 WLM and WLM-1

The Local Group of galaxies includes three large spiral galaxies – the Milky Way, the Andromeda Galaxy, and the Triangulum Spiral – and more than two dozen smaller systems. Two of the toughest to see make up this next challenge object found in the constellation Cetus.

Let's begin with IC 1613, discovered in 1906 by German astronomer Max Wolf on photographs taken with the Bruce 16-inch refractor at the Astrophysical Observatory in Heidelberg. Using data from the Hubble Space Telescope in 1999, Andrew Cole and colleagues confirmed that IC 1613 lies 2.3 million light years away. This places it a bit closer to us than M31 and its cadre of satellites. Like that galactic family, IC 1613 is also approaching the Milky Way, in this case at a rate of 234 km/s. While that is comparable to NGC 147, one of M31's companions, IC 1613 is not gravitational kin to Andromeda. In fact, it is nearly as far away from M31 as it is from the Milky Way.

Despite being relatively nearby, IC 1613 suffers from the same problem that plagues many other Local Group members, such as monster-scope Challenges 158 (Leo I) and 159 (Leo II). Like those systems, the feeble light from IC 1613 is spread across such a large area that its surface brightness drops precipitously. Although it is listed as 9th magnitude, remember that this is its *integrated* brightness, that is, how bright it would appear if it were compressed to a stellar point. The fact that the galaxy is spread across an area half the apparent diameter of the Full Moon lowers its surface brightness to only magnitude 15.5.

Photographs show that IC 1613's internal structure is both loosely and poorly organized. Outwardly, it resembles Barnard's Galaxy (NGC 6822; small-scope Challenge 60), although its total mass is much less. IC 1613 has an estimated mass equal to just 80 million Suns. By comparison, NGC 6822 has a mass equal to 130 million Suns. Like that summertime object, IC 1613 is designated as a dwarf barred irregular, a strangely bloated mishmash highlighted by an axial "bar" across its core. There is also the subtle suggestion of spirality, with a gently curving extension tapering away from a jumbled core.

Although it can be a bear to see, IC 1613 is not hard to locate thanks to its proximity to 26 Ceti. To get there, aim your finderscope along the rope attached to the southern Fish's tail at Epsilon (ε) Piscium. Look $2\frac{1}{2}°$ south-southeast of Epsilon for an equilateral triangle of stars made up of 73, 77, and 80 Piscium. From here,

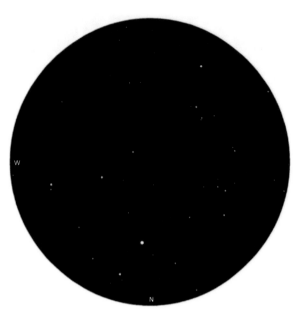

Figure 7.22 IC 1613

move about 4° south to 26 Ceti. Be careful not to mistake it for 29 Ceti, which is more to the east-northeast. IC 1613 is 47′ north of 26 Ceti and just 11′ south of an orangish 7th-magnitude sun, SAO 109653.

The galaxy's large apparent size conspires with its low surface brightness to make this a very difficult catch. In order to squeeze it all into a single eyepiece field, we need to use low power. But low power also lowers image contrast. Higher magnifications boost contrast but reduce the field size too much. It's a vicious circle that requires experimentation to solve. In my 18-inch, I have found the best compromise to be a 22-mm Tele Vue Panoptic eyepiece that yields 94× and a real field measuring 42′ in diameter. Figure 7.22 shows that view.

With that combination, I can make out that the northeastern section is slightly brighter than the region toward the southwest, but beyond that I see no distinct boundaries. Instead, the galaxy just diffuses softly into the background.

The brighter northeastern segment corresponds to a region populated with blue-giant stars distributed among more than two dozen stellar associations. The galaxy also contains more than two dozen very small star clusters, as well as a dozen small nebulae. Under extraordinary conditions the largest scopes may be able to just resolve some of these distant deep-sky objects, as well as several background galaxies shining through IC 1613 itself.

Don't get too full of yourself if you see IC 1613 through your *big* telescope, however. Arizona astronomer Brian Skiff has seen it through his 2.8-inch refractor, while super-eyed Steve O'Meara reports an observation of IC 1613 through his 4-inch refractor from a high-altitude site in Hawaii.

If you thought IC 1613 was just too *easy*, try your luck with another member of the Local Group that is also within Cetus. The Wolf–Lundmark–Mellotte Dwarf, or WLM, for short, was discovered by Max Wolf three years after IC 1613. The true extragalactic nature of this dwarf irregular system remained unconfirmed, however, until 1926, when it was studied by Knut Lundmark and Philibert Melotte.[17] As a result, this little galaxy carries three names.

WLM, also cross-listed as MCG-3-1-15 in the *Morphological Catalogue of Galaxies* and as PGC 143 in the *Principal Galaxy Catalog*, is a loner. Distance estimates vary greatly, but the most recently published values place it an average of 3.4 million light years from the Milky Way. If true, WLM is at the edge of the Local Group and more than a million light years from IC 1613, which is its nearest physical neighbor.

Structurally, the WLM Dwarf appears to be quite elongated in photos taken with the European Organization for Astronomical Research in the Southern Hemisphere's (informally the European Southern Observatory, ESO) 3.5-meter New Technology Telescope (NTT). These high-resolution images clearly show individual stars across the galaxy's

---

[17] P. J. Melotte, "New Nebulae Shown on Franklin-Adams Chart Plates," *Monthly Notices of the Royal Astronomical Society*, Vol. 86 (1926), pp. 636–8.

full disk as well as a halo of very old stars surrounding that disk.

The fact that this halo exists is quite surprising for two reasons. First, it shows that the galaxy is older than previously thought, perhaps as old as the Milky Way itself. It also proves that a halo can form around a dwarf galaxy in the first place. More often than not, such halos are reserved for far more massive galaxies. Adding to the mystery is the fact that the stars in the WLM halo appear much redder than the ones in the galaxy's central disk. That would seem to indicate that they are also much older. How is it that stars in a galaxy's halo formed before those in the galaxy itself?

The WLM Dwarf is 2.2° due west of 4.9-magnitude 6 Ceti and 56′ northeast of 6.3-magnitude 1 Ceti. Starhoppers should use 2nd-magnitude Deneb Kaitos [Beta (β) Ceti] as a guide to find 6 Ceti; it lies 8° to 6 Ceti's southeast. Look for the galaxy's faint glow just west of the halfway point between two widely separated 9th-magnitude stars. SAO 147053 is 27′ to the galaxy's northeast, while SAO 147052 is 25′ to its southeast. The stars' north–south orientation parallels the galaxy's long, slender disk, which through my 18-inch offers little more than a soft, featureless glow. An arc of 12th- to 14th-magnitude stars appears to cradle the galaxy's western edge. Like IC 1613, WLM is going to require some eyepiece experimentation before an ideal combination is found. In my scope, a 17-mm eyepiece producing 121× and a 3.8-mm exit pupil worked best.

Finally, if you could see WLM, then you just might be able to nab its solitary globular cluster. Can you spot two faint points just west of the center of the galaxy's elongated disk? The northern star shines at 15th magnitude, while the southern "star" is actually the 16th-magnitude globular. It's a daunting test for even the largest amateur apertures.

181   Palomar 1

Target	Type	RA	Dec.	Constellation	Magnitude	Size	Chart
Palomar 1	Globular cluster	03 33.3	+79 35.0	Cepheus	13.2	2.8′	7.24

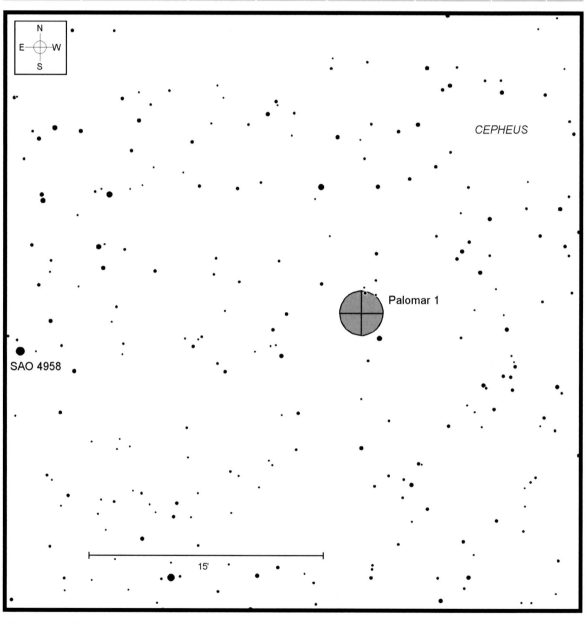

Chart 7.24 Palomar 1

By now, some of the 15 Palomar globular clusters should be old friends. If you have tried some of Chapter 6's large-scope challenges, then you have already met Palomar 11 and 13 (large-scope Challenges 142 and 145, respectively). Palomar 4 was also profiled earlier in this chapter as monster-scope Challenge 161. Now, we come to the lead member of that hit list, which is also one of the toughest to see. Palomar 1 lies in northern Cepheus and never sets for most northern hemisphere observers. Given how difficult it is, however, it is best to wait for it to reach culmination. Even then, it will take a special night to see this distant swarm of stars.

Not only does Palomar 1 hold the distinction of being the most northerly globular cluster visible in amateur telescopes, it is also one of the youngest. A 1998 study conducted by a team of astronomers led by Alfred Rosenberg of the Telescopio Nazionale Galileo in Padova, Italy, concluded that Palomar 1 is between 6.3 and 8 billion years old.[18] If true, Palomar 1 is half as old as most Milky Way globulars, which are believed to be as many as 13 billion years old. The study found that, since the mass of Palomar 1 also seems to differ significantly from most other Milky Way globulars, it "might be an interesting indication of a different origin for this object. This fact, coupled with its young age, might be further evidence that Palomar 1 should be considered a member of a different globular cluster population, with a different formation process and time."

Finding Palomar 1 is tough because of its sparse surroundings. As a general guide, it lies on the opposite side of the celestial pole from the Little Dipper asterism. Draw a line from Delta (δ) Ursae Minoris through Polaris [Alpha (α) Ursae Minoris] and continue for 9° to an obtuse triangle of 5th- and 6th-magnitude stars – SAO 650, 670, and 691. Follow the triangle's wide base roughly 2° to the west, passing the very red variable star SS Cephei along the way, to the orangish 8th-magnitude star SAO 4958. Palomar 1 is 22′ farther west, just 2′ northeast of an 11th-magnitude field star (Figure 7.23).

Figure 7.23 Palomar 1

Even veteran observers using large telescopes under pristine skies have trouble capturing Palomar 1. Superobserver Barbara Wilson from Houston, Texas, comments about her experiences.

I have suspected its very faint glow on three separate occasions, and have positively seen it once after seeing had settled down. On each of those four occasions, I could see or suspect something in exactly the same position. There were two stars to the south, and two fainter stars at the very edge to the north.

Part of the predicament posed by Palomar 1 is its closeness to those stars mentioned by Wilson, which is compounded by the fact that it appears a magnitude or more fainter than its published value. The brightness discrepancy is caused in part by the cluster's very loose stellar structure. On the standard, 12-point Shapley–Sawyer globular cluster concentration scale, with 12 being the most loosely structured, Palomar 1 rates a 12. Such a poor central concentration lowers the overall surface brightness, which further masks an already demanding target.

Even if you do manage to see it, don't expect to resolve its stars. The brightest shines at magnitude 16.5.

[18] A. Rosenberg, I. Saviane, G. Piotto, A. Aparicio, and S. R. Zaggia, "Palomar 1 – Another Young Galactic Halo Globular Cluster," *Astronomical Journal*, Vol. 115 (1998), p. 648.

182   UGC 2838

Target	Type	RA	Dec.	Constellation	Magnitude	Size	Chart
UGC 2838	Galaxy	03 43.8	+24 03.6	Taurus	17.9	1.6′×0.2′	7.25

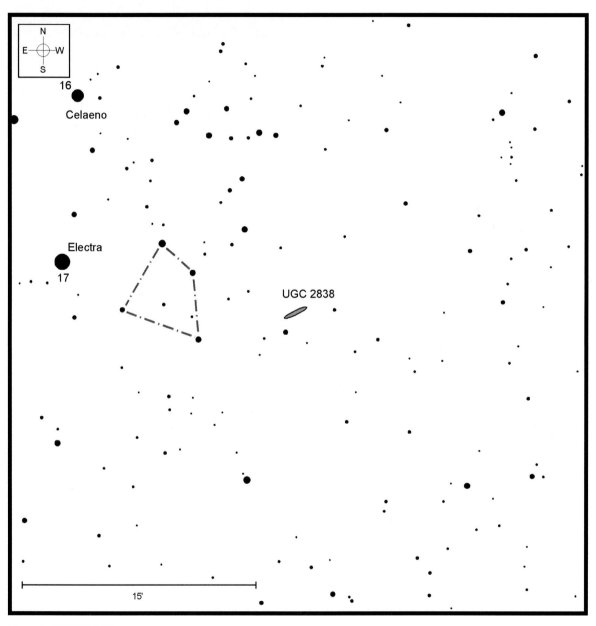

Chart 7.25 UGC 2838

The Pleiades is one of everyone's favorite winter sights and was the basis for naked-eye Challenge 13 in Chapter 1 and small-scope Challenge 69 in Chapter 3. We return one more time here to look for buried treasure. Did you know there was a small galaxy hidden behind the Pleiades? Few people do. Its faint disk was not seen by the Herschels or known to John Dreyer when he assembled the *New General Catalogue* and supplemental *Index Catalogues*. Instead, this little treasure was first included in the UGC, the *Uppsala General Catalogue of Galaxies*, in 1973.

Observing UGC galaxies is great sport for advanced deep-sky observers using very large telescopes. Truth be told, however, most UGC galaxies are so faint that the best we can hope for are very dim glimmers just barely perceptible with averted vision. But this little galaxy is different because of its prominent location. UGC 2838 appears just 16′ west of Electra (17 Tauri). In reality, it lies 200 million light years behind the western edge of the Pleiades.

UGC 2838's position is both a blessing and a curse. It's a blessing in that we don't have to starhop to some far-off place in a starless field. It's a curse for the very same reason. Not only do the cluster stars dazzle our eyes when looking for the faint starlight from this distant spiral galaxy, the cosmic dust littering the cluster also dulls its appearance. Taking all that into account and considering the galaxy's listed magnitude is 17.9, can anyone possibly see it visually?

To find out, let's first locate UGC 2838's field by centering on Electra, the southwestern star in the Pleiades "bowl." Viewing with your "other" eye (i.e., the eye you will not be using to search for the galaxy, in order to maintain its full dilation), shift your gaze 7′ westward from Electra to a crooked trapezoid of four stars that reminds me of Corvus the Crow. Now, look for a lone 10.6-magnitude star another 6′ west of the "Crow's" southwestern corner. That's GSC 1799:791 in the Hubble Guide Star catalog, and your destination.

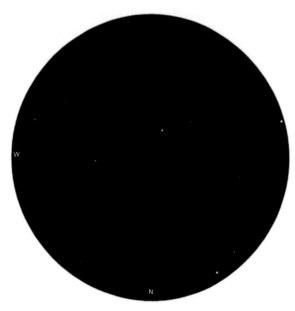

Figure 7.24 UGC 2838

Switch to an eyepiece with a narrow enough field to move the trapezoid's brightest star out of view. Offset GSC 1799:791 1.4′ to the southeast of center, and take a look.

Even with the galaxy's field isolated from the cluster's stars, the background sky is still brightened by the glow of M45's nebulosity. Averted vision and lightly tapping the side of the telescope tube to impart motion should help to overcome that handicap, but only if the sky itself is dark and transparent.

I have seen UGC 2838 through my 18-inch reflector using 7- and 10-mm eyepieces from a site with a naked-eye limiting magnitude of 6+. The 7-mm eyepiece (294×, 13′ field) proved the best (Figure 7.24), but only after I had spotted the galaxy initially with the 10-mm (206×, 15′ field). Trying a 5-mm eyepiece (411×, 10′ field), however, dimmed the images so much that the galaxy faded completely from view.

183   Palomar 2

Target	Type	RA	Dec.	Constellation	Magnitude	Size	Chart
Palomar 2	Globular cluster	04 46.1	+31 22.9	Auriga	13.0	2′	7.26

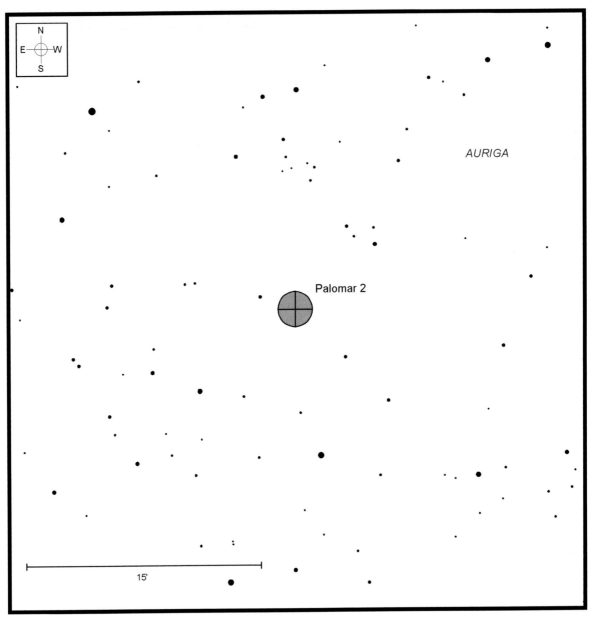

AURIGA

Palomar 2

15′

Chart 7.26 Palomar 2

We now come to the fifth and final Palomar globular cluster to be discussed in this book, Palomar 2. Palomar 2 is a particularly nasty cluster to spot because it is partially concealed behind a dark nebula in Auriga. Barnard 221 spans $\frac{3}{4}°$ and is centered just 32′ northwest of Palomar 2. The dark nebula may not be readily evident by eye, but it has a stifling influence on Palomar 2.

Most globular clusters associated with the Milky Way are positioned around the galactic nucleus, and so are referred to as "inner-halo globulars." There is a second family, however, whose members lie far beyond the Galaxy's center and so are known as "outer-halo globulars." Of all the outer-halo globulars known, Palomar 2 is the most extreme, located almost directly opposite the Galactic Center in Sagittarius, separated by 85,400 light years. A 1997 analysis conducted by Professor William E. Harris and colleagues from McMaster University in Hamilton, Ontario, found that Palomar 2 is one of the brightest and most massive clusters in the outer halo.[19]

That distinction goes completely unappreciated through our telescopes, however, because Palomar 2 is also one of the most heavily obscured. As a result, we see only a vague inkling of the cluster's true self. To find it, scan 2° southwest of Iota (ι) Aurigae, the southwestern star in the Auriga pentagon. Keep an eye out for a triangle of three 6th- to 7th-magnitude stars. The brightest and westernmost star in the triangle, SAO 57441, is 40′ due east of the cluster. A close-set pair of 12th- and 13th-magnitude stars lying halfway between the two makes a useful checkpoint along the way. Another close stellar duo, shining faintly at magnitudes 13 and 14.5, is just 2.3′ northeast of the cluster.

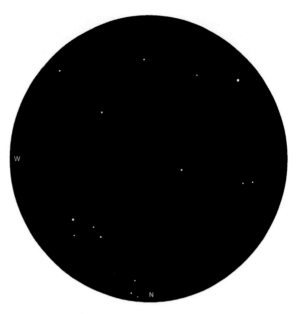

Figure 7.25 Palomar 2

Palomar 2 has a listed magnitude of 13.0, and so may be visible through smaller telescopes than those in this chapter if attempted under very dark skies. For most readers, however, it will take the firepower of at least a 15-inch scope to pull out its tiny glow from the surroundings. Through my 18-inch under suburban skies, Palomar 2 is an extremely dim, but doable, smudge measuring no more than 1′ across (Figure 7.25). Averted vision is a must no matter which eyepiece I use, although I have found the best view was at 171×. The overall impression is more reminiscent of a distant galaxy than a globular cluster. Indeed, it was this vague appearance that led Boris Vorontsov-Velyaminov and V. P. Arkhipova to misclassify Palomar 2 as a galaxy in their *Morphological Catalogue of Galaxies* (MCG) compiled in the 1960s. Even today some references cross-list Palomar 2 as MCG +05-12-1.

[19] W. E. Harris, P. R. Durrell, G. R. Petitpas, T. M. Webb, and S. C. Woodworth, "Unveiling Palomar 2: The Most Obscure Globular Cluster in the Outer Halo," *Astronomical Journal*, Vol. 114 (1997), pp. 1043–50.

184 Simeis 147

Target	Type	RA	Dec.	Constellation	Magnitude	Size	Chart
Simeis 147	Supernova remnant	05 39.0	+28 00.0	Taurus	–	200′	7.27

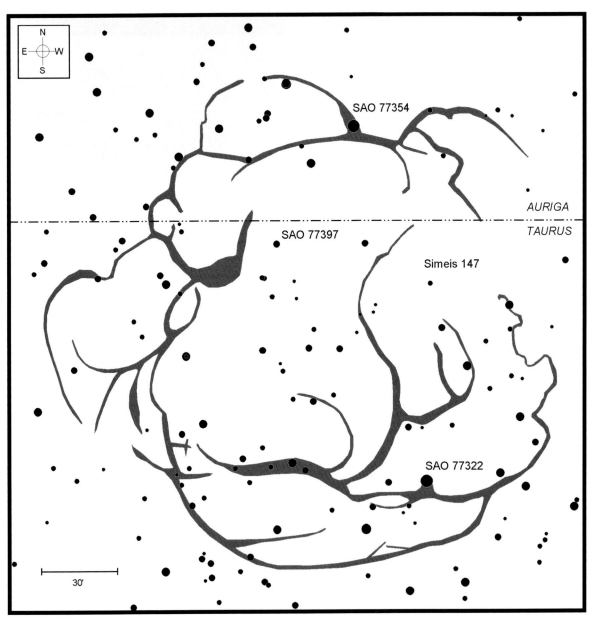

Chart 7.27 Simeis 147

The year 1054 must have been an active one for
stargazers. That was the year that the famous Crab
Nebula supernova blasted forth, shining brightly
enough for Chinese and Native American skywatchers
to note a "new star" blazing near the tip of one of what
we now call Taurus the Bull's two horns. The 1054
supernova was so bright that it was visible in broad
daylight during the summer of that year, and remained
visible to the naked eye for nearly a year. Today, we
know the fading gaseous remnant of that all-consuming
event as the Crab Nebula, M1 (binocular Challenge
41).

Some 99,000 years earlier, another massive star in
the Bull, just 7° north of the Crab, underwent a similar
detonation. Our earliest cave-dwelling ancestors could
have witnessed the explosive devastation first-hand if it
was bright enough to reach naked-eye visibility, but
there are no records to confirm that. The only evidence
we have of that once mighty star is a filamentary debris
field that continues to expand around ground
zero.

None of the classical observers from the eighteenth
and nineteenth centuries ever spotted those remains.
Instead, the discovery of this ruptured bubble of
starsplatter was made only recently, in 1952 by G. A.
Shajn and V. E. Hase at the Crimean Astrophysical
Observatory at Simeis, Russia. The tendrils of
expanding nebulosity are so faint and so broad that it
took the light-gathering ability of a 25″ Schmidt camera
to record their whisper. Today, we know it as Simeis
147, or S147. Some references, however, prefer to
identify it as Sh2-240, for its entry in Stewart Sharpless's
1959 catalog of nebulae.

Call it what you will, this is one tough target. The
biggest problem with seeing Simeis 147 visually is its
huge size, more than 3° across. Can your 15-inch scope
cram 3° into a single field of view? Probably not. That's
why we need a strategy for hunting down this tough
object.

Most observers who have seen it report success after
dividing the nebula into four or more regions, and then
looking for those specific sections. Even with that
approach, glimpsing a small portion of Simeis 147
takes a concentrated effort. Dark skies are also needed
to see the full breadth of the clouds, but I have been
able to spot the brightest section through my 18-inch
telescope under naked-eye limiting magnitude 5.0 skies
using a low-power, wide-field eyepiece coupled with an
oxygen-III filter (Figure 7.26).

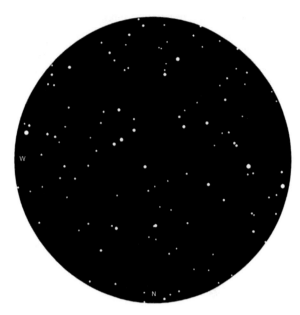

Figure 7.26 *Simeis 147*

To zero in on Simeis 147, scan 3° due east of
Elnath [Beta (β) Tauri], the northern horn. That puts
you right in the thick of things, but there is nothing to
see at that exact location. The brightest section, and the
one that I spotted under 5th-magnitude skies, lies about
a degree due south. Place 6th-magnitude SAO 77322
along the western edge of your eyepiece field and scan
slowly to the east. With averted vision, look for a thin
lane of nebulosity threading across the center of the
field. The effect looks like a fainter, less structured
version of the Veil Nebula in Cygnus. You might also
see a hint of the two branches that veer off to the
north.

Another "bright" section of nebulosity is centered at
right ascension 05h 43m, declination +28° 16′, or
about half a degree southeast of 8th-magnitude SAO
77397. You are in the right area if you see a square of
four 8th- to 10th-magnitude stars surrounded by
several fainter suns. The stars appear embedded in
nebulosity, with a tuft extending about 10′ further to
the northwest of the square. Although it is invisible
from my backyard, this section is fairly obvious through
the same scope under darker skies, again with the
oxygen-III filter firmly in place. Without the filter, all
bets are off.

A third segment, more intricate, but also more
challenging than the other two, lies at right ascension
05h 44.5m, declination +28° 58′. Though only visible

from truly dark sites, this portion hints at the true complexity of the entire cloud that we marvel at in photographs.

Finally, a fourth subdivision of the Simeis 147 complex lies across the border in southern Gemini, centered at right ascension 05h 39m, declination +29° 08′, near 6th-magnitude SAO 77354. All we can hope for here is the faintest hint of an east–west lane of nebulosity, curving slightly to the southeast as it passes the star.

Few amateurs have seen, or perhaps have even tried to see, Simeis 147. But with a little patience, top-notch optics, and a good eye for fine detail, you just might be surprised at not just seeing a dim hint of this once mighty star, but a patchwork of gossamer clouds interwoven throughout a starry backdrop.

185 NGC 2363 and NGC 2366

★
★★
★

	Target	Type	RA	Dec.	Constellation	Magnitude	Size	Chart
185a	NGC 2363	Galaxy	07 28.5	+69 11.6	Camelopardalis	13.0	1.8′×1.0′	7.28
185b	NGC 2366	Galaxy	07 28.9	+69 12.7	Camelopardalis	11.5b	8.2′×3.3′	7.28

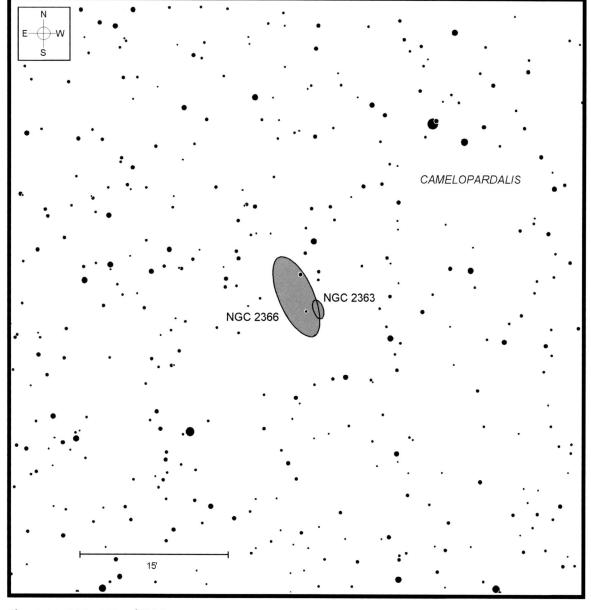

Chart 7.28 NGC 2363 and NGC 2366

Will the real NGC 2363 please stand up? For years, there has been an ongoing debate over the true identity of the 2,363rd entry in the *New General Catalogue*. Many references cite it as a huge area of ionized hydrogen (an H II region) within NGC 2366, a dim irregular galaxy.

That was the explanation in the press release that accompanied a magnificent shot of the galaxy taken with the Hubble Space Telescope in 1996. In part, the release said, "Clusters of stars and a fishhook-shaped cloud of luminescent gases glow brilliantly in NGC 2363, a giant star-forming region in the galaxy NGC 2366." The press release went on to describe how the Hubble image revealed that the brightest individual star in the region is a rare example of an erupting Luminous Blue Variable. This star is thought to be between 30 and 60 times more massive than our Sun and is currently enduring a very unstable, eruptive phase of its life. The same Hubble image also shows two dense clusters of massive stars. The older cluster is about a tenth of the age of our Solar System, while the other is probably less than half as old, judging by how much remnant gas and dust remains.

Recently, however, some historians have suggested that William Herschel, who is credited with discovering both NGC 2363 and NGC 2366, was describing the H II region and the galaxy collectively when he recorded a circular patch of light with a dim protruding extension. According to Dr. Harold G. Corwin, Jr. on the NGC/IC Project's website (www.ngcic.com), the catalog number NGC 2366 refers to both the brighter H II region as well as its faint home galaxy.

If that's the case, then what is NGC 2363? Corwin's research points to an even fainter galaxy just to the southwest, which is identified as UGC 3847 in the *Uppsala General Catalogue of Galaxies*. He contends that this second galaxy is actually NGC 2363. UGC 3847, also an irregular galaxy, shines at 13th magnitude.

Fortunately, these targets are not terribly difficult to pinpoint, as they lie 4° due north of the bright galaxy NGC 2403. To get there, begin at Omicron (o) Ursae Majoris, the 3rd-magnitude star marking the tip of the Great Bear's nose. Head 4° north to a triangular asterism created by Pi-1 (π1), Pi-2 (π2), and 2 Ursae

Figure 7.27 NGC 2363 and NGC 2366

Majoris, and then due west 5° to 51 Camelopardalis. NGC 2403 is just a degree west of the star and is always worth a stopover. Then, it's off to the north for 4° to a 6th-magnitude field sun and our targets, which lie just a bit farther north still.

When I turned my 18-inch reflector toward this area recently, I could see the extragalactic H II cloud directly using a 12-mm eyepiece (171×; Figure 7.27). I estimated it to be about 2 arc-minutes across, perhaps 12th magnitude, and with a fairly bright stellar core. Spotting the faint disk of its home galaxy, however, proved more difficult. I had to use averted vision to catch even a passing glimpse of its extended disk, which measures about 4′×2′. Together, they reminded me of a faint comet, with the H II region serving as the coma, and the disk of the irregular galaxy forming a dim tail extending toward the north.

That same night, try as I might, I saw no sign of the second, smaller galaxy to the south of the H II region. If the former is indeed NGC 2363, then it's a challenge that may only be met with the largest backyard telescopes.

186 NGC 2474 and NGC 2475

★★
★★★

	Target	Type	RA	Dec.	Constellation	Magnitude	Size	Chart
186a	NGC 2474	Galaxy pair	07 58.0	+52 51.4	Lynx	14.3	0.6′	7.29
186b	NGC 2475	Galaxy pair	07 58.0	+52 51.8	Lynx	14.0b	0.7′	7.29

187 PK 164+31.1

★★★
★★★
★

Target	Type	RA	Dec.	Constellation	Magnitude	Size	Chart
PK 164+31.1	Planetary nebula	07 57.9	+53 25.3	Lynx	14.0p	6.3′	7.29

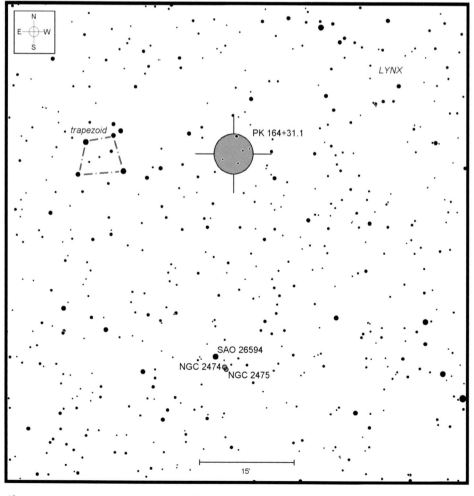

Chart 7.29 NGC 2474, NGC 2475 and PK 164+31.1

There once was a mystery in Lynx. The story opened in 1790 when William Herschel discovered a small, nebulous glow about $2\frac{1}{2}°$ northwest of 27 Lyncis. He later added it as number 830 in his list of "very faint nebulae" (abbreviated H-III-830) and apparently moved on without noticing a second, fainter blur of light just to the northeast. That second object was discovered 66 years later by William Parsons, 3rd Earl of Rosse, through his 72-inch "Leviathan" reflector. Both were later incorporated into John Dreyer's *New General Catalogue*. NGC 2474 is described as "faint, pretty small, extended?, brighter middle, very small star?, large star north following." NGC 2475 is simply noted as "makes a double nebula with" NGC 2474.

The mystery of NGC 2474 and 2475 was hatched in 1939, when Rebecca Jones and Richard Emberson, astronomers at Harvard Observatory, discovered a planetary nebula on patrol photographs that seemed to lie close to the original NGC 2474/2475 coordinates. Their nebula was a large ring-type cloud with two distinctly brighter east–west lobes.

Jones and Emberson announced their discovery in the August 1939 issue of *Harvard College Observatory Bulletin*:

On a recent photographic plate, a faint nebular ring has been detected joining two condensations, NGC 2474, observed by Sir John Herschel, and NGC 2475. The latter was discovered by Lord Rosse who described it as forming a double nebula with Herschel's object.[20]

Oops. Not only did they incorrectly credit John Herschel as the discoverer of NGC 2474, they also misidentified NGC 2474 and NGC 2475 as a single planetary nebula.

Their error did not become apparent for more than 40 years, however. That was more than enough time for NGC 2474/2475 to be misclassified as a planetary nebula in the original Perek–Kohoutek (PK) catalog of planetary nebulae and several other reliable references.

After more than four decades of confusion, thanks in large part to the research of Nancy and Ronald Buta of McDonald Observatory, University of Texas, we now know that NGC 2474 and NGC 2475 are a close pair of elliptical galaxies discovered by Herschel and Parsons, respectively. Look for them just 2.4′ southwest of the

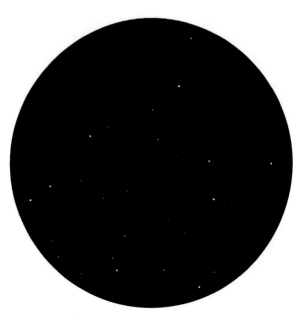

Figure 7.28 PK 164+31.1

9th-magnitude golden star SAO 26594 (the "large star'' in the NGC description).

While their identity is now certain, there is still some discrepancy between the listed magnitudes of these two objects and their visual appearances. Although both are listed at photographic magnitude 14, NGC 2474 is just bright enough to be seen in 8-inch instruments. NGC 2475, however, requires at least a 12-incher. Look for them just 2.4′ southwest of 10th-magnitude SAO 26594.

So, if those are the galaxies, what did Jones and Emberson discover? The answer to the "Mystery of the Missing Lynx" lies $\frac{1}{2}°$ further north. There we find a planetary nebula, a *real* planetary nebula that today is identified correctly as Jones–Emberson 1 and cross-listed in the revised Perek–Kohoutek listing as PK 164+31.1.

It's easy for us to criticize Jones and Emberson for their goof, especially since they realized that the planetary did not match the original location of NGC 2474. But one look at the planetary through a giant backyard scope and you will see, well, a double nebula just like the NGC said you would. As luck would have it, PK 164+31.1 is an unusual "ring nebula" with two brighter lobes connected by opposing arcs of faint nebulosity. The visual resemblance to a pair of faint galaxies is undeniable.

[20] R. B. Jones, and R. Emberson, "A Large New Planetary Nebula," *Harvard College Observatory Bulletin*, No. 911 (1939), pp. 11–13.

Although it is visible in smaller scopes, PK 164+31.1 needs as much aperture as you can throw at it to reveal the full ring. That is why I chose to include it in this chapter. You will find it 21′ due west of a distinctive trapezoid of 9th- to 11th-magnitude stars. Use between 100× and 150× to see the full span, but avoid going too much higher, since more magnification actually works against the planetary. As to filters, a narrowband nebula filter will help improve the odds of spotting the full 360° ring, but an oxygen-III filter will actually smother it. With averted vision, I have seen the full ring through my 18-inch at 121× (Figure 7.28). The two lobes were seen directly, the southern knot being the brighter of the two. I only caught fleeting glimpses of the full ring, however, and then only with averted vision.

★
★★★ **188 Pluto/Charon**
★

Target	Type	Magnitude	Separation
Pluto/Charon	Double dwarf planet	~14 (Pluto) ~16 (Charon)	0.9″

Almost every amateur astronomer is familiar with the story of Clyde Tombaugh's discovery of the dwarf planet Pluto in February 1930 after spending many grueling months photographing the night sky with the 13-inch photographic telescope at Lowell Observatory in Flagstaff, Arizona. We also learned how he would study pairs of photos taken weeks apart through a special microscope equipped with a blink comparator to show any objects that shifted position from one plate to the next. Talk about a diligent effort!

It would be another 48 years before we discovered that Pluto was not alone out there in the depths of the outer Solar System. Instead, it had a friend, a small orbiting satellite named Charon. Charon was found by James Christy on June 22, 1978, as he was examining highly magnified photographs of Pluto taken at the United States Naval Observatory.

This discovery changed our view of Pluto forever. Astronomers recalculated the mass and size of the system, since now they were looking at two objects rather than one. Based on this and later observations, Pluto's diameter is believed to be about 1,420 miles (2,320 km), while Charon's diameter is about 750 miles (1,207 km). The large size of Charon relative to Pluto has led some astronomers to call it a dwarf double planet, as both orbit a common center of gravity, called a *barycenter*, located in the empty gap between both worlds. Two smaller moons of Pluto, named Nix and Hydra, were discovered with the Hubble Space Telescope in 2005.

Small-scope Challenge 90 was to find Pluto itself. That can be a formidable task through any telescope, large or small, because of its dim, stellar appearance. As if that isn't hard enough, can you possibly spot Charon?

As Pluto and Charon orbit around their common gravitational center, completing an orbit once every 6.39 days, they remain separated by 12,150 miles (19,570 km). From Earth, some 3.7 billion miles

(5.9 billion km) away on average, that translates to an angular separation of less than 1″.

Still, some talented observers have successfully split Charon from Pluto. One of the first published reports appeared in *Sky & Telescope* magazine in January 1993. In their article "Exotic Worlds," coauthors William Sheehan and Stephen O'Meara recall a special night at the Pic Du Midi Observatory in southern France.[21] Both observers were afforded the unprecedented opportunity to view through the facility's 41.7-inch (106 cm) *f*/16 Cassegrain reflector on four nights in the summer of 1992. On their final night, they successfully split Charon and Pluto using magnifications between 800× and 1,200×. They recall, "The observation is at the utmost limits of perception, but our independent sketches agree reasonably well with predictions."

Does it take a 41.7-inch telescope at one of the world's leading astronomical observatories to see Charon? Not necessarily. In 1993, Arizona amateur telescope-maker Steven Aggas reports that he split Pluto and Charon through his homemade 20-inch *f*/4.2 reflector.

I used 509× by combining a 4.8mm eyepiece with a Paracorr Coma Corrector to split Charon from Pluto at the SMURFS (Southern Michigan Unorganized Regional Festival of Stargazers) Star Party near Midland, Michigan, on July 17, 1993. Charon was near maximum separation from Pluto with reasonably good elevation above the horizon at midnight that summer's night. The night itself had good transparency and very good seeing, as often happens in Michigan at that time of year.

Has Charon been seen through even smaller apertures?

It stands to reason that steady seeing and dark, transparent skies are needed even to consider this

---
[21] W. Sheehan, and S. O'Meara, "Exotic Worlds," *Sky & Telescope*, Vol. 85, No. 1 (1993), pp. 20–4.

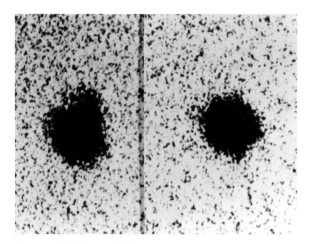

Figure 7.29 Pluto and Charon

ultimate challenge. As with large-scope Challenges 155 and 156, which dared you to spot the larger satellites of Uranus and Neptune, respectively, the Pluto–Charon test is best attempted when they are at or near opposition. That will occur in late June or early July throughout the twenty-first century.

Pluto's orbit takes it as close as 2.8 billion miles (4.4 billion km) to the Sun at perihelion and as far as 4.6 billion miles (7.4 billion km) at aphelion. Right now, Pluto is on the way out, having passed perihelion

in 1989. Back then, Pluto shone at magnitude 13.7, brighter than at any other point in its orbit. Right now, Pluto is on the wane at about 14th magnitude. Over the next century, it will slowly fade on its way to aphelion in 2113, when it will shine at only 16th magnitude. Charon, being smaller and with a lower albedo (surface reflectivity) than Pluto, is always about two magnitudes fainter.

With Pluto and Charon so dim and so close to each other, it is critical to know where one will appear with respect to the other to confirm the sighting. Several popular computer programs, such as Project Pluto's *Guide*, plot the locations of Charon and Pluto for any date and time. To remain perfectly impartial, however, you should not look at that prediction before making your attempt. That way, you go into it unbiased.

Unfortunately, at the time that this book is released, Pluto is transiting star-filled Sagittarius, which will only confound attempts at sighting Charon. As the years progress, Pluto will continue to move slowly eastward, away from the plane of the Milky Way. By 2017, it will have shifted beyond the Sagittarius teapot, which should make this challenge a little easier. Perhaps if seeing is steady enough to allow high magnifications like Sheehan, O'Meara, and Aggas used, then it just might be possible to duplicate their amazing feat by spotting tiny Charon.

Epilogue

The edge of imagination

Trying to draw up a finite list of challenging objects scattered through the infinite universe is an almost insurmountable task. That became all too apparent as this book evolved. In fact, when I wrote the "Challenge Object" column in *Deep Sky* magazine back the 1980s, targets that I thought were difficult tests back then have subsequently proven to become old friends that I visit often. Others, however, proved to be so difficult back then that they jumped up an aperture class, if they are included in this book.

It would be a very easy task to assemble a volume 2 to this book simply from the leftovers that I chose not to include here for one reason or another. Indeed, the *Cosmic Challenge* could become a multi-volume set. But, rather than do that, I leave the task of developing volume 2 to you, the reader.

As you explore the objects discussed here, you will undoubtedly bump into other targets that are just as challenging, and perhaps even more so. Prompted by those, began to compile your own favorite list of cosmic challenges.

Perhaps your challenges favor Solar System targets. If so, begin with the Lunar 100 list of targets. Created by Charles Wood and first appearing in the April 2004 issue of *Sky & Telescope* magazine, the Lunar 100 offers, in Wood's own words, "a selection of the Moon's 100 most interesting regions, craters, basins, mountains, rilles, and domes; to provide Moon lovers with something akin to what deep-sky observers enjoy with the Messier catalog." And he certainly has done just that. The Lunar 100 has given this devout deep-sky observer something new and enjoyable to do on those moonlit nights that otherwise would have been spent in front of the television.

You may prefer to specialize in one particular type of deep-sky object, such as tightly spaced binary stars. If those are your thing, begin your cosmic challenge with

the 100 binaries listed in Appendix B. One of my favorites is seeing the "greenish" companion of Antares in summer's Scorpius. That's a great challenge for medium and large instruments. To determine which of the binary stars in the appendix will be resolvable through your telescope, review the discussion of Dawes' Limit and the Rayleigh Criteria in Chapter 1. But use those only as guides, not as absolutes.

If you enjoy conquering lists of cosmic challenges, there are many other notable and notorious inventories available. One of my favorites is the Royal Astronomical Society of Canada's "Deep-Sky Challenging Objects," which is included their annual *Observer's Handbook*.

Another great reference is the "Adventures in Deep Space: Challenging Observing Projects for Amateur Astronomers of All Ages" found on Astronomy-mall.com. Here, you will find not one, but several compendia with such intriguing titles as 100 Peculiar Galaxies; Off the Beaten Path; Observing Galaxy Clusters; and, my personal favorite, the AINTNO 100.

The AINTNO 100, created by the devious minds of Texas amateur astronomers Barbara Wilson and Larry Mitchell, is purportedly presented by the Association of Invisible Nebulae and Things Nobody Observes (AINTNO). Here, we find 100 fantastically and fancifully intriguing targets, including observing astronaut footprints on the Moon and the arrow that points to Pluto. Great stuff to imagine doing on cloudy nights. And, incidentally, looking at photos of these through your telescope doesn't count – besides, I had already thought of that. These must be direct observations.

Whether you decide to pursue one of these observing projects or create one of your own, I hope that you will continue to push the limits of your telescope, your observing site, and your talents as an observer. Trying to see the unseeable not only improves your skills as an observer, it also unlocks a part of the universe that most amateurs ignore. You will see targets that others only read about.

I would love to hear about your triumphs and your failures as you take on the Cosmic Challenge. Consider joining the "Cosmic Challenge" Internet mailing list and posting your observations, comments, and questions. You will find a link to the mail list from the Cosmic Challenge section of my website, www .philharrington.net.

Appendix A
The cosmic challenge

Challenge	Instrument	Season	Target	Type	RA	Dec.	Constellation	Magnitude	Size/separation
1	Naked eye	Spring	M81	Galaxy	09 55.6	+69 03.8	Ursa Major	7.9b	27.1′×14.2′
2	Naked eye	Spring	Melotte 111	Open cluster	12 25	+26	Coma Berenices	1.8	275′
3	Naked eye	Spring	Alcor and Mizar	Double star (optical)	13 23.9	+54 55.5	Ursa Major	2.2, 4.0	11.8′
4	Naked eye	Spring	M3	Globular cluster	13 42.2	+28 22.5	Canes Venatici	6.3	18′
5	Naked eye	Summer	Stars in Ursa Minor	Asterism	15 40	+80	Ursa Minor	–	–
6	Naked eye	Summer	M13	Globular cluster	16 41.7	+36 27.6	Hercules	5.8	20′
7	Naked eye	Summer	Great Dark Horse Nebula	Dark nebula	17 10	−27	Sagittarius	–	~480′
8	Naked eye	Summer	NGC 7000	Emission nebula	20 58.0	+44 20.0	Cygnus	–	120′
9	Naked eye	Autumn	M2	Globular cluster	21 33.5	−00 49.2	Aquarius	6.6	16′
10	Naked eye	Autumn	Stars in the Great Square	Asterism	23 40	+21 30	Pegasus	–	–
11	Naked eye	Autumn	Extent of M31	Galaxy	00 42.7	+41 16.1	Andromeda	4.4b	192.4′×62.2′
12	Naked eye	Autumn	M33	Galaxy	01 33.8	+30 39.6	Triangulum	6.3b	65.6′×38.0′
13	Naked eye	Winter	Pleiades	Open cluster	03 47.5	+24 06.3	Taurus	1.2	110′
14	Naked eye	Winter	Barnard's Loop	Supernova remnant	05 54	−01 00	Orion	–	420′×60′
15	Naked eye	Winter	M35	Open cluster	06 09.0	+24 21.0	Gemini	5.1	28.0′
16	Naked eye	Winter	M41	Open cluster	06 46.0	−20 45.3	Canis Major	4.5	38.0′

Challenge	Instrument	Season	Target	Type	RA	Dec.	Constellation	Magnitude	Size/ separation
17	Naked eye	Year-round	Moon: Young Moon/old Moon	Solar System					
18	Naked eye	Year-round	Spotting Mercury	Solar System					
19	Naked eye	Year-round	Spotting Uranus	Solar System					
20	Naked eye	Year-round	Glimpsing Vesta	Solar System					
21	Naked eye	Year-round	Zodiacal light and Gegenshein	Solar System					
22	Binoculars	Spring	M82	Galaxy	09 55.9	+69 41.0	Ursa Major	9.3b	11.3′×4.2′
23a	Binoculars	Spring	M84	Galaxy	12 25.1	+12 53.2	Virgo	10.1b	6.4′×5.5′
23b	Binoculars	Spring	M86	Galaxy	12 26.2	+12 56.8	Virgo	9.8b	8.9′×5.7′
24	Binoculars	Spring	M104	Galaxy	12 40.0	−11 37.6	Virgo	9.0b	8.8′×3.5′
25	Binoculars	Spring	NGC 5128	Galaxy	13 25.5	−43 01.0	Centaurus	7.8b	25.8′×20.0′
26a	Binoculars	Spring	M65	Galaxy	11 18.9	+13 05.6	Leo	10.3b	9.8′×2.8′
26b	Binoculars	Spring	M66	Galaxy	11 20.3	+12 59.0	Leo	9.7b	9.1′×4.1′
26c	Binoculars	Spring	NGC 3628	Galaxy	11 20.3	+13 35.4	Leo	10.3b	14.8′×2.9′
27	Binoculars	Spring	M101	Galaxy	14 03.2	+54 20.9	Ursa Major	8.3b	28.9′×26.9′
28a	Binoculars	Summer	16+17 Dra	Binary star	16 36.2	+52 54.0	Draco	5.5, 5.6	84″
28b	Binoculars	Summer	Nu Dra	Binary star	17 32.2	+55 11.1	Draco	4.9, 4.9	60″
28c	Binoculars	Summer	Psi Dra	Binary star	17 41.9	+72 08.9	Draco	4.6, 5.8	30″
29	Binoculars	Summer	M20	Reflection nebula	18 02.4	−22 59.0	Sagittarius	9	17′×12′
30a	Binoculars	Summer	M57	Planetary nebula	18 53.6	+33 01.7	Lyra	9.7	1.8′×1.4′
30b	Binoculars	Summer	M11	Open cluster	18 51.1	−06 16.0	Scutum	5.8	13′
30c	Binoculars	Summer	M27	Planetary nebula	19 59.6	+22 43.2	Vulpecula	7.6	6.7′
31	Binoculars	Summer	Barnard's E (B142–143)	Dark nebula	19 41.1	+10 54.3	Aquila	–	40′×23′
32a	Binoculars	Summer	Veil Nebula (NGC 6960)	Supernova remnant	20 45.9	+30 43.0	Cygnus	–	60.0′×9.0′
32b	Binoculars	Summer	Veil Nebula (NGC 6992)	Supernova remnant	20 57.0	+31 30.0	Cygnus	–	80.0′×26.0′
33	Binoculars	Summer	Pelican Nebula (IC 5067/5070)	Emission nebula	20 51.0	+44 00.0	Cygnus	–	60.0′×50.0′
34	Binoculars	Autumn	NGC 7293	Planetary nebula	22 29.6	−20 50.2	Aquarius	7.3	16′

(cont.)

Challenge	Instrument	Season	Target	Type	RA	Dec.	Constellation	Magnitude	Size/separation
35a	Binoculars	Autumn	M110	Galaxy	00 40.4	+41 41.4	Andromeda	8.9b	21.9′×10.9′
35b	Binoculars	Autumn	M32	Galaxy	00 42.7	+40 51.9	Andromeda	9.0b	8.7′×6.4′
36a	Binoculars	Autumn	NGC 247	Galaxy	00 47.1	−20 45.6	Sculptor	9.1	21.4′×6.0′
36b	Binoculars	Autumn	NGC 253	Galaxy	00 47.5	−25 17.3	Sculptor	8.0b	27.7′×6.7′
37	Binoculars	Autumn	NGC 288	Globular cluster	00 52.8	−26 34.9	Sculptor	8.1	13′
38a	Binoculars	Autumn	Psi-1 Psc	Double star	01 05.7	+21 28.4	Pisces	5.3, 5.6	29″
38b	Binoculars	Autumn	Zeta Psc	Double star	01 13.7	+07 34.5	Pisces	4.9, 6.3	23″
38c	Binoculars	Autumn	Lambda Ari	Double star	01 57.9	+23 35.8	Aries	4.8, 7.3	36″
38d	Binoculars	Autumn	30 Ari	Double star	02 37.0	+24 38.8	Aries	6.5, 7.4	38″
39	Binoculars	Autumn	M74	Galaxy	01 36.7	+15 47.0	Pisces	10.0b	10.5′×9.5′
40	Binoculars	Winter	NGC 1499	Emission nebula	04 00.5	+36 33.0	Perseus	–	160′×42′
41	Binoculars	Winter	M1	Supernova remnant	05 34.5	+22 01.0	Taurus	8.4	6.0′×4.0′
42	Binoculars	Winter	M78	Reflection nebula	05 46.8	+00 03.5	Orion	8.3	8.4′×7.8′
43	Binoculars	Winter	NGC 2158	Open cluster	06 07.4	+24 05.8	Gemini	8.6	5.0′
44	Binoculars	Winter	NGC 2237	Emission nebula	06 31.7	+05 04.0	Monoceros	–	80′×60′
45	Binoculars	Winter	NGC 2403	Galaxy	07 36.9	+65 36.2	Camelopardalis	8.9b	22.1′×12.4′
46	Binoculars	Year-round	Moon: Apollo landing sites	Solar System					
47a	Small scopes	Spring	NGC 2959	Galaxy	09 45.1	+68 35.7	Ursa Major	13.6p	1.3′×1.2′
47b	Small scopes	Spring	NGC 2976	Galaxy	09 47.3	+67 55.1	Ursa Major	10.8b	5.9′×2.6′
47c	Small scopes	Spring	NGC 3077	Galaxy	10 03.4	+68 44.0	Ursa Major	9.9	5.5′×4.0′
48	Small scopes	Spring	M109	Galaxy	11 57.6	+53 22.5	Ursa Major	10.6b	7.6′×4.6′
49a	Small scopes	Spring	NGC 4284	Galaxy	12 20.2	+58 05.6	Ursa Major	14.3p	2.5′×1.1′
49b	Small scopes	Spring	NGC 4290	Galaxy	12 20.8	+58 05.6	Ursa Major	12.7p	2.3′×1.5′
49c	Small scopes	Spring	M40	Double star	12 22.2	+58 05.0	Ursa Major	9.0, 9.3	50″
50	Small scopes	Spring	Markarian's Chain	Galaxy group	12 28	+13	Coma–Virgo	–	1.5°
51	Small scopes	Spring	3C 273	Quasar	12 29.1	+02 03.2	Virgo	13.0	stellar

Challenge	Instrument	Season	Target	Type	RA	Dec.	Constellation	Magnitude	Size/separation
52	Small scopes	Spring	NGC 5195	Galaxy	13 30.0	+47 16.4	Canes Venatici	10.5b	5.8′×4.6′
53	Small scopes	Spring	Izar [Epsilon (ε) Boötis]	Binary star	14 45.0	+27 04.5	Boötes	2.5, 5.0	2.8″
54	Small scopes	Summer	NGC 6369	Planetary nebula	17 29.3	−23 45.6	Ophiuchus	12.9p	38″
55	Small scopes	Summer	Barnard's Star	Star with high proper motion	17 57.6	+04 41.6	Ophiuchus	9.5	stellar
56	Small scopes	Summer	NGC 6781	Planetary nebula	19 18.4	+06 32.3	Aquila	11.8p	108″
57	Small scopes	Summer	NGC 6803	Planetary nebula	19 31.3	+10 03.3	Aquila	11.3p	6″
58	Small scopes	Summer	NGC 6804	Planetary nebula	19 31.6	+09 13.5	Aquila	12.2p	35″
59	Small scopes	Summer	Campbell's Hydrogen Star	Planetary nebula	19 34.8	+30 31.0	Cygnus	9.6p	35″
60	Small scopes	Summer	NGC 6822	Galaxy	19 44.9	−14 48.2	Sagittarius	9.3b	15.6′×13.5′
61	Small scopes	Summer	61 Cygni	Binary star	21 06.9	+38 44.8	Cygnus	5.2, 6.0	16″
62a	Small scopes	Autumn	Barnard 168	Dark nebula	21 53.3	+47 16	Cygnus	–	100′×10′
62b	Small scopes	Autumn	IC 5146	Emission nebula	21 53.5	+47 15.7	Cygnus	–	11.0′×10.0′
63a	Small scopes	Autumn	NGC 147	Galaxy	00 33.2	+48 30.5	Cassiopeia	10.5b	13.2′×7.7′
63b	Small scopes	Autumn	NGC 185	Galaxy	00 39.0	+48 20.2	Cassiopeia	10.1b	11.9′×10.1′
64	Small scopes	Autumn	NGC 188	Open cluster	00 47.5	+85 14.5	Cepheus	8.1	13′
65	Small scopes	Autumn	NGC 300	Galaxy	00 54.9	−37 41.0	Sculptor	8.7b	22.1′×16.6′
66	Small scopes	Autumn	NGC 404	Galaxy	01 09.5	+35 43.1	Andromeda	11.2b	3.4′×3.4′
67	Small scopes	Autumn	NGC 604	Emission nebula/ stellar association in galaxy	01 34.5	+30 47.0	Triangulum	11.5b	1.5′
68	Small scopes	Winter	NGC 1360	Planetary nebula	03 33.3	−25 52.2	Fornax	9.6p	6.4′
69	Small scopes	Winter	Pleiades nebulosity	Reflection nebula	03 46	+24 10	Taurus	n/a	70′×60′
70	Small scopes	Winter	NGC 1535	Planetary nebula	04 14.3	−12 44.4	Eridanus	9.6p	60.0″

(cont.)

Challenge	Instrument	Season	Target	Type	RA	Dec.	Constellation	Magnitude	Size/ separation
71	Small scopes	Winter	NGC 1851	Globular cluster	05 14.1	−40 02.8	Columba	7.1	12.0′
72	Small scopes	Winter	Horsehead Nebula	Dark nebula	05 41.0	−02 27.7	Orion	n/a	4′
73	Small scopes	Winter	Jonckheere 900	Planetary nebula	06 26.0	+17 47.4	Gemini	12.4p	9.0″
74	Small scopes	Winter	NGC 2419	Globular cluster	07 38.1	+38 52.9	Lynx	10.3	4.6′
75	Small scopes	Year-round	Moon: Bailly	Solar System			Days 13 and 23–24		
76	Small scopes	Year-round	Moon: Clavius craterlets	Solar System			Days 8 and 21		
77	Small scopes	Year-round	Moon: Hesiodus A	Solar System			Days 8 and 22		
78	Small scopes	Year-round	Moon: Lambert R	Solar System			Days 9 and 22		
79	Small scopes	Year-round	Moon: Lunar X and Lunar V	Solar System			Day 7		
80	Small scopes	Year-round	Moon: Mare Marginis, Mare Smythii	Solar System			Days 2–4 and 15		
81	Small scopes	Year-round	Moon: Mare Orientale	Solar System			Days 13 and 26		
82	Small scopes	Year-round	Moon: Messier/ Messier A	Solar System			Days 4–8 and 15–16		
83	Small scopes	Year-round	Moon: Rupes Recta, Birt, and Rima Birt	Solar System			Days 8 and 22		
84	Small scopes	Year-round	Moon: Schröter's Valley	Solar System			Days 11 and 24		
85	Small scopes	Year-round	Moon: Serpentine Ridge (Dorsae Lister & Smirnov)	Solar System			Days 6 and 19		
86	Small scopes	Year-round	Venus: Ashen Light	Solar System					
87	Small scopes	Year-round	Eye of Mars	Solar System					
88	Small scopes	Year-round	Jupiter's Great Red Spot	Solar System					
89	Small scopes	Year-round	Cassini's Division	Solar System					

Challenge	Instrument	Season	Target	Type	RA	Dec.	Constellation	Magnitude	Size/ separation
90	Small scopes	Year-round	Finding Pluto	Solar System					
91	Medium scopes	Spring	Zeta Cancri	Quadruple star	08 12.2	+17 38.9	Cancer	5.6/6.0/ 6.1/10.0	1″/0.3″
92	Medium scopes	Spring	Leo Trio 2	Galaxy group	09 43.2	+31 55.7	Leo	n/a	~11′
93	Medium scopes	Spring	The Antennae	Galaxy pair	12 01.9	−18 52.8	Corvus	n/a	~4′
94	Medium scopes	Spring	NGC 4361	Planetary nebula	12 24.5	−18 47.0	Corvus	10.3p	118″
95	Medium scopes	Spring	NGC 5053	Globular cluster	13 16.4	+17 41.9	Coma Berenices	9.0	10.0′
96	Medium scopes	Spring	M51 spiral structure	Galaxy structure	13 29.9	+47 11.8	Canes Venatici	9.0b	10.3′×8.1′
97	Medium scopes	Summer	NGC 6445	Planetary nebula	17 49.3	−20 00.6	Sagittarius	13.2p	44″×30″
98	Medium scopes	Summer	NGC 6453	Globular cluster	17 50.9	−34 35.9	Scorpius	10.2	7.6′
99	Medium scopes	Summer	NGC 6517	Globular cluster	18 01.8	−08 57.5	Ophiuchus	10.1	4.0′
100	Medium scopes	Summer	NGC 6539	Globular cluster	18 04.7	−07 35.2	Serpens	8.9	7.9′
101	Medium scopes	Summer	Barnard 86	Dark nebula	18 03.0	−27 52.0	Sagittarius	Opacity 5	5.0′
102	Medium scopes	Summer	NGC 6886	Planetary nebula	20 12.7	+19 59.3	Sagitta	12.2p	6″
103	Medium scopes	Summer	IC 4997	Planetary nebula	20 20.1	+16 43.9	Sagitta	11.6p	2″
104	Medium scopes	Summer	NGC 6905	Planetary nebula	20 22.4	+20 06.3	Delphinus	11.9p	72″×37″
105	Medium scopes	Autumn	IC 5217	Planetary nebula	22 23.9	+50 58.0	Lacerta	12.6p	7″
106	Medium scopes	Autumn	Deer Lick Group	Galaxy group	22 37.3	+34 26.1	Pegasus	–	~6′
107	Medium scopes	Autumn	NGC 7354	Planetary nebula	22 40.3	+61 17.1	Cepheus	12.9p	36″
108a	Medium scopes	Autumn	NGC 7537	Galaxy	23 14.6	+04 29.9	Pisces	13.9b	2.2′×0.5′
108b	Medium scopes	Autumn	NGC 7541	Galaxy	23 14.7	+04 32.0	Pisces	12.4b	3.5′×1.2′
109	Medium scopes	Autumn	STF 3057	Binary star	00 04.9	+58 32.0	Cassiopeia	6.7/9.3	3.9″
110	Medium scopes	Autumn	STF 3062	Binary star	00 06.3	+58 26.2	Cassiopeia	6.4/7.3	1.3″
111	Medium scopes	Autumn	Lambda Cas	Binary star	00 31.8	+54 31.3	Cassiopeia	5.3/5.6	0.3″
									(cont.)

Challenge	Instrument	Season	Target	Type	RA	Dec.	Constellation	Magnitude	Size/separation
112	Medium scopes	Autumn	IC 10	Galaxy	00 20.3	+59 18.1	Cassiopeia	11.8b	6.3′×5.1′
113	Medium scopes	Autumn	Sculptor Dwarf Galaxy	Galaxy	01 00.2	−33 43.3	Sculptor	10.5b	39.8′×30.8′
114	Medium scopes	Winter	NGC 1343	Galaxy	03 37.8	+72 34.3	Cassiopeia	13.5p	2.5′×1.5′
115	Medium scopes	Winter	AGCS 373	Galaxy cluster	03 38.5	−35 27.0	Fornax	–	180′
116	Medium scopes	Winter	Jonckheere 320	Planetary nebula	05 05.6	+10 42.4	Orion	12.9p	26″×14″
117	Medium scopes	Winter	IC 418	Planetary nebula	05 27.5	−12 41.8	Lepus	10.7p	12″
118	Medium scopes	Winter	NGC 1924	Galaxy	05 28.0	−05 18.6	Orion	13.3b	1.5′×1.1′
119	Medium scopes	Winter	PK 189+7.1	Planetary nebula	06 37.3	+24 00.6	Gemini	13.4p	32″×15″
120	Medium scopes	Winter	Sirius and the Pup	Binary star	06 45.1	−16 42.8	Canis Major	−1.46, 8.30	Varies
121	Medium scopes	Winter	NGC 2298	Globular cluster	06 49.0	−36 00.3	Puppis	9.3	5.0′
122	Medium scopes	Winter	NGC 2371-72	Planetary nebula	07 25.6	+29 29.4	Gemini	13.0p	55″
123	Medium scopes	Winter	NGC 2438	Planetary nebula	07 41.8	−14 44.1	Puppis	10.1p	64″
124	Medium scopes	Year-round	Moon: Alpine Valley rille	Solar System			Days 7 and 20–21		
125	Medium scopes	Year-round	Moon: Armstrong, Aldrin, and Collins	Solar System			Days 5 and 19		
126	Medium scopes	Year-round	Moon: Hortensius dome field	Solar System			Days 9 and 22		
127	Medium scopes	Year-round	Moon: Mons Hadley and Rima Hadley	Solar System			Days 7 and 20–21		
128	Medium scopes	Year-round	Moon: Plato's craterlets	Solar System			Days 8 and 21		
129	Medium scopes	Year-round	Mercury: Surface markings	Solar System					
130	Medium scopes	Year-round	Mars: Phobos and Deimos	Solar System					

Challenge	Instrument	Season	Target	Type	RA	Dec.	Constellation	Magnitude	Size/separation
131	Medium scopes	Year-round	Saturn: Encke's Division	Solar System					
132	Large scopes	Spring	Abell 33	Planetary nebula	09 39.2	−02 48.5	Hydra	13.4p	270″
133	Large scopes	Spring	Hickson Compact Galaxy Group 44	Galaxy group	10 18.0	+21 49.3	Leo	−	16′
134	Large scopes	Spring	AGC 1060	Galaxy cluster	10 36.9	−27 31.0	Hydra	−	168′
135	Large scopes	Spring	NGC 3172	Galaxy	11 47.2	+89 05.6	Ursa Minor	14.8	1.2′×1.1′
136	Large scopes	Spring	Abell 36	Planetary nebula	13 40.7	−19 53.0	Virgo	13.0p	480″×300″
137	Large scopes	Spring	M101 nebulae	Emission nebulae	14 03.2	+54 20.9	Ursa Major	−	−
138	Large scopes	Summer	M13 propellers	Globular cluster	16 41.7	+36 27.6	Hercules	5.8	20.0′
139a	Large scopes	Summer	NGC 6194	Galaxy	16 36.6	+36 12.0	Hercules	14.6	1.0′×0.8′
139b	Large scopes	Summer	IC 4614	Galaxy	16 37.8	+36 06.9	Hercules	15.4	0.8′×0.8′
139c	Large scopes	Summer	NGC 6196	Galaxy	16 37.9	+36 04.4	Hercules	13.9b	2.0′×1.1′
139d	Large scopes	Summer	IC 4616	Galaxy	16 38.0	+35 59.8	Hercules	15.4	0.6′×0.3′
139e	Large scopes	Summer	IC 4617	Galaxy	16 42.1	+36 41.0	Hercules	15.5	1.2′×0.4′
139f	Large scopes	Summer	NGC 6207	Galaxy	16 43.1	+36 50.0	Hercules	12.2b	3.3′×1.7′
140	Large scopes	Summer	NGC 6578	Planetary nebula	18 16.3	−20 27.1	Sagittarius	13.1p	9″
141	Large scopes	Summer	IC 4732	Planetary nebula	18 33.9	−22 38.7	Sagittarius	13.3p	10″
142	Large scopes	Summer	Palomar 11	Globular cluster	19 45.2	−08 00.4	Aquila	9.8	10.0′
143	Large scopes	Summer	Abell 70	Planetary nebula	20 31.6	−07 05.3	Aquila	14.3p	42″
144	Large scopes	Autumn	Stephan's Quintet	Galaxy group	22 36.0	+33 57.0	Pegasus	−	~3′
145	Large scopes	Autumn	Palomar 13	Globular cluster	23 06.7	+12 46.3	Pegasus	13.8	0.7′
146a	Large scopes	Autumn	NGC 1	Galaxy	00 07.3	+27 42.5	Pegasus	12.8	1.8′×1.1′
146b	Large scopes	Autumn	NGC 2	Galaxy	00 07.3	+27 40.7	Pegasus	14.1	1.2′×0.7′
									(*cont.*)

Challenge	Instrument	Season	Target	Type	RA	Dec.	Constellation	Magnitude	Size/separation
147	Large scopes	Autumn	M33 nebulae	Emission nebulae	01 33.8	+30 39.6	Triangulum	–	–
148a	Large scopes	Autumn	Fornax 1	Globular cluster	02 37.0	−34 11.0	Fornax	15.6	0.9′
148b	Large scopes	Autumn	Fornax 2	Globular cluster	02 38.7	−34 48.6	Fornax	13.5	0.8′
148c	Large scopes	Autumn	NGC 1049	Globular cluster	02 39.8	−34 15.4	Fornax	12.6	0.8′
148d	Large scopes	Autumn	Fornax 6	Globular cluster	02 40.1	−34 25.2	Fornax	–	0.6′
148e	Large scopes	Autumn	Fornax 4	Globular cluster	02 40.1	−34 32.2	Fornax	13.6	0.8′
148f	Large scopes	Autumn	Fornax 5	Globular cluster	02 42.4	−34 06.2	Fornax	13.4	1.7′
149	Large scopes	Autumn	Arp 77	Galaxy pair	02 46.3	−30 16.4	Fornax	10.2b	12.7′×9.4′
150	Large scopes	Winter	Arp 41	Galaxy pair	03 09.8	−20 34.9	Eridanus	10.5b	7.4′×6.4′
151	Large scopes	Winter	AGC 426	Galaxy cluster	03 18.6	+41 30.0	Perseus	–	190′
152	Large scopes	Winter	Abell 12	Planetary nebula	06 02.3	+09 39.3	Orion	13.9p	37″
153	Large scopes	Winter	Sharpless 2-301	Emission nebula	07 09.8	−18 29.8	Canis Major	–	9.0′×8.0′
154	Large scopes	Winter	Arp 82	Galaxy pair	08 11.2	+25 12.4	Cancer	13.3	3.3′×1.8′
155	Large scopes	Year-round	Major satellites of Uranus	Solar System					
156	Large scopes	Year-round	Triton: satellite of Neptune	Solar System					
157	Monster scopes	Spring	Galaxies beyond M44	Galaxy collection	08 40.4	+19 40	Cancer	–	66′
158	Monster scopes	Spring	Leo I	Galaxy	10 08.5	+12 18.5	Leo	9.8	9.8′×7.4′
159	Monster scopes	Spring	Leo II	Galaxy	11 13.5	+22 09.2	Leo	11.9	10.1′×9.0′
160	Monster scopes	Spring	Hickson Galaxy Group 50	Galaxy group	11 17.1	+54 55.0	Ursa Major	–	<1′
161	Monster scopes	Spring	Palomar 4	Globular cluster	11 29.3	+28 58.4	Ursa Major	14.2	1′
162	Monster scopes	Spring	Copeland's Septet	Galaxy group	11 37.9	+21 58.9	Leo	–	5′
163	Monster scopes	Spring	AGC 1367	Galaxy cluster	11 44.5	+19 50	Leo	–	100.8′

Challenge	Instrument	Season	Target	Type	RA	Dec.	Constellation	Magnitude	Size/ separation
164	Monster scopes	Spring	AGC 1656	Galaxy cluster	12 59.8	+27 58	Coma Berenices	–	224′
165	Monster scopes	Summer	AGC 2065	Galaxy cluster	15 22.7	+27 43	Corona Borealis	–	22.4′
166	Monster scopes	Summer	Seyfert's Sextet	Galaxy group	15 59.2	+20 45	Serpens	–	1′
167	Monster scopes	Summer	AGC 2151	Galaxy cluster	16 05.2	+17 44.0	Hercules	–	56′
168	Monster scopes	Summer	UKS 1	Globular cluster	17 54.5	−24 08.7	Sagittarius	17.3	2′
169	Monster scopes	Summer	PK 9-7.1	Planetary nebula	18 36.4	−23 55.3	Sagittarius	15.0p	9″
170	Monster scopes	Summer	IC 1296	Galaxy	18 53.3	+33 04.0	Lyra	14.8p	0.9′×0.5′
171	Monster scopes	Summer	M57 central star	Central star	18 53.6	+33 01.7	Lyra	15.2	n/a
172	Monster scopes	Summer	Terzan 7	Globular cluster	19 17.7	−34 39.5	Sagittarius	12.0	1.2′
173	Monster scopes	Summer	Simeis 57	Emission nebula	20 16.2	+43 41.2	Cygnus	–	23′×4′
174	Monster scopes	Autumn	Pease 1	Planetary nebula	21 30.0	+12 10.4	Pegasus	14.9p	1″
175	Monster scopes	Autumn	Elephant's Trunk Nebula (IC 1396A)	Reflection nebula	21 39.0	+57 30.0	Cepheus	–	13′×5′
176	Monster scopes	Autumn	Einstein's Cross	Gravitatio-nal lens	22 40.5	+03 21.5	Pegasus	15.1p	1.1′×0.5′
177	Monster scopes	Autumn	NGC 51 group	Galaxy group	00 14.2	+48 12.1	Andromeda	–	10′
178	Monster scopes	Autumn	M31 globulars	Globular clusters	00 42.7	+41 16.1	Andromeda	n/a	n/a
179	Monster scopes	Autumn	IC 1613	Galaxy	01 04.8	+02 17.1	Cetus	9.2	16.3′×14.5′
180a	Monster scopes	Autumn	WLM	Galaxy	00 01.9	−15 27.8	Cetus	10.6	9.5′×3.0′
180b	Monster scopes	Autumn	WLM-1	Globular cluster	00 01.8	−15 27.6	Cetus	16.1	–
181	Monster scopes	Winter	Palomar 1	Globular cluster	03 33.3	+79 35.0	Cepheus	13.2	2.8′
182	Monster scopes	Winter	UGC 2838	Galaxy	03 43.8	+24 03.6	Taurus	17.9	1.6′×0.2′
183	Monster scopes	Winter	Palomar 2	Globular cluster	04 46.1	+31 22.9	Auriga	13.0	2′

(cont.)

Challenge	Instrument	Season	Target	Type	RA	Dec.	Constellation	Magnitude	Size/separation
184	Monster scopes	Winter	Simeis 147	Supernova remnant	05 39.0	+28 00.0	Taurus	–	200′
185a	Monster scopes	Winter	NGC 2363	Galaxy	07 28.5	+69 11.6	Camelopardalis	13.0	1.8′×1.0′
185b	Monster scopes	Winter	NGC 2366	Galaxy	07 28.9	+69 12.7	Camelopardalis	11.5b	8.2′×3.3′
186a	Monster scopes	Winter	NGC 2474	Galaxy pair	07 58.0	+52 51.4	Lynx	14.3	0.6′
186b	Monster scopes	Winter	NGC 2475	Galaxy pair	07 58.0	+52 51.8	Lynx	14.0b	0.7′
187	Monster scopes	Winter	PK 164+31.1	Planetary nebula	07 57.9	+53 25.3	Lynx	14.0p	6.3′
188	Monster scopes	Year-round	Pluto and charon	Solar System					

Appendix B
Suggested further reading

BOOKS

Burnham, R., Jr., *Burnham's Celestial Handbook*, volumes 1, 2, and 3 (Dover, 1978)

Cragin, M. and Bonnano, E., *Uranometria 2000.0 Atlas Volume 3: Deep Sky Field Guide* (Willmann-Bell, 2001)

Dickinson, T. and Dyer, A., *Backyard Astronomer's Guide* (Camden House, 2002)

Harrington, P., *Star Ware*, fourth edition (John Wiley & Sons, 2007)

Harrington, P., *Touring the Universe through Binoculars* (John Wiley and Sons, 1990)

Hickson, P., *Atlas of Compact Groups of Galaxies* (CRC Press, 1994)

Houston, W. and O'Meara, S., *Deep-Sky Wonders* (Sky Publishing, 1998)

Kepple, G. and Sanner, G., *Night Sky Observer's Guide*, volumes 1 and 2 (Willmann-Bell, 1998)

Mollise, R., *The Urban Astronomer's Guide* (Springer, 2006)

Mullaney, J. and Tirion, W., *The Cambridge Double Star Atlas* (Cambridge University Press, 2009)

O'Meara, S., *Deep-Sky Companions: Hidden Treasures* (Cambridge University Press, 2006)

O'Meara, S., *Deep-Sky Companions: The Caldwell Objects* (Cambridge University Press, 2003)

O'Meara, S., *Deep-Sky Companions: The Messier Objects* (Cambridge University Press, 2000)

Pennington, H., *Year-Round Messier Marathon Field Guide* (Willmann-Bell, 1998)

Stoyan, R., *Atlas of the Messier Objects* (Cambridge University Press, 2008)

Tirion, W., *Sky Atlas 2000.0*, second edition (Sky Publishing, 1998)

Tirion, W. and Cragin, M., *Uranometria 2000.0*, volumes 1 and 2 (Willmann-Bell, 2001)

SOFTWARE

Guide (CD-ROM, Windows), Project Pluto, 168 Ridge Road, Bowdoinham, ME 04008; www.projectpluto.com

Megastar (CD-ROM, Windows), Willmann-Bell, PO Box 35025, Richmond, VA 23235; www.willbell.com

SkyMap Pro (CD-ROM, Windows), World Wide Software Publishing, PO Box 326, Elk River, MN 55330; www.skymap.com

Starry Night (CD-ROM, Windows or Macintosh), Space.com, 284 Richmond Street East, Toronto, ON, Canada M5A 1P4; www.starrynight.com

The Sky (CD-ROM, Windows or Macintosh), Software Bisque, 912 12th Street, Golden, CO; www.bisque.com

Virtual Moon Atlas (Windows), Christian Legrand and Patrick Chevalley, available for free download at ap-i.net/avl/en/start

WEBSITES

Abell Planetaries: www.astronomy-mall.com/Adventures.In.Deep.Space/abellpn.htm

Adventures in Deep Space: www.astronomy-mall.com/Adventures.In.Deep.Space/

American Association of Variable Star Observers: www
.aavso.org

Association of Lunar and Planetary Observers: www.lpl
.arizona.edu/alpo.com

Astronomical Society of the Pacific: www.astrosociety
.org

British Astronomical Association: www.ast.cam.ac.uk/~
baa.com

Consolidated Lunar Atlas: www.lpi.usra.edu/resources/
cla

Edward Emerson Barnard's Photographic Atlas of
Selected Regions of the Milky Way: www.library
.gatech.edu/barnard

Environment Canada: www.weatheroffice.gc.ca

Earth System Scientist Network, Atmospheric Aerosols:
cesse.terc.edu/essn/projects/proj_brooks01/
resrc_01.cfm

Huey, Alvin; Abell Planetary Observer's Guide: www
.faintfuzzies.com

Huey, Alvin; Hickson Group Observer's Guide: www
.faintfuzzies.com

HyperLeda galactic database: leda.univ-lyon1.fr

Lunar Calculator: www3.telus.net/public/aling/
lunarcal/lunarcal.htm

Millennium Galaxy Catalogue: www.eso.org/~jliske/
mgc

NASA-Astrophysics Data System (ADS): adswww
.harvard.edu

NASA/IPAC Extragalactic Database (NED): nedwww
.ipac.caltech.edu/forms/byname.html

NGC/IC Project: www.ngcic.org

Perek–Kohoutek Catalogue of Galactic Planetary
Nebulae: www.hs.uni-hamburg.de/DE/Ins/Per/
Kohoutek/kohoutek/WEBcgpn2/text2/index.html

Saguaro Astronomy Club (SAC): www.saguaroastro.org

SEDS (Students for the Exploration and Development
of Space): www.seds.org

Sharpless Catalog: galaxymap.org/cgi-bin/
sharpless.py?s=1

SIMBAD Astronomical Database: simbad.u-strasbg.fr

Sloan Digital Sky Survey: astronomerica.awardspace
.com

Washington Double Star Catalog: ad.usno.navy.mil/
wds

Wood, Charles; Lunar 100 observing list: www
.skyandtelescope.com/observing/objects/moon/
3308811.html

Appendix C
100 challenging double stars

Observing double and multiple stars is a fun activity that can be enjoyed by every amateur astronomer under just about any clear-sky condition. Since they are point sources (as opposed to extended, diffuse targets, like most traditional deep-sky objects), double stars punch right through light pollution, moonlight, and even haze. They make great targets on nights that might not otherwise be usable for other types of observing.

While several challenging double stars are mentioned in earlier chapters, here is a list of 100 challenging targets to try your luck. Some are difficult because they lie close to one another, while others are tricky because of the discrepancy between their comparative brightnesses.

Star	Constellation	RA	Dec.	Components	Mag A	Mag B	Separation (arc-seconds)	Position angle
STF 3056	Andromeda	00 04.7	+34 16	AB	8.0	8.0	0.6	148
STF 3062	Cassiopeia	00 06.3	+58 26		6.4	7.2	1.2	235
Kappa-1 (κ1) Scl	Sculptor	00 09.3	−27 59	AB	6.1	6.2	1.4	265
STF 13	Cepheus	00 16.2	+76 57		7.0	7.3	0.9	51
AC 1	Andromeda	00 20.9	+32 59		7.0	7.5	1.7	288
36 And	Andromeda	00 55.0	+23 38		6.0	6.4	0.9	313
Burnham 303	Pisces	01 09.7	+23 48		7.3	7.5	0.7	290
STF 138	Pisces	01 36.1	+07 39		7.6	7.6	1.7	56
STF 162	Perseus	01 49.3	+47 54	AB	5.8	6.8	2.0	203
1 Ari	Aries	01 50.1	+22 17		6.2	7.2	2.8	166
STF 186	Cetus	01 55.9	+01 51		6.8	6.8	1.1	51
Alpha (α) Psc	Pisces	02 02.0	+02 46		4.2	5.1	1.8	267
STF 228	Andromeda	02 14.0	+47 29		6.6	7.1	1.0	283
STF 314	Perseus	02 52.9	+53 00	ABxC	6.5	7.1	1.6	308
Epsilon (ε) Ari	Aries	02 59.2	+21 20	AB	5.2	5.5	1.5	208
7 Tau	Taurus	03 34.4	+24 28		6.6	6.7	0.7	55
STF 425	Perseus	03 40.1	+34 07		7.6	7.6	1.8	76
Burnham 184	Eridanus	04 27.9	−21 30		7.3	7.7	1.5	252
14 Ori	Orion	05 07.9	+08 30		5.8	6.5	0.8	71
STF 644	Auriga	05 10.3	+37 18		6.8	7.0	1.6	219
STF 749	Taurus	05 37.2	+26 55	AB	6.5	6.6	1.1	325
								(cont.)

463

Star	Constellation	RA	Dec.	Components	Mag A	Mag B	Separation (arc-seconds)	Position angle
STF 757	Orion	05 38.1	−00 11	AB	8.0	8.0	1.6	237
52 Ori	Orion	05 48.0	+06 27		6.1	6.1	1.6	210
STF 3115	Camelopardalis	05 49.1	+62 49		6.5	7.6	1.0	5
Burnham 568	Canis Major	06 23.8	−19 47		7.2	7.5	0.8	155
STF 932	Gemini	06 34.4	+14 45		8.0	8.0	1.8	317
12 Lyn	Lynx	06 46.2	+59 26	AB	5.5	6.0	1.8	73
Burnham 324	Canis Major	06 49.7	−24 05	AB	6.3	7.6	1.8	206
STF 987	Monocerotis	06 54.1	−05 51		7.1	7.3	1.4	175
15 Lyn	Lynx	06 57.3	+58 25	AB	4.8	5.9	0.9	33
Burnham 328	Canis Major	07 06.7	−11 18	AB	5.7	6.9	0.6	116
STF 1037	Gemini	07 12.8	+27 13		7.0	7.0	1.2	329
STF 1057	Canis Major	07 14.8	−15 29	AB	8.0	8.0	0.7	75
STT 170	Canis Minor	07 17.6	+09 18		7.6	7.9	0.7	74
Alpha (α) Gem	Gemini	07 34.6	+31 53		1.9	2.9	2.2	164
Alpha (α) CMi	Canis Minor	07 39.3	+05 13.6		0.4	10.8	2.3	221
STF 1126	Canis Minor	07 40.1	+05 14	AB	6.4	6.7	0.9	132
STF 1216	Hydra	08 21.3	−01 36		7.1	7.4	0.7	287
Epsilon (ε) Hya	Hydra	08 46.8	+06 25	AB	3.8	4.7	2.7	113
57 Cnc	Cancer	08 54.2	+30 35	AB	6.0	6.5	1.4	316
Iota-2 (ι2) Cnc	Cancer	08 54.2	+30 35	AB	6.0	6.5	1.4	316
STF 1291	Cancer	08 54.3	+30 34		6.1	6.4	1.5	312
STF 1333	Lynx	09 18.4	+35 22		6.4	6.7	1.6	49
STF 1348	Hydra	09 24.5	+06 21		7.5	7.6	1.9	317
STT 215	Leo	10 16.3	+17 44		7.2	7.5	1.5	184
Xi (ξ) UMa	Ursa Major	11 18.2	+31 32		4.3	4.8	1.6	273
Gamma (γ) Vir	Virgo	12 41.7	−01 27		3.5	3.5	1.0	35
48 Vir	Virgo	13 03.9	−03 40		7.2	7.5	0.8	205
STF 1734	Virgo	13 20.7	+02 57		6.8	7.5	1.0	183
STF 1816	Boötes	14 13.9	+29 06		7.5	7.6	1.1	88
STF 1819	Virgo	14 15.3	+03 08		7.8	7.9	0.8	189
Zeta (ζ) Boo	Boötes	14 41.1	+13 44		4.5	4.6	1.0	307
STF 1871	Boötes	14 41.7	+51 24		7.0	7.0	2.0	306
Epsilon (ε) Boo	Boötes	14 45.0	+27 04		2.5	4.9	2.8	339
Mu (μ) Lib	Libra	14 49.3	−14 09		5.8	6.7	1.8	355
39 Boo	Boötes	14 49.8	+48 43		6.0	6.5	2.9	45

Star	Constellation	RA	Dec.	Components	Mag A	Mag B	Separation (arc-seconds)	Position angle
STT 288	Boötes	14 53.5	+15 42		6.5	7.0	1.5	175
59 Hya	Hydra	14 58.7	−27 39		6.3	6.6	0.8	335
44 Boo	Boötes	15 03.8	+47 39		5.3	6.2	2.4	56
STF 1932	Corona Borealis	15 18.3	+26 50		7.3	7.4	1.6	258
Eta (η) CrB	Corona Borealis	15 23.2	+30 17	AB	5.6	5.9	0.5	128
STT 298	Boötes	15 36.1	+39 48		7.5	7.5	1.0	191
Burnham 122	Libra	15 39.9	−19 46		7.7	7.7	1.8	226
Xi (ξ) Sco	Scorpius	16 04.4	−11 22	AB	4.8	5.1	0.5	358
STF 2054	Draco	16 23.8	+61 42		6.0	7.2	1.0	355
Alpha (α) Sco	Scorpius	16 29.4	−26 26		1.0	5.4	2.5	274
Zeta (ζ) Her	Hercules	16 41.3	+31 36		2.9	5.5	1.1	196
20 Dra	Draco	16 56.4	+65 02		7.1	7.3	1.4	67
24 Oph	Ophiuchus	16 56.8	−23 09		6.2	6.5	0.8	294
Mu (μ) Dra	Draco	17 05.3	+54 28		5.7	5.7	2.3	10
Eta (η) Oph	Ophiuchus	17 10.4	−15 43		3.0	3.5	1.0	325
STT 338	Hercules	17 52.0	+15 20		6.8	7.1	0.7	357
Tau (τ) Oph	Ophiuchus	18 03.1	−08 11		5.2	5.9	1.6	286
STF 2289	Hercules	18 10.1	+16 29		6.5	7.2	1.2	224
Burnham 132	Sagittarius	18 11.2	−19 51		6.9	7.3	1.0	203
Burnham 133	Sagittarius	18 27.8	−26 38		7.0	7.2	1.3	247
STT 359	Hercules	18 35.5	+23 36		6.3	6.5	0.7	6
STT 358	Hercules	18 35.9	+16 59	AC	6.8	7.0	1.5	149
Epsilon-2 (ε2) Lyr	Lyra	18 44.3	+39 40	CD	5.2	5.5	2.3	94
Epsilon-1 (ε1) Lyr	Lyra	18 44.3	+39 40	AB	5.0	6.1	2.6	357
STF 2375	Serpens	18 45.5	+05 30		5.8	5.8	2.6	119
STF 2438	Draco	18 57.5	+58 14		7.0	7.4	0.8	358
STT 371	Lyra	19 16.0	+27 27	AB	7.0	7.3	0.9	161
Pi (π) Aql	Aquila	19 48.7	+11 49		6.1	6.9	1.4	110
STT 387	Cygnus	19 48.7	+35 19		7.0	7.6	0.6	152
16 Vul	Vulpecula	20 02.2	+24 56		5.8	6.2	0.8	115
STF 2644	Aquila	20 12.6	+00 52		6.9	7.1	2.7	206
STT 403	Cygnus	20 14.4	+42 06		7.4	7.6	0.8	173
STT 410	Cygnus	20 39.6	+40 35	AB	6.8	7.1	0.8	10
								(*cont.*)

Star	Constellation	RA	Dec.	Components	Mag A	Mag B	Separation (arc-seconds)	Position angle
4 Aqr	Aquarius	20 51.4	−05 38	AB	6.4	7.2	0.7	27
Epsilon (ε) Equ	Equuleus	20 59.1	+04 18	AB	6.0	6.3	0.9	285
STF 2751	Cepheus	21 02.2	+56 40		6.2	7.0	1.6	354
12 Aqr	Aquarius	21 04.1	−05 49		5.9	7.3	2.8	192
STT 437	Cygnus	21 20.8	+32 27	AB	6.2	6.9	2.1	28
STF 2799	Pegasus	21 29.0	+11 06	AB	7.3	7.4	1.8	265
STF 2843	Cepheus	21 51.6	+65 45		7.1	7.3	1.5	147
Eta (η) PsA	Piscis Austrinus	22 00.8	−28 27		5.8	6.8	1.7	115
STF 2854	Pegasus	22 04.5	+13 39		7.6	7.9	2.0	83
Zeta (ζ) Aqr	Aquarius	22 28.8	−00 01		4.3	4.5	2.3	183
STF 3050	Andromeda	23 59.5	+33 43		6.5	6.7	2.1	329

Notes:
1. Star: This column lists each star's Bayer or Flamsteed designation, or the entry listing in its discoverer's catalog.
   a. AC: Alvan Clark
   b. Burnham: S. W. Burnham
   c. STF: F. G. W. Struve
   d. STT: Otto Struve
2. Components: An entry in this column indicates that the target is a multiple-star system. The corresponding values cited in the following columns apply to the specified stars. For instance, the star STF 314 is listed as ABxC. This indicates that the data apply to the system's third component in relation to the combined pair of brightest stars in this multiple star, specified as AB.
3. Position angle: Specifies the angular location of the companion star with respect to the primary star. Due north corresponds to a position angle of 0°, east is 90°, south is 180°, and west is 270°.

Index